Deposition, Diagenesis, and Geochemistry of Carbonate Sequences

Deposition, Diagenesis, and Geochemistry of Carbonate Sequences

Editors

Hamzeh Mehrabi
Vahid Tavakoli

MDPI

Basel • Beijing • Wuhan • Barcelona • Belgrade • Novi Sad • Cluj • Manchester

Editors

Hamzeh Mehrabi
Geology
University of Tehran
Tehran
Iran

Vahid Tavakoli
Geology
University of Tehran
Tehran
Iran

Editorial Office
MDPI
St. Alban-Anlage 66
4052 Basel, Switzerland

This is a reprint of articles from the Special Issue published online in the open access journal *Minerals* (ISSN 2075-163X) (available at: www.mdpi.com/journal/minerals/special_issues/4RSIJ25U92).

For citation purposes, cite each article independently as indicated on the article page online and as indicated below:

Lastname, A.A.; Lastname, B.B. Article Title. *Journal Name* **Year**, *Volume Number*, Page Range.

ISBN 978-3-7258-0688-1 (Hbk)
ISBN 978-3-7258-0687-4 (PDF)
doi.org/10.3390/books978-3-7258-0687-4

Contents

About the Editors

Hamzeh Mehrabi

Dr. Hamzeh Mehrabi is an Assistant Professor at the School of Geology, College of Science, University of Tehran. He completed his PhD studies in Sedimentary Petrology and Geochemistry at the University of Tehran and has been awarded PhD student of the year (2016), the best young professor in basic sciences of Iran (2021), and the best geology researcher of the year (2019). He recently led projects on geological, geochemical, and petrophysical studies of carbonate hydrocarbon reservoirs in the Zagros Basin and the Persian Gulf. He has published more than 40 papers in various WOS and ISI journals and serves as an Editorial Board Member and Associate Editor for *Minerals*, the *Journal of Petroleum Science and Technology*, and *Frontiers in Earth Science*. His teaching disciplines include carbonate reservoir characterization, sedimentary petrology and geochemistry, reservoir modeling, flow in porous media, seismic stratigraphy, and chemo-stratigraphy. His interests include the geology of carbonate hydrocarbon reservoirs, sedimentary geochemistry, and paleoclimate studies.

Vahid Tavakoli

Vahid Tavakoli, Ph.D., is an Associate Professor at the University of Tehran, Iran. He obtained his bachelor's degree in geology from the University of Tehran and completed his master's at the same institution. In 2011, he earned a Ph.D. in petroleum geology from the University of Tehran. Since 2003, he has been actively engaged in the field of reservoir characterization of hydrocarbon formations utilizing core and wireline log data. Vahid has supervised over fifty graduate students. In recognition of his contributions to the field, Vahid was honored with the Young Geologist Award by the Iranian Academy of Sciences in 2018. Additionally, in 2023, he was selected as the top geology researcher of the country by the Geological Society of Iran. Vahid has authored three international books, edited two journal Special Issues, and authored approximately 150 scientific papers. Currently, he serves as a Reviewer and Editor for high-impact journals worldwide, focusing primarily on the geology of carbonate reservoirs.

Preface

The intricate interplay of geological processes has long captivated the minds of scientists and enthusiasts alike. Within the realm of sedimentary geology, the evolution of carbonate sequences stands as a testament to the dynamic forces that have shaped our planet over millennia. It is within this realm that the narrative of "Deposition, Diagenesis, and Geochemistry of Carbonate Sequences" unfolds.

This reprint endeavors to delve deep into the complexities of carbonate sedimentology, offering a comprehensive exploration of the deposition, diagenesis, and geochemistry inherent within these formations. From the meticulous processes of sediment deposition to the transformative effects of diagenetic alteration, each facet is meticulously dissected to unravel the geological story encoded within carbonate sequences.

In the pursuit of understanding the intricacies of carbonate systems, this text synthesizes a wealth of research findings, theoretical frameworks, and practical applications. Through an interdisciplinary approach, drawing upon insights from sedimentology, geochemistry, and stratigraphy, a holistic perspective emerges, illuminating the multifaceted nature of carbonate deposition and evolution.

Furthermore, this reprint serves as a guide for both seasoned geologists and aspiring scholars, offering a roadmap through the labyrinthine world of carbonate geology. From fundamental principles to advanced methodologies, each chapter is crafted to facilitate comprehension and foster a deeper appreciation for the complexities inherent within carbonate sequences.

It is our hope that "Deposition, Diagenesis, and Geochemistry of Carbonate Sequences" will not only serve as a valuable resource for academia but also inspire curiosity and spark further inquiry into the intricate tapestry of Earth's geological history. As we embark on this intellectual journey, may the pages that follow serve as a beacon, guiding readers through the mesmerizing landscape of carbonate sedimentology.

Hamzeh Mehrabi and Vahid Tavakoli
Editors

minerals MDPI

Editorial

Editorial for the Special Issue: Deposition, Diagenesis, and Geochemistry of Carbonate Sequences

Hamzeh Mehrabi * and Vahid Tavakoli

School of Geology, College of Science, University of Tehran, Tehran 14155-6455, Iran; vtavakoli@ut.ac.ir or vahid6105@yahoo.com
* Correspondence: mehrabi.hamze@ut.ac.ir or hmehrabi.ut@gmail.com

Citation: Mehrabi, H.; Tavakoli, V. Editorial for the Special Issue: Deposition, Diagenesis, and Geochemistry of Carbonate Sequences. *Minerals* **2024**, *14*, 269. https://doi.org/10.3390/min14030269

Received: 21 February 2024
Accepted: 29 February 2024
Published: 3 March 2024

The study of carbonate sequences encompasses a multifaceted exploration of depositional characteristics, diagenetic alterations, and geochemical variations that shape Earth's geological history. From ancient reef systems to modern marine environments, carbonate formations serve as archives of past climates, tectonic events, and biological evolution. Understanding the intricate interplay between depositional processes, diagenetic transformations, and geochemical signatures within carbonate sequences is essential for unraveling their geological significance and economic potential.

Depositional characteristics provide invaluable insights into the environmental conditions and sedimentary processes that govern the formation of carbonate rocks. Researchers can reconstruct past depositional environments by deciphering sedimentary facies, bedding patterns, and fossil assemblages and elucidate the factors driving carbonate deposition. Furthermore, diagenetic alterations play a pivotal role in modifying carbonate rocks' original mineralogy, texture, and porosity, thereby influencing their reservoir properties and hydrocarbon potential. Investigating diagenetic pathways, such as dolomitization, cementation, and dissolution, offers crucial information for reservoir characterization and exploration strategies.

Moreover, geochemical variations within carbonate sequences offer a window into past environmental changes, fluid–rock interactions, and diagenetic processes. Isotopic compositions, elemental concentrations, and trace metal distributions provide valuable proxies for reconstructing paleoenvironments, tracing fluid migration pathways, and quantifying diagenetic reactions. By integrating geochemical data with petrological and sedimentological observations, researchers can unravel the complex history of carbonate sequences and discern the driving forces behind their evolution.

In this special issue of the *Minerals* journal, we present a collection of articles that delve into the deposition, diagenesis, and geochemistry of carbonate sequences worldwide. These contributions offer diverse perspectives and cutting-edge methodologies for unraveling the complexities of carbonate geology. From the early dolomitization mechanisms in the Tarim Basin to the reservoir characterization of the Sarvak Formation in the Zagros Basin, each article contributes to our understanding of carbonate systems and highlights their geological significance.

Through the exploration of depositional characteristics, diagenetic alterations, and geochemical variations within carbonate sequences, we aim to advance our knowledge of Earth's geological history and inform sustainable resource management practices. We invite readers to delve into the wealth of insights presented in this Special Issue and join us in the ongoing exploration of carbonate geology.

Our first article by Pan Tang et al. offers an in-depth exploration of early dolomitization mechanisms, employing a multi-proxy approach involving petrography, rare earth elements, and clumped isotope analysis. This study provides valuable insights into the complex pathways of dolomite formation, shedding light on the diagenetic history of carbonate reservoirs.

Moving forward, Hamzeh Mehrabi and colleagues present a comprehensive analysis of pore-type characterization and reservoir zonation within the Sarvak Formation, contributing to enhanced reservoir management strategies in the Abadan Plain. Saif Ur Rehman et al. further elucidate the diagenetic evolution of Upper Cretaceous carbonates in the Attock Hazara Fold and Thrust Belt of Pakistan, unraveling the intricate interplay of diagenetic processes in sedimentary basins.

The diverse topics covered in this issue extend to the petrological, geochemical, and chronological characteristics of dolomites in the Permian Maokou Formation of the Central Sichuan Basin, as investigated by Xuejing Bai and colleagues. Additionally, Alireza Vaezi and Razyeh Lak present an ecological risk assessment of sedimentary deposits in the Northeast Persian Gulf, underscoring the importance of geochemical variability in marine environments.

Furthermore, studies exploring the facies, depositional environment, and reservoir quality of Early Cambrian carbonates in the Tarim Basin (Yongjin Zhu et al.), the genesis of dolomite reservoirs in the Ediacaran Chigbrak Formation of the Tarim Basin (Jianfeng Zheng et al.), and the epigenetic changes of Famennian–Tournaisian carbonate rocks in the Northern Marginal Shear Zone of the Caspian Syneclise (Valeriy Korobkin et al.) offer valuable contributions to our understanding of carbonate systems worldwide.

Lastly, Zhenwu Liu et al. provide insights into early diagenesis in lacustrine ostracods from the Songliao Basin, highlighting the significance of paleoenvironmental reconstructions in deciphering geological histories. Meanwhile, Hamzeh Mehrabi offers a detailed examination of deposition, diagenesis, and geochemistry within Upper Cretaceous carbonates of the Zagros Basin and the Persian Gulf, underscoring regional variations in carbonate evolution.

In conclusion, this Special Issue has comprehensively explored deposition, diagenesis, and geochemistry within carbonate sequences, spanning diverse geological settings and research methodologies. Through the collective efforts of esteemed researchers, we have gained valuable insights into the complex interplay of sedimentary processes, diagenetic transformations, and geochemical signatures shaping carbonate geology.

From the early dolomitization mechanisms of the Upper Ediacaran Qigebrak Formation in the Tarim Basin to the reservoir characterization of the Sarvak Formation in the Zagros Basin, each article has contributed to our understanding of carbonate systems worldwide. By elucidating the depositional characteristics, diagenetic alterations, and geochemical variations within carbonate sequences, these studies have advanced our knowledge of Earth's geological history and informed sustainable resource management practices.

As we reflect on the wealth of insights presented in this Special Issue, it is evident that the study of carbonate sequences remains a dynamic and multifaceted field ripe for further exploration. By continuing to unravel the complexities of carbonate geology and addressing key knowledge gaps, we can better understand past environmental changes, decipher geological histories, and inform future research endeavors.

Acknowledgments: We extend our sincere gratitude to the authors, reviewers, and editorial team for their contributions to this Special Issue. It is our hope that the articles presented herein will inspire further collaboration and innovation in the field of carbonate geology, paving the way for continued exploration and discovery.

Conflicts of Interest: The author declares no conflict of interest.

List of Contributions

1. Early Dolomitization Mechanism of the Upper Ediacaran Qigebrak Formation, Northwestern Tarim Basin: Evidence from Petrography, Rare Earth Elements, and Clumped Isotope. https://doi.org/10.3390/min14010035. Authors: Pan Tang, Daizhao Chen, Shuangjian Li, Yuanzheng Wang, Moataz El-Shafeiy, and Bo Yang;

2. Pore-Type Characterization and Reservoir Zonation of the Sarvak Formation in the Abadan Plain, Zagros Basin, Iran. https://doi.org/10.3390/min13121464. Authors: Hamzeh Mehrabi, Fatemeh Karami, Nafiseh Fakhar-Shahreza, and Javad Honarmand;
3. Diagenetic Evolution of Upper Cretaceous Kawagarh Carbonates from Attock Hazara Fold and Thrust Belt, Pakistan. https://doi.org/10.3390/min13111438. Authors: Saif Ur Rehman, Muhammad Jawad Munawar, Mumtaz Muhammad Shah, Naveed Ahsan, Muhammad Kashif, Hammad Tariq Janjuhah, Vasiliki Lianou, and George Kontakiotis;
4. Petrological, Geochemical and Chronological Characteristics of Dolomites in the Permian Maokou Formation and Constraints to the Reservoir Genesis, Central Sichuan Basin, China. https://doi.org/10.3390/min13101336. Authors: Xuejing Bai, Jianfeng Zheng, Kun Dai, Shuxin Hong, Junmao Duan, and Yunmiao Liu;
5. Sediment Texture, Geochemical Variation, and Ecological Risk Assessment of Major Elements and Trace Metals in the Sediments of the Northeast Persian Gulf. https://doi.org/10.3390/min13070850. Authors: Alireza Vaezi and Razyeh Lak;
6. Facies, Depositional Environment and Reservoir Quality of an Early Cambrian Carbonate Ramp in the Tarim Basin, NW China. https://doi.org/10.3390/min13060791. Authors: Yongjin Zhu, Jianfeng Zheng, Jiankun Zhang, Xinsheng Luo, Guang Yu, Jun Li, Fangjie Hu, and Guo Yang;
7. Genesis of Dolomite Reservoir in Ediacaran Chigbrak Formation of Tarim Basin, NW China: Evidence from U–Pb Dating, Isotope and Element Geochemistry. https://doi.org/10.3390/min13060725. Authors: Jianfeng Zheng, Hui Wang, Anjiang Shen, Xianying Luo, Zhao Cheng, and Kun Dai;
8. Results of the Study of Epigenetic Changes of Famennian–Tournaisian Carbonate Rocks of the Northern Marginal Shear Zone of the Caspian Syneclise (Kazakhstan). https://doi.org/10.3390/min13020249. Authors: Valeriy Korobkin, Akhan Chaklikov, Zhamal Tulemissova, Iskander Samatov, and Yelena Dobrovolskaya;
9. Early Diagenesis in the Lacustrine Ostracods from the Songliao Basin 91.35 Million Years Ago and Its Geological Implications. https://doi.org/10.3390/min13010005. Authors: Zhenwu Liu, Yuke Liu, Xuejia Du, Dan Lyu, Huaichun Wu, and Huajian Wang;
10. Deposition, Diagenesis, and Geochemistry of Upper Cretaceous Carbonates (Sarvak Formation) in the Zagros Basin and the Persian Gulf, Iran. https://doi.org/10.3390/min13081078. Author: Hamzeh Mehrabi.

minerals

MDPI

Article

Early Dolomitization Mechanism of the Upper Ediacaran Qigebrak Formation, Northwestern Tarim Basin: Evidence from Petrography, Rare Earth Elements, and Clumped Isotope

Pan Tang [1], Daizhao Chen [2,3,*], Shuangjian Li [4], Yuanzheng Wang [4], Moataz El-Shafeiy [2] and Bo Yang [2]

1 Research Institute of Mud Logging Technology and Engineering, Yangtze University, Jingzhou 434023, China; tangpan2013@126.com
2 Institute of Geology and Geophysics, Chinese Academy of Sciences, Beijing 100029, China; motazadel80@hotmail.com (M.E.-S.); hyangbo21@163.com (B.Y.)
3 University of Chinese Academy of Sciences, Beijing 100049, China
4 Petroleum Exploration and Production Research Institute, SINOPEC, Beijing 100083, China; lishuangjian.syky@sinopec.com (S.L.); wyz2023.syky@sinopec.com (Y.W.)
* Correspondence: dzh-chen@mail.iggcas.ac.cn

Citation: Tang, P.; Chen, D.; Li, S.; Wang, Y.; El-Shafeiy, M.; Yang, B. Early Dolomitization Mechanism of the Upper Ediacaran Qigebrak Formation, Northwestern Tarim Basin: Evidence from Petrography, Rare Earth Elements, and Clumped Isotope. *Minerals* **2024**, *14*, 35. https://doi.org/10.3390/min14010035

Academic Editors: Hamzeh Mehrabi and Vahid Tavakoli

Received: 8 November 2023
Revised: 13 December 2023
Accepted: 26 December 2023
Published: 28 December 2023

Abstract: The deeply buried Upper Ediacaran Qigebrak Formation dolostones in the Tarim Basin are promising future hydrocarbon exploration targets in China. However, the origin of these pervasive matrix dolomites is not well understood, which hampers further hydrocarbon exploration. In this study, petrographic, isotopic (C, O, and Sr), rare earth element (REE), and clumped isotope analyses were performed to unravel the mechanisms of early dolomitization. Petrographic investigations indicate that the Qigebrak Formation carbonates were completely replaced by three distinct types of dolomites: (1) dolomicrite (MD-1), (2) fabric-preserving dolomite (MD-2), and (3) fabric-destructive dolomite (MD-3). Despite different crystal textures, these three dolomite types have a narrow range of $\delta^{13}C$ and $^{87}Sr/^{86}Sr$ values similar to those of coeval seawater. Furthermore, their seawater-normalized REE compositions display a seawater-like REE pattern with positive Ce anomalies. These findings suggest that the dolomitization fluids were seawater derived. From the clumped isotope temperature ($T_{\Delta 47} \approx 60$ °C) and the $\delta^{18}O$ water values of the dolomitization fluids, it can be inferred that the main mechanism for the formation of matrix dolomites was seepage-reflux dolomitization by mesosaline to penesaline seawater in the evaporative environment. MD-1 and MD-2 precipitated from mesosaline to penesaline seawater in slightly evaporated settings. MD-3 was likely formed via recrystallization of MD-1 and/or MD-2 at a greater depth. This study provides an insight into early dolomitization processes related to mesosaline to penesaline seawater, which may make the origins of dolomite reservoirs with similar geological backgrounds better understood.

Keywords: Upper Ediacaran; Qigebrak Formation; seepage-reflux dolomitization; Tarim Basin

1. Introduction

Dolomitization has continued to attract interest because dolomites comprise a large number of hydrocarbon reservoirs worldwide [1]. This diagenetic process is an important controlling factor affecting the physical properties of carbonate rock reservoirs and thus affects the development of high-quality reservoirs [2]. Dolomitization models, such as sabkha dolomitization [3], seepage-reflux dolomitization [4], hydrothermal dolomitization [5], and seawater thermal convective dolomitization [6] have been reported to interpret the formation of large-scale dolomitization in various geological circumstances. Seepage-reflux dolomitization by hypersaline brines is the main mechanism proposed for the formation of shallow marine platform dolomites, and this process usually involves the precipitation of evaporite minerals [7,8]. In contrast, dolomitization via the reflux of mesosaline to penesaline brines is generally characterized by the absence of the precipitation of evaporite minerals [9,10]. This process can also cause large-scale dolomitization, which is conducive to the formation of dolomite reservoirs. For example, the development of large-scale high-quality reservoirs in the Cambrian Longwangmiao Formation and Triassic Feixianguan Formation in the Sichuan Basin was closely related to this type of dolomitization [11,12].

Precambrian dolomite reservoirs are currently promising exploration targets for oil and gas in China. Deeper exploration in the Tarim Basin led to a significant breakthrough in 2020 (well Luntan-1) with the discovery of dolomite reservoirs in the Upper Ediacaran Qigebrak Formation, indicating a broad exploration potential [13], and the pervasive early-formed dolomites are well developed [14]. However, there are still different views on the large-scale dolomitization mechanism of the Qigebrak Formation, and the focus of the controversy is whether the formation of these matrix dolomites occurred via primary precipitation [15,16] or secondary replacement [17]. Studying the origin of these early-formed dolomites can not only provide an insight into the early dolomitization process during fluid–rock interactions but can also lead to a better understanding of the genesis of related dolomite reservoirs and hence further improve exploration strategies.

Here, based on comprehensive analysis of the petrography, rare earth elements, carbon-oxygen-strontium isotopes, and clumped isotopes, detailed analyses have been performed on the matrix dolomites in the Qigebrak Formation of the NW Tarim Basin. This study aims to (1) document the petrographic and geochemical characteristics of the matrix dolomites; (2) trace the source of dolomitization fluids; and (3) interpret the mechanisms that were responsible for the early dolomitization. The results are expected to help guide oil and gas explorations of dolomite reservoirs in the Qigebrak Formation of the study area and provide an improved understanding of the formation of such reservoirs.

2. Geological Setting

The Tarim Basin is a superimposed hydrocarbon-bearing basin, which is made of the Paleozoic Craton basin and the Mesozoic-Cenozoic foreland basin, with a rhombic shape [18] (Figure 1a). From the Cryogenian to the Early Ediacaran, due to the continuous extension process, a continental rift basin developed in the interior and margin of the Tarim Basin, which resulted in clastic deposits thousands of meters thick interbedded with volcanic rocks [19]. In the late Ediacaran, the continuous extension of the rift led to subsidence of the ocean basin, forming a depression basin characterized by the occurrence of shallow marine carbonates [20]. The Neoproterozoic strata in the NW Tarim Basin comprise the Qiaoenbrak, Youermeinake, Sugaitebrak, and Qigebrak Formations in ascending order (Figure 1b). During the depositional period of the Qigebrak Formation, the study area was the inner ramp of a carbonate platform (Figure 1c) and characterized by relatively high salinity seawater and a dry climate [21]. The main lithofacies in the Qigebrak Formation, in order of decreasing volume percentage, include dolomicrobialites, dolowackstone/dolomudstone, dolograinstone, and minor sandstone (Figure 2). By the end of the Ediacaran period and owing to either regional tectonic movement or a large-scale sea level fall [22,23], the uppermost Qigebrak Formation was substantially eroded and karstified, forming an unconformity (Figure 2) associated with karst breccias.

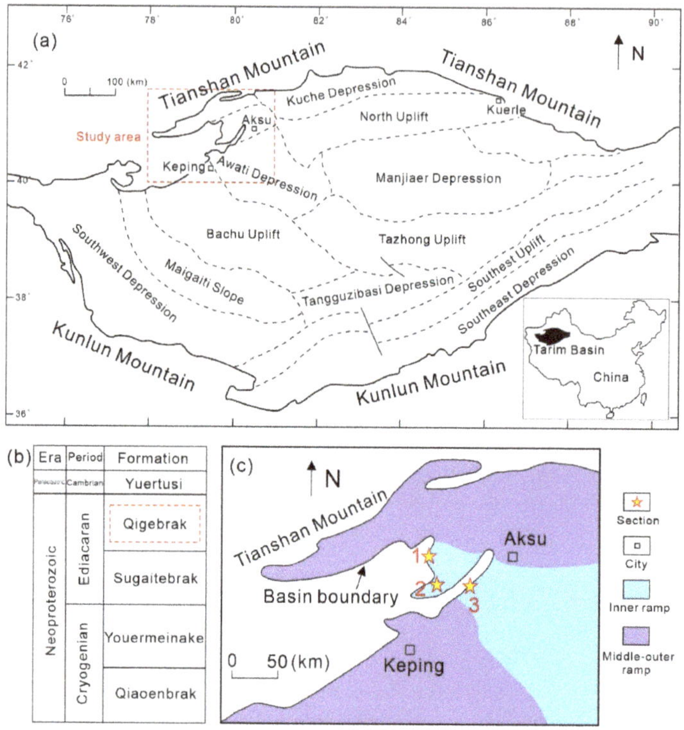

Figure 1. Geological setting map of the study area: (**a**) location and structural feathers of the Tarim Basin; (**b**) simplified stratigraphic column of the Neoproterozoic to Early Cambrian in the Aksu area; (**c**) paleogeography of the study area in the late Ediacaran. The number 1, 2 and 3 show the studied outcrop sections.

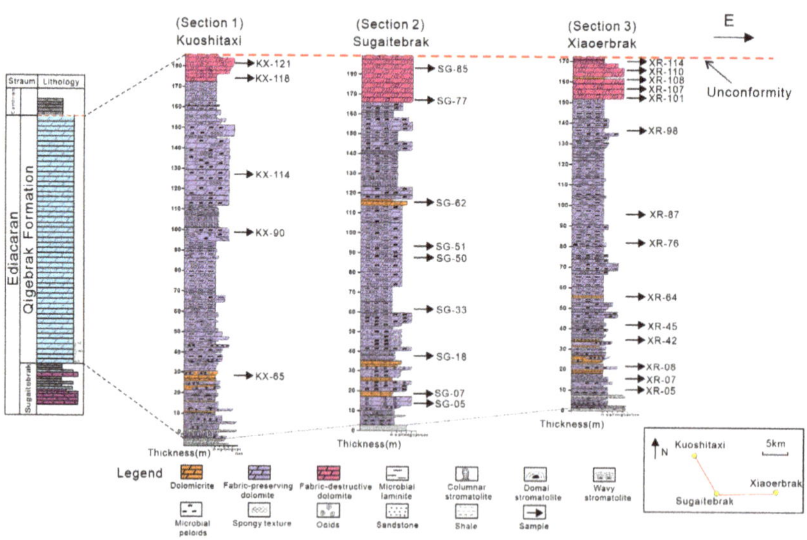

Figure 2. The stratigraphic column shows the distribution of different matrix dolomites in the Qigebrak Formation.

3. Samples and Methods

The samples analyzed in this study were collected from three sections in the Aksu area of the NW Tarim Basin (Figure 2). Thereafter, geochemical analyses ($\delta^{13}C$, $\delta^{18}O$, $^{87}Sr/^{86}Sr$, REE, and clumped isotopes) were conducted on part of representative samples.

Petrographical analyses were conducted at the Institute of Geology and Geophysics, Chinese Academy of Sciences (IGGCAS). A total of 45 thin sections were prepared and examined under both a transmitted light optical microscope and a cathodoluminescence (CL) apparatus with a Reliotron, model Relion III, and voltage as well as gun current ranged 5–8 kV and 300–400 μA, respectively. Additionally, 10 samples were chosen for the examination of mineral morphological characteristics and spacing using a Nova NanoSEM 450.

Carbon and oxygen isotopic measurements were conducted at IGGCAS. For the $\delta^{13}C$ and $\delta^{18}O$ analyses (n = 23), 10 mg of powdered samples were obtained using a micro-drilling apparatus on the surface of polished slabs as well as hand specimens. Dolomite powders were acidified by adding H_3PO_4 at 25 °C and left for 72 h [24]. The analysis of the product CO_2 was performed by the Finnigan MAT-252 mass spectrometer. Results were documented as a per-mil difference from the Vienna Peedee Belemnite (VPDB) international standard [24]. To monitor the precision of the measurement, the international standard NBS-19 was also utilized. The typical standard deviation yields 1σ values < 0.15‰ for $\delta^{13}C$ and <0.20‰ for $\delta^{18}O$.

$^{87}Sr/^{86}Sr$ measurements (n = 12) were performed at IGGCAS. A total of 70–80 mg of powdered samples was dissolved by adding 2.5 N HCl on a hotplate at 90 °C. A cation exchange resin AG50Wx12 of 200–400 mesh was used for the Sr separation from the sample matrix. The NBS-987 was used as a reference standard to correct $^{87}Sr/^{86}Sr$ ratios. The two-standard error for all $^{87}Sr/^{86}Sr$ measurements varied from 0.0000010 to 0.000013.

Furthermore, 20 samples were selected for the measurement of trace elements and REEs at the Beijing Research Institute of Uranium Geology. Detailed operating procedures to obtain solution for measurement were presented by Du's [25] research. The results are reported with a precision of ±4%. To calculate the Ce, Eu, and Pr anomalies, respectively, the equations listed below were used: $(Ce/Ce^*)_{SN} = 2 \times [Ce]_{SN}/([La]_{SN} + [Pr]_{SN})$ [26], $(Eu/Eu^*)_{SN} = [Eu]_{SN}/(2/3 \times [Sm]_{SN} + 1/3 \times [Tb]_{SN})$ [27], and $(Pr/Pr^*)_{SN} = 2 \times [Pr]_{SN}/([Ce]_{SN} + [Nd]_{SN})$ [26], SN represents seawater-normalized.

In order to accurately determine the temperature of the dolomitization fluid, clumped isotope tests were carried out on ten obtained samples. A 10–12 mg sample of dolomite powder was weighed out, an Imperial Batch Extraction (IBEX) gas automatic extraction and purification device was utilized, and the sample was reacted with 1.90 g/mL phosphoric acid to release the carbon dioxide gas. Using a MAT-253 stable isotope mass spectrometer, automatic online analysis of Δ47 was realized. For detailed instrument operation steps, please refer to Wang's [28] research. In this test, standard ETH1-ETH4 was used to perform nonlinear correction of the original data. NB-4 (marble) and P1 (coral) were selected as standard samples to correct the related errors in the sample preparation and testing process. The sample test results were converted to the Δ47 value of the absolute reference system (carbon dioxide equilibrated scale, CDES) [29]. The Δ47 values (CDES 25 °C) of NB-4 and P1 were 0.476 (±0.002)‰ and 0.702 (±0.002)‰, respectively. For the relationship between the formation temperature of the dolomite minerals and the Δ47 value, refer to the formula: $\Delta 47 = 0.0449\ (\pm 0.001) \times 10^6/T^2 + 0.167\ (\pm 0.01)$, where T is the temperature (K) [30].

4. Results

4.1. Dlomite Petrology

Following the description terminology of different dolomite textures [31,32], the matrix dolomites in the Qigebrak Formation can be divided into three types in the field: (1) dolomicrite (MD-1; Figure 3a); (2) fabric-preserving dolomite (MD-2; Figure 3b,c); and (3) fabric-destructive dolomite (MD-3; Figure 3d).

Figure 3. Outcrop features and occurrence of the matrix dolomites from the Qigebrak Formation in the studied area: (**a**) thin bedded dolomicrites, Xiaoerbrak section; (**b**) fabric-preserving dolomites, characterized by the occurrence of stromatolites (yellow arrows), Xiaoerbrak section; (**c**) fabric-preserving dolomites (below dotted line) developed vertically with dolomicrites (above dottedline), Xiaoerbrak section; (**d**) fabric-destructive dolomite, Sugaitebrak section.

MD-1 was very rarely observed in the studied sections. Volumetrically, it accounts for less than 5% of the dolostones in the Qigebrak Formation (Figure 2). Microscopically, MD-1 dolomite crystals are less than 5 μm in size and generally show planar-s to nonplanar-a textures (Figure 4a,b). Under cathodoluminescence, this type of dolomite shows nonluminescence (Figure 4c).

MD-2 is the predominant dolomite type in volume (more than 80%) of the study area (Figure 2). In the thin sections, MD-2 dolomite crystals are 5–20 μm in size (Figure 4d). Under SEM, MD-2 includes euhedral to subhedral crystals that formed with relatively smooth but uncorroded surfaces (Figure 4e). Under CL, this type has a very dull red color (Figure 4f).

MD-3 constitutes approximately 15% of the total dolostones (Figure 2). This type is characterized by nonplanar-a dolomite crystals, and the crystal sizes vary between 50 μm and 200 μm (Figure 4g). This type is characterized by the obliteration of primary fabrics (Figure 4g), and some MD-3 dolomite crystals have undulatory extinction (Figure 4h). Under CL, this type shows dull red cathodoluminescence (Figure 4i).

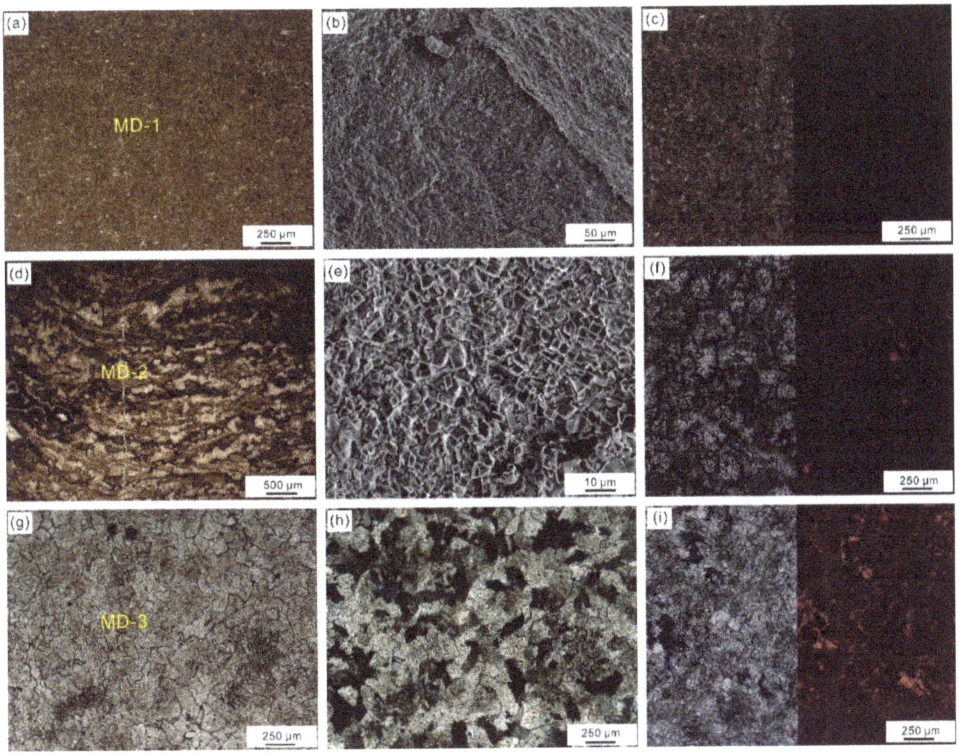

Figure 4. Thin section, CL, and SEM photomicrographs of matrix dolomites in the Qigebrak Formation of the NW Tarim Basin: (**a**) Dolomicrite (MD-1) with tight texture, in plane-polarized light (PPL), Xiaoerbrak section. (**b**) SEM image showing planar-s to nonplanar-a dolomite crystals in MD-1, Xiaoerbrak section. (**c**) MD-1 (PPL, left) shows nonluminescence (right), Xiaoerbrak section. (**d**) Very finely crystalline dolomite (MD-2), with well-preserved stromatolitic laminae, PPL, Xiaoerbrak section. (**e**) SEM image showing planar-e to planar-s dolomite crystals in MD-2, Xiaoerbrak section. (**f**) MD-2 with well-preserved spongy fabrics (PPL, left) shows very dull red luminescence (right), Xiaoerbrak section. (**g**) Fine to medium crystalline dolomite (MD-3) in which the primary depositional fabrics have been obliterated, PPL, Sugaitebrak section. (**h**) Thin section photomicrograph under cross-polarized light for (**g**); MD-3 shows nonplanar-a dolomite crystals with undulating extinction. (**i**) MD-3 with intercrystalline pores (PPL, left) shows dull red luminescence (right), Kuoshitaxi section.

4.2. Geochemical Characteristics

4.2.1. Carbon and Oxygen Isotopes

The carbon and oxygen isotope analysis results for the different carbonate minerals are presented in Figure 5 and Table 1. The $\delta^{13}C$ values of MD-1 range from 2.79‰ to 5.16‰ VPDB (average of 3.63‰; n = 4), and the $\delta^{18}O$ values range from −3.89‰ to −0.04‰ VPDB (average of −2.36‰). MD-2 has $\delta^{13}C$ values of 1.79‰ to 4.79‰ VPDB (average of 2.89‰; n = 13) and $\delta^{18}O$ values of −5.72‰ to −0.72‰ VPDB (average of −3.07‰). MD-3 has lighter $\delta^{13}C$ and $\delta^{18}O$ values. Its $\delta^{13}C$ values are 1.79‰ to 3.34‰ VPDB (average of 2.37‰; n = 6), and its $\delta^{18}O$ values are −6.78‰ to −2.35‰ VPDB (average of −3.07‰). In sum, the $\delta^{13}C$ values of MD-1, MD-2, and MD-3 are similar to those of Ediaracan seawater range [33].

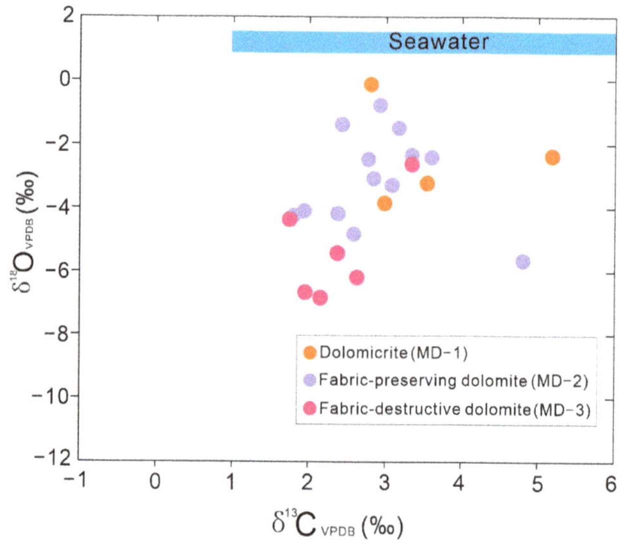

Figure 5. Crossplot of δ^{13}C and δ^{18}O values for different types of matrix dolomites in the Qigrbrak Formation.

Table 1. Carbon, oxygen, and strontium isotopic data of matrix dolomites in the Qigebrak Formation, NW Tarim Basin.

Sample	Dolomite Type	$\delta^{13}C_{VPDB}$	$\delta^{18}O_{VPDB}$	$^{87}Sr/^{86}Sr$	Error (2s)
XR-42	MD-1	5.16	−2.2	0.708835	0.000013
XR-64	MD-1	2.98	−3.89	0.708499	0.000012
SG-62	MD-1	3.57	−3.31	0.708612	0.000011
KX-65	MD-1	2.79	−0.04	-	-
XR-05	MD-2	3.34	−2.35	-	-
XR-07	MD-2	1.88	−4.05	0.708745	0.000012
XR-08	MD-2	1.79	−4.12	-	-
XR-45	MD-2	2.77	−2.51	-	-
XR-87	MD-2	2.85	−3.07	0.708477	0.000013
XR-98	MD-2	3.08	−3.28	0.708480	0.000013
KX-90	MD-2	2.36	−4.14	-	-
KX-114	MD-2	2.93	−0.72	0.708559	0.000011
KX-99	MD-2	3.59	−2.36	-	-
SG-18	MD-2	2.57	−4.79	-	-
SG-33	MD-2	3.18	−1.48	-	-
SG-50	MD-2	4.79	−5.72	0.708929	0.000011
SG-05	MD-2	2.42	−1.38	-	-
XR-114	MD-2	1.79	−4.12	-	-
XR-110	MD-3	2.09	−6.78	0.708828	0.000012
XR-107	MD-3	2.63	−6.05	-	-
XR-101	MD-3	3.34	−2.35	0.708466	0.000014
SG-85	MD-3	1.93	−6.54	0.708675	0.000012
KX-121	MD-3	2.41	−5.49	0.708819	0.000014

4.2.2. Strontium Isotopes

The three types of matrix dolomite have relatively narrow ranges of $^{87}Sr/^{86}Sr$ values (Figure 6, Table 1). The $^{87}Sr/^{86}Sr$ values of MD-1 are 0.708499–0.708835 (n = 3). The $^{87}Sr/^{86}Sr$ values of MD-2 are 0.708477–0.708929 (n = 5). The $^{87}Sr/^{86}Sr$ values of MD-3 are 0.708466–0.708828 (n = 4). Overall, there is no significant difference in the strontium isotope compositions of the matrix dolomites, and these $^{87}Sr/^{86}Sr$ ratios largely overlap with those of the late Ediacaran seawater range [33].

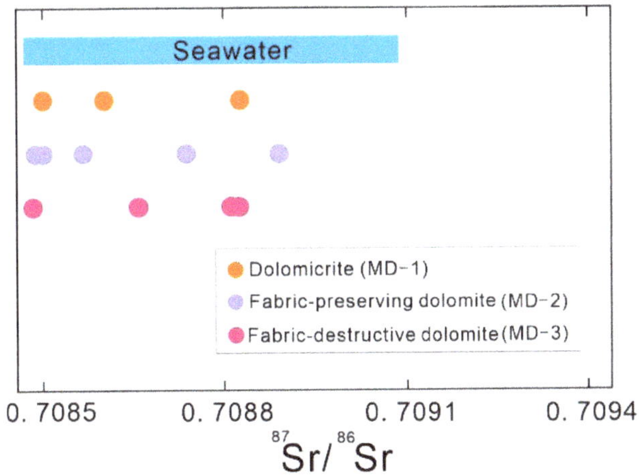

Figure 6. The $^{87}Sr/^{86}Sr$ ratios of studied matrix dolomites.

4.2.3. REE Geochemistry

Seawater-normalized REE data for the studied samples are shown in Figure 7, and Tables 2 and 3 display REE concentrations in the Qigebrak Formation matrix dolomites and REE concentrations normalized to seawater, respectively. Table 4 exhibits the concentrations of several trace elements for each dolomite type.

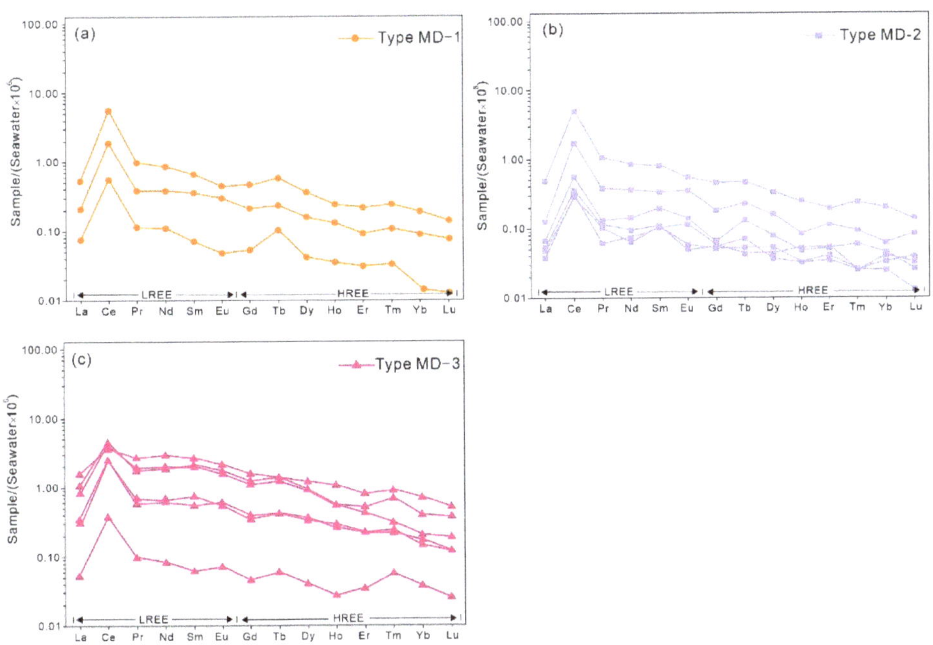

Figure 7. Seawater-normalized REE patterns of matrix dolomites from the Qigebrak Formation in the study area. (**a**) REE pattern of MD-1; (**b**) REE pattern of MD-2; (**c**) REE pattern of MD-3.

Table 2. REE concentrations of matrix dolomites from the Qigebrak Formation, NW Tarim Basin.

Sample	Dolomite Type	La	Ce	Pr	Nd	Sm	Eu	Gd	Tb	Dy	Ho	Er	Tm	Yb	Lu	ΣLREE	ΣHREE	ΣREE	ΣLREE /ΣHREE
XR-42	MD-1	1.940	4.460	0.506	2.040	0.307	0.056	0.334	0.067	0.312	0.061	0.181	0.029	0.155	0.022	9.309	1.161	10.47	8.018
XR-64	MD-1	0.273	0.447	0.059	0.260	0.033	0.006	0.038	0.012	0.036	0.009	0.026	0.004	0.008	0.002	1.078	0.135	1.213	7.985
SG-07	MD-1	3.330	7.570	0.966	3.680	0.682	0.132	0.581	0.105	0.597	0.118	0.340	0.059	0.271	0.045	16.360	2.116	18.476	7.732
SG-62	MD-1	0.758	1.510	0.198	0.904	0.165	0.037	0.151	0.027	0.137	0.033	0.077	0.013	0.074	0.012	3.572	0.524	4.096	6.817
KX-65	MD-1	5.640	14.50	1.500	5.950	1.210	0.266	1.070	0.187	1.100	0.245	0.680	0.121	0.577	0.113	29.066	4.093	33.159	7.101
XR-98	MD-2	0.451	1.390	0.195	0.841	0.152	0.043	0.127	0.026	0.137	0.021	0.093	0.011	0.050	0.013	3.072	0.478	3.55	6.427
XR-45	MD-2	1.210	3.700	0.462	1.620	0.315	0.064	0.294	0.051	0.263	0.062	0.154	0.028	0.161	0.022	7.371	1.035	8.406	7.122
XR-87	MD-2	0.189	0.276	0.057	0.221	0.049	0.006	0.038	0.008	0.031	0.008	0.028	0.003	0.033	0.005	0.798	0.154	0.952	5.182
SG-18	MD-2	1.720	3.880	0.544	2.030	0.369	0.065	0.316	0.052	0.282	0.060	0.157	0.027	0.148	0.024	8.608	1.066	9.674	8.075
SG-33	MD-2	2.310	5.010	0.529	2.150	0.438	0.088	0.456	0.075	0.394	0.073	0.213	0.029	0.171	0.030	10.525	1.441	11.966	7.304
SG-51	MD-2	0.171	0.224	0.052	0.147	0.049	0.007	0.036	0.006	0.045	0.008	0.033	0.003	0.020	0.002	0.65	0.153	0.803	4.248
KX-90	MD-2	0.232	0.436	0.066	0.330	0.088	0.017	0.045	0.015	0.067	0.012	0.043	0.007	0.036	0.004	1.169	0.229	1.398	5.105
KX-114	MD-2	0.134	0.249	0.031	0.171	0.046	0.014	0.040	0.005	0.037	0.013	0.044	0.003	0.027	0.006	0.645	0.175	0.82	3.686
XR-114	MD-3	12.80	7.550	3.410	16.20	3.090	0.562	2.150	0.364	1.950	0.357	1.010	0.155	0.847	0.125	43.612	6.958	50.57	6.268
XR-110	MD-3	3.000	3.570	0.936	4.520	0.955	0.197	0.777	0.144	0.790	0.140	0.429	0.084	0.328	0.059	13.178	2.751	15.929	4.790
XR-107	MD-3	3.830	3.550	0.891	4.380	0.977	0.214	0.875	0.161	1.050	0.272	0.679	0.109	0.586	0.082	13.842	3.814	17.656	3.629
XR-101	MD-3	0.192	0.306	0.051	0.200	0.029	0.009	0.033	0.007	0.035	0.007	0.029	0.007	0.031	0.004	0.787	0.153	0.94	5.144
SG-85	MD-3	5.600	2.890	1.370	6.910	1.230	0.267	1.130	0.161	0.808	0.142	0.360	0.038	0.169	0.030	18.267	2.838	21.105	6.437
KX-121	MD-3	1.170	1.980	0.344	1.530	0.345	0.068	0.252	0.049	0.293	0.074	0.189	0.027	0.143	0.019	5.437	1.046	6.483	5.198
KX-118	MD-3	1.110	1.960	0.300	1.450	0.253	0.075	0.279	0.048	0.318	0.067	0.185	0.029	0.119	0.019	5.148	1.064	6.212	4.838

(ppm).

Table 3. Seawater-normalized REEs of matrix dolomites from the Qigebrak Formation, NW Tarim Basin.

Sample	Dolomite Type	La_{SN}	Ce_{SN}	Pr_{SN}	Nd_{SN}	Sm_{SN}	Eu_{SN}	Gd_{SN}	Tb_{SN}	Dy_{SN}	Ho_{SN}	Er_{SN}	Tm_{SN}	Yb_{SN}	Lu_{SN}	$(Eu/Eu^*)_{SN}$	$(Ce/Ce^*)_{SN}$
XR-42	MD-1	0.543	5.624	0.992	0.873	0.672	0.455	0.473	0.583	0.363	0.242	0.219	0.242	0.189	0.140	0.709	5.062
XR-64	MD-1	0.076	0.564	0.116	0.111	0.072	0.049	0.054	0.104	0.042	0.036	0.031	0.033	0.010	0.013	0.588	4.695
SG-07	MD-1	0.933	9.546	1.894	1.575	1.492	1.073	0.823	0.913	0.694	0.468	0.411	0.492	0.331	0.287	0.826	4.314
SG-62	MD-1	0.212	1.904	0.388	0.387	0.361	0.301	0.214	0.235	0.159	0.131	0.093	0.108	0.090	0.076	0.943	4.889
KX-65	MD-1	1.580	18.285	2.941	2.547	2.648	2.163	1.516	1.626	1.279	0.972	0.821	1.008	0.705	0.720	0.937	5.482
XR-98	MD-2	0.126	1.753	0.382	0.360	0.333	0.350	0.180	0.226	0.159	0.083	0.112	0.092	0.061	0.083	1.177	4.331
XR-45	MD-2	0.339	4.666	0.906	0.693	0.689	0.520	0.416	0.443	0.306	0.246	0.186	0.233	0.197	0.140	0.857	4.172
XR-87	MD-2	0.053	0.348	0.112	0.095	0.107	0.049	0.054	0.070	0.036	0.032	0.034	0.025	0.040	0.032	0.515	2.700
SG-18	MD-2	0.482	4.893	1.067	0.869	0.807	0.528	0.448	0.452	0.328	0.238	0.190	0.225	0.181	0.153	0.767	3.870
SG-33	MD-2	0.647	6.318	1.037	0.920	0.958	0.715	0.646	0.652	0.458	0.290	0.257	0.242	0.209	0.191	0.835	5.474
SG-51	MD-2	0.048	0.282	0.102	0.063	0.107	0.057	0.051	0.052	0.052	0.032	0.040	0.025	0.024	0.013	0.640	2.003
KX-90	MD-2	0.065	0.550	0.129	0.141	0.193	0.138	0.064	0.130	0.078	0.048	0.052	0.058	0.044	0.025	0.804	4.677
KX-114	MD-2	0.038	0.314	0.061	0.073	0.101	0.114	0.057	0.043	0.043	0.052	0.053	0.025	0.033	0.038	1.395	6.492
XR-114	MD-3	3.585	9.521	6.686	6.935	6.761	4.569	3.045	3.165	2.267	1.417	1.220	1.292	1.035	0.796	0.821	1.479
XR-110	MD-3	0.840	4.502	1.835	1.935	2.090	1.602	1.101	1.252	0.919	0.556	0.518	0.700	0.401	0.376	0.885	2.594
XR-107	MD-3	1.073	4.477	1.747	1.875	2.138	1.740	1.239	1.400	1.221	1.079	0.820	0.908	0.716	0.522	0.920	2.765
XR-101	MD-3	0.054	0.386	0.100	0.086	0.063	0.073	0.047	0.061	0.041	0.028	0.035	0.058	0.038	0.025	1.169	3.374
SG-85	MD-3	1.569	3.644	2.686	2.958	2.691	2.171	1.601	1.400	0.940	0.563	0.435	0.317	0.207	0.191	0.960	1.509
KX-121	MD-3	0.328	2.497	0.675	0.655	0.755	0.553	0.357	0.426	0.341	0.294	0.228	0.225	0.175	0.121	0.857	3.597
KX-118	MD-3	0.311	2.472	0.588	0.621	0.554	0.610	0.395	0.417	0.370	0.266	0.223	0.242	0.145	0.121	1.200	4.447

$La_{SN} = La/3.57$; $Ce_{SN} = Ce/0.793$; $Pr_{SN} = Pr/0.510$; $Nd_{SN} = Nd/2.336$; $Sm_{SN} = Sm/0.457$; $Eu_{SN} = Eu/0.123$; $Gd_{SN} = Gd/0.706$; $Tb_{SN} = Tb/0.115$; $Dy_{SN} = Dy/0.860$; $Ho_{SN} = Ho/0.252$; $Er_{SN} = Er/0.828$; $Tm_{SN} = Tm/0.120$; $Yb_{SN} = Yb/0.818$; $Lu_{SN} = Lu/0.157$.

Table 4. Concentrations of some trace elements of matrix dolomites from the Qigebrak Formation, NW Tarim Basin.

Sample	Dolomite Type	Zr	Th	Ni	Cu	Pb	Sc	Y	Ho	Y/Ho
XR-42	MD-1	2.75	0.258	18.8	1.26	1.16	1.15	1.96	0.061	32.131
XR-64	MD-1	0.444	0.018	17.1	0.573	0.334	0.324	0.349	0.009	38.778
SG-62	MD-1	2.46	0.101	16.4	1.03	0.278	0.529	1.28	0.033	38.788
XR-98	MD-2	0.747	0.064	18.2	0.634	0.824	0.569	0.779	0.021	37.095
XR-87	MD-2	1.61	0.005	18.1	0.445	0.257	0.301	0.513	0.008	64.125
SG-18	MD-2	2.9	0.201	15.5	1.36	0.964	0.732	1.86	0.06	31.000
SG-33	MD-2	3.88	0.33	20.5	5.64	8.43	0.86	2.29	0.073	31.370
SG-51	MD-2	1.72	0.013	16.1	0.65	0.342	0.336	0.429	0.008	53.625
KX-90	MD-2	0.431	0.039	16.1	0.942	0.563	0.531	0.729	0.012	60.750
KX-114	MD-2	2.18	0.012	17.1	0.616	0.823	0.392	0.717	0.013	55.154
XR-110	MD-3	3.5	0.18	17.3	1.23	0.571	0.715	8.54	0.14	61.000
XR-107	MD-3	3.47	0.296	19	0.858	0.493	0.884	13.1	0.272	48.162
XR-101	MD-3	1.68	0.017	18.7	0.74	0.562	0.463	0.468	0.007	66.857
SG-85	MD-3	3.74	0.091	16.1	3.09	0.438	0.396	7.16	0.142	50.423
KX-121	MD-3	0.992	0.024	16.1	0.695	0.666	0.29	3.95	0.074	53.378
KX-118	MD-3	1.93	0.046	14.7	0.72	0.62	0.329	3.6	0.067	53.731

(ppm).

The total REE (ΣREE) content of the MD-1 ranges from 1.213 ppm to 33.159 ppm, with an average of 13.483 ppm, which is similar to that of MD-3 (average ΣREE = 16.985 ppm) (Table 2). In contrast, MD-2 has the lowest ΣREE, with an average of 4.696 ppm (Table 2). In general, all the studied dolomite types display right-leaning REE patterns, with light REEs (LREEs) enriched and heavy REEs (HREEs) depleted (Figure 7). In addition, the average LREE/HREE ratios of the three types of dolomite range from 5.186 to 7.531, which also indicates LREE enrichment and HREE depletion. The average $(La/Sm)_{SN}$ values are low, ranging from 0.474 to 0.736, displaying the low differentiation among LREEs (Table 3). The average $(Gd/Yb)_{SN}$ ranges from 2.151 to 3.022, showing a high degree of differentiation among HREEs (Table 3).

All of the dolomite samples exhibit positive Ce anomalies. The average $(Ce/Ce^*)_{SN}$ values of MD-1, MD-2, and MD-3 are 4.882, 4.221, and 3.048, respectively (Table 3). In contrast, the Eu anomaly values of the three types are less than 1, with their average values ranging from 0.747 to 0.999 (Table 3).

4.2.4. Clumped Isotopes

Compared with fluid inclusion thermometry, clumped isotope thermometry could more accurately determine the temperatures of carbonate minerals in the early diagenetic stages. The clumped isotope results for the three types of matrix dolomite are presented in Figure 8 and Table 5. The Δ47 values of MD-1 and MD-2 range from 0.562‰ to 0.594‰, and the calculated clumped temperatures ($T_{\Delta47}$) range from 51.06 °C to 63.86 °C, with an average of 59.03 °C (n = 4) (Figure 8). The $\delta^{18}O_{water}$ values of the corresponding dolomitization fluids are from 3.09‰ to 4.90‰ VSMOW. The clumped temperatures ($T_{\Delta47}$) of MD-1 and MD-2 are very close to those of the fibrous dolomites in the Qigebrak Formation (60.0–63.0 °C) [34]. In contrast, MD-3 display higher clumped temperatures. Its Δ47 values range from 0.533‰ to 0.566‰, and its $T_{\Delta47}$ values range from 62.28 °C to 75.81 °C, with an average value of 70.47 °C (n = 6). The calculated $\delta^{18}O_{water}$ of the diagenetic fluids range from 0.99‰ to 2.88‰ VSMOW (Figure 8).

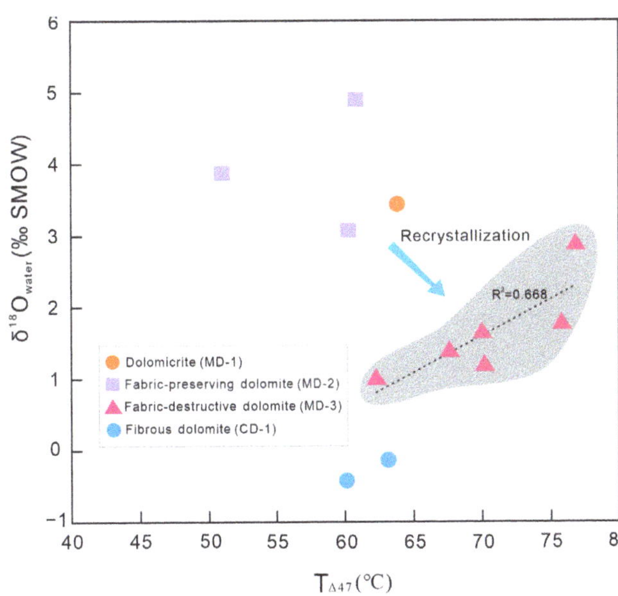

Figure 8. Relationship between clumped temperatures and $\delta^{18}O_{water}$ of the diagenetic fluids from matrix dolomites in the Qigebrak Formation.

Table 5. Clumped isotopes and related carbon/oxygen isotopes of Sinian depositional to early-diagenetic dolomites in the NW Tarim Basin.

Sample	Dolomite Type	$\delta^{13}C$ (VPDB)	$\delta^{18}O$ (VPDB)	$\delta^{18}O_{water}$ (VSMOW)	Δ47 (‰)	$T_{\Delta47}$ (°C)
SG-62	MD-1	3.57	−3.31	3.44	0.562	63.86
XR-76	MD-2	4.34	−3.08	3.09	0.571	60.29
SG-51	MD-2	2.42	−1.38	4.90	0.569	60.90
KX-114	MD-2	2.93	−0.72	3.87	0.594	51.06
XR-110	MD-3	2.09	−6.78	1.76	0.536	75.81
XR-107	MD-3	2.63	−6.05	1.64	0.548	69.98
SG-85	MD-3	1.93	−6.54	1.18	0.548	70.24
KX-121	MD-3	2.41	−5.49	0.99	0.566	62.28
SG-77	MD-3	1.93	−5.94	1.39	0.554	67.63
XR-108	MD-3	2.41	−5.83	2.88	0.533	76.90
Q-58-1-2 *	Fibrous dolomite	2.16	−6.73	-0.14	-	63.00
Q-76-1 *	Fibrous dolomite	2.03	−6.57	-0.47	-	60.00

* Data are cited from [34].

5. Discussion

5.1. Dolomitization Fluid Properties

5.1.1. Feasibility of Seawater-Normalized REEs

The REE concentrations of dolomites are commonly normalized to North America Shale Composite (NASC), Post Archean Australian Shale (PAAS), and C1 chondrite [35,36]. However, these dolomites do not have an inherent genetic relationship with shales or chondrites. In recent years, studies have shown that in the absence of terrestrial contaminants, the REE composition of seawater has remained relatively stable throughout geological history [37]. Moreover, numerous studies show the REE compositions of dolomites formed in marine environments are usually closely related to fluids driven by seawater, exhibiting a strong affinity [8,10–12,14,25]. Therefore, it seems more reasonable to utilize seawater as the standard for normalization [38], and some researchers have chosen seawater for a standardized treatment to study the genesis of marine dolomite [39,40]. In this study, the average REE concentrations of Pacific surface water were used as the standard for normalization [41]. Due to the very low content of REEs in seawater, the REE concentrations in seawater were magnified 10^6 times (Figure 7).

5.1.2. Validity Assessment of REE Data

Terrigenous clastic materials commonly have high REE contents, and even a small proportion (1–2%) of terrigenous clastics (such as shale) can significantly affect the REE concentrations of marine carbonates [42]. To minimize the risk of contamination by terrigenous components, samples with shale intercalations were avoided during field sampling, and a millimeter microdrill was used to extract powder from fresh hand specimens.

Some trace metals are enriched in clay minerals (e.g., Zr and Th), oxides (e.g., Ni and Cu), and sulfides (e.g., Pb and Sc). Therefore, the cross-plot between Y/Ho (as the seawater signal) and the concentrations of the above elements can assess the degree of contamination [43,44]. Figure 9 implies that none of the aforementioned elements have an anti-correlation with Y/Ho, excluding the likelihood of contamination by silicates, oxides, and sulfides. This is further supported by the $^{87}Sr/^{86}Sr$ values of the above dolomite samples (MD-1, MD-2, and MD-3), which fall within the coeval seawater range (Figure 6), suggesting that the terrigenous components were not absorbed into the dolomitization fluids. Therefore, the above data suggest that the REE patterns of the Qigebrak dolomite samples do not reflect contamination by terrigenous components and could provide useful information for tracing the source of dolomitization fluids.

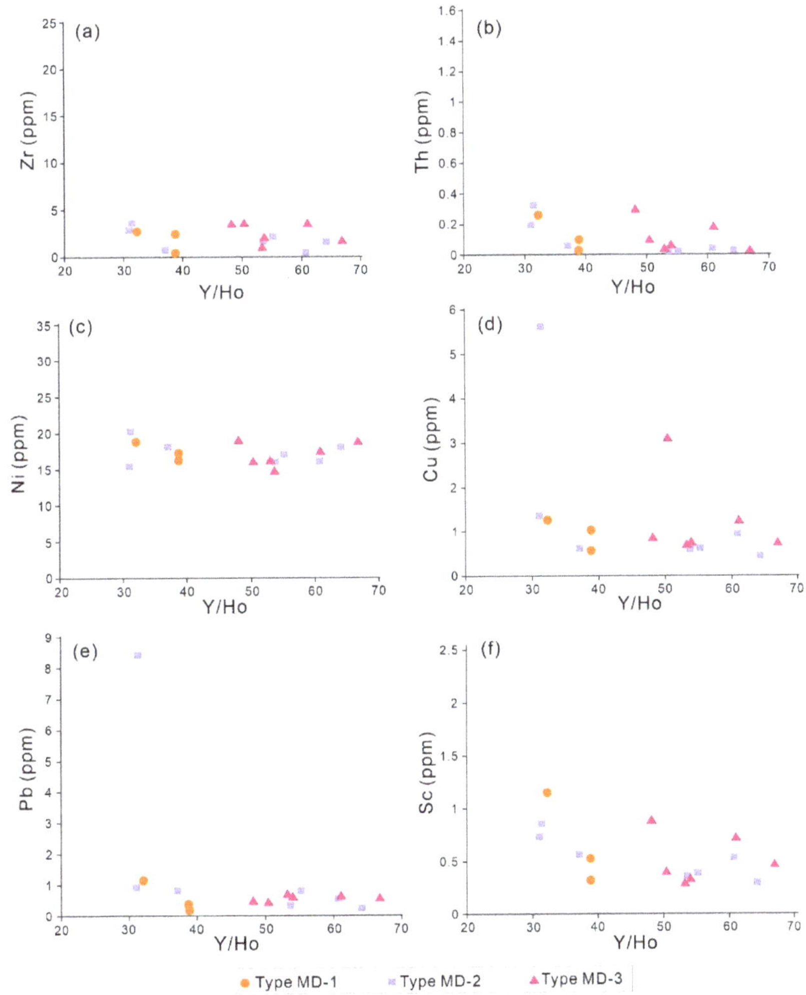

Figure 9. Binary plots showing relationships of Y/Ho values to contents of (**a**) Zr, (**b**) Th, (**c**) Ni, (**d**) Cu, (**e**) Pb, and (**f**) Sc in matrix dolomites from the Qigebrak Formation.

5.1.3. Ce Anomaly

Cerium is very sensitive to redox conditions and can be regarded as an indicator of marine paleoredox changes [45]. In oxidizing conditions, Ce^{3+} in seawater can be oxidized to Ce^{4+}, and the latter is less mobile and more easily absorbed into sediment particles [46], leading to Ce depletion of the seawater. Hence, when the REE compositions are normalized to PAAS, NASC, or C1 chondrite, seawater usually has a negative Ce anomaly [47]. However, the REE compositions of all dolomite samples in the present study demonstrate positive Ce anomalies after seawater normalization (Figure 7). The element Ce occurs as CeO_2 and is easily absorbed into carbonate particles during the sedimentary differentiation process; hence, Ce is more likely to be enriched in minerals than the two adjacent rare earth elements (i.e., Pr and La) [48]. Therefore, the fact that the samples exhibited positive Ce anomalies after seawater normalization, indicates that they reflect the seawater signal.

Unfortunately, the interpretation of Ce anomalies can also be complicated by the possible anomalous abundance of La in marine precipitates. Since the concentrations of La in seawater are relatively high, the Ce anomaly values may only reflect La anomalies rather than true Ce anomalies. In comparison, the chemical properties of Pr and Nd are stable, exhibiting no obvious anomalies in seawater, and the true Ce anomalies may lead to Pr anomalies. If the $(Pr/Pr^*)_{SN}$ value is greater than 1, it represents a negative Ce anomaly, whereas if the value is less than 1, it represents a positive Ce anomaly. Moreover, if $(Ce/Ce^*)_{SN} < 1$ and $(Pr/Pr^*)_{SN} \approx 1$, then this would represent a positive La anomaly rather than a Ce anomaly [49]. In this study, all studied samples display $(Ce/Ce^*)_{SN} > 1$ and $(Pr/Pr^*)_{SN} < 1$, thus representing true Ce anomalies. The MD-1, MD-2, and MD-3 samples all exhibited significant positive Ce anomalies, possibly indicating that the formation of these dolomites was seawater derived.

5.1.4. Eu Anomaly

Europium usually occurs as Eu^{3+} in seawater, whereby fractionation is not as obvious as cerium fractionation in dissolution-precipitation processes [50]. Therefore, seawater-normalized Eu anomalies do not differ much from those resulting from PAAS, NASC, or C1 chondrite normalization. However, in a reducing environment, part of the Eu^{3+} will be reduced to Eu^{2+}, causing the radius of the Eu ions to increase; thus, Eu^{2+} is more easily incorporated into the carbonate lattice and enriched in carbonates [51,52]. Studies have shown that the positive Eu anomaly is commonly related to reducing hydrothermal fluids with temperatures higher than 200 °C [53], so the positive Eu anomalies in marine carbonate rocks may be closely related to high-temperature hydrothermal processes [54].

The average values of Eu anomalies for MD-1, MD-2, and MD-3 are 0.747, 0.876, and 0.999, respectively (Table 3), and there were no obvious positive Eu anomalies. This shows that the three types of dolomite were not reformed by obvious high-temperature hydrothermal fluids. Although the Aksu area underwent multiple hydrothermal events [55,56], they did not have a significant impact on the REE compositions of the matrix dolomites in the Qigebrak Formation. From MD-1 to MD-3, the Eu anomaly increased gradually, which may correspond to the gradual rising temperature during dolomitization processes [40], consistent with the observed size increase of the dolomite crystals.

5.1.5. Accuracy Assessment of Clumped Temperature ($T_{\Delta 47}$) Data

As an emerging geological thermometer, clumped isotopes have great potential for tracing the temperature of dolomitization fluids [57]. However, for ancient dolomite samples, the ^{13}C-^{18}O bonds within carbonate crystals may be reordered through solid-state exchange reactions during the diagenetic processes [58]. As a result, the temperature and oxygen isotope composition calculated based on the measured Δ47 values cannot represent the information when the dolomites were formed. Laboratory measurements [59] and numerical simulations [60] have shown that calcite is prone to solid-state reordering if it is buried at a temperature exceeding 100 °C. In contrast, dolomite usually does not undergo obvious solid-state reordering at temperatures below 300 °C and has a stronger resistance to reordering.

Although the Qigebrak Formation in the NW Tarim has undergone complex diagenetic phases, the primary Δ47 signatures have not been affected by significant solid-state reordering. The evidence is as follows. Firstly, the $\delta^{13}C$ and $^{87}Sr/^{86}Sr$ values of these matrix dolomite samples are very close to those of the coeval seawater, indicating the late diagenesis did not significantly modify the original geochemical signals. Secondly, assuming that the surface temperature in the late Ediacaran was 25 °C, based on the burial history and geothermal gradient, it could be estimated that the Qigebrak Formation reached its maximum burial depth in the Triassic, and the highest burial temperature was about 192 °C [34], which is lower than the temperature (300 °C) at which dolomite undergoes obvious solid-state reordering. Thirdly, the clumped temperature ranges ($T_{\Delta 47}$) of MD-1 and MD-2 are very close to the formation temperature of fibrous dolomite cements

(Figure 8), of which the fibrous dolomites commonly indicate the paleoenvironment during the sedimentary-early diagenetic period [61]. This further indicates the original geochemical signals about MD-1 and MD-2 have not been modified. It should be noted that the clumped temperature ($T_{\Delta 47}$) of MD-3 has a positive correlation with the $\delta^{18}O_{water}$ value of the diagenetic fluid (Figure 8), which may reflect the influence of recrystallization on the $T_{\Delta 47}$ values. Thus, the $T_{\Delta 47}$ values of MD-3 possibly represent the lowest temperature at which MD-3 was formed via recrystallization [62,63].

5.2. Origin of the Different Types of Matrix Dolomites

5.2.1. Origin of MD-1 and MD-2

The development of MD-1 was very limited. The crystal size (<5 μm) and planar-s to nonplanar-a texture of MD-1 suggest that it was formed from a relatively low-temperature syndepositional to near-surface environment [2]. Even though the nonplanar-a texture dolomite crystals are commonly formed above the critical roughening temperature (50–60 °C), when the dolomitization fluids have high supersaturation levels, this type of dolomite can also precipitate below temperatures less than 50 °C [31,64]. This dolomite type represents the first stage of dolomitization in early diagenesis, with an associated burial depth less than 500 m [65] (Figure 10). As for MD-2, the very fine dolomite crystals (10–20 μm) with planar-e(s) textures indicate that the formation temperature was lower than the critical roughness temperature [31]. This was further supported by the $T_{\Delta 47}$ of the MD-2 (51.06–60.29 °C). Assuming that the surface temperature was 25 °C and the average geothermal gradient was 35 °C/km [34], its formation depth can be further restricted within the range of 600–1000 m [66] (Figure 10), indicating that the dolomitization process occurred in a shallow burial environment [67].

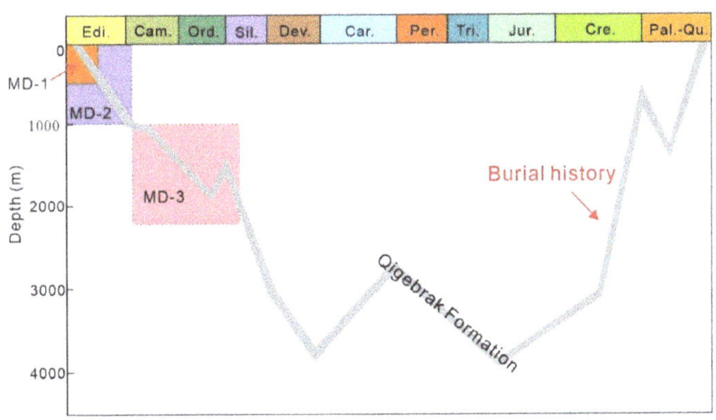

Figure 10. Possible formation timing and depth of various matrix dolomites from the Qigebrak Formation.

The REE patterns of MD-1 and MD-2 are similar, characterized by LREE enrichment, slight HREE depletion, and positive Ce anomalies, indicating that the dolomitization fluids for the formation of MD-1/MD-2 were closely related to original seawater. This explanation is supported by the $\delta^{13}C$ and $^{87}Sr/^{86}Sr$ values of MD-1/MD-2 lying within the scope of Ediacaran seawater (Figure 5).

The $\delta^{18}O$ in dolomite is governed by the formation temperature and the $\delta^{18}O$ of dolomitization fluid [68], and the oxygen isotope value of the dolomitization fluid can be calculated [69]. According to the $T_{\Delta 47}$ values, we calculated that the $\delta^{18}O_{water}$ values of the dolomitization fluid were 3.09–4.90‰ VSMOW when MD-1 and MD-2 were formed. This is about 8–10‰ VSMOW higher than the $\delta^{18}O$ of Neoproterozoic seawater (−5‰ VSMOW) [70] and is higher than the $\delta^{18}O_{water}$ (0‰ VSMOW) of current seawater. This suggests that the dolomitization fluids that formed MD-1 and MD-2 were associated with seawater subjected to evaporation.

5.2.2. Origin of MD-3

The crystal size for MD-3 is much larger (50 to 200 μm) than that of the other two types, exhibiting planar-s to nonplanar-a texture (Figure 4g,h), implying a faster growth rate and higher temperature (>50–60 °C) [31]. The cathodoluminescence color of MD-3 is dull red, and the $\delta^{18}O$ values are depleted, further indicating that MD-3 was formed in the weakly reducing environment at higher temperatures. The similarities in MD-3's REE patterns and Sr isotope range with the other two dolomite types indicate that its dolomitization fluid was also derived from seawater.

Chen's [71] research showed that the late-stage hydrothermal pore-filling saddle dolomites were formed mainly between 100 °C and 130 °C. Generally, the formation temperature of MD-3 appears to be lower than that of saddle dolomite; thus, it can be assumed that the formation temperature of MD-3 was 60–100 °C. The $T_{\Delta 47}$ range of MD-3 (62.28–75.81 °C) also supports the above inference. Similarly, according to the surface temperature and geothermal gradient, it can be estimated that the burial depth of MD-3 was 1000–2143 m (Figure 10), indicating a shallow–medium burial environment [72]. MD-3 is mostly found in intervals where hydrothermal fluids developed [56], but MD-3 does not have Eu anomalies, indicating that the hydrothermal fluids did not significantly modify the geochemical signals of MD-3. The temperature of MD-3 is about 10 °C higher than that of MD-1 and MD-2, which is the result of the greater burial depth rather than the influence of hydrothermal fluids. In addition, the calculated $\delta^{18}O_{water}$ values of the MD-3 dolomitization fluid is 0.99–2.88‰ VSMOW, so the oxygen isotope composition of the dolomitization fluids still exhibited the characteristics of evaporated seawater.

5.3. *Early Dolomitization Mechanism*

5.3.1. Primary or Secondary Dolomitization?

In previous studies, microbial dolomitization [16] and seepage-reflux dolomitization [17] have been proposed to interpret the pervasive early-formed matrix dolomites in the Qigebrak Formation. Some researchers believe that in the late Ediacaran there existed "dolomite seas" in which primary dolomite was easily precipitated [73]. Thus, coeval seawater, microorganisms, and sedimentary organic matter jointly provided Mg^{2+}, and large-scale dolomitization occurred via microbial induction [16]. In this study, MD-2 consists of a large number of well-preserved microbial fabrics, and its formation may have been related to the early activities of microbial mats [74]. Therefore, this study does not reject the occurrence of microbial dolomitization. However, it should be noted that how to accurately assess the scale of microbial dolomitization still deserves further research. You [75] proposed that dolomite with a spherical, dumbbell shape can be considered to be of microbial origin. Although this special crystal morphology is not the unique product of microbial activities [76,77], dolomites with the above-mentioned morphology were not found by SEM observations in the Qigebrak Formation. This suggests that the formation of large-scale matrix dolomites in the study area were not formed by microbial dolomitization. Furthermore, the gradient of the biochemical environment inside the microbial mat changes strongly, and there are dozens of corresponding mineralization products [78]. Dolomite is only one of the products induced by microorganisms, and it is still unclear whether it can be formed on a large scale. Therefore, it is more reasonable to explain the early dolomitization mechanism of the Qigebrak Formation from the view of secondary replacement.

5.3.2. Dolomitization Model

Seepage-reflux dolomitization has been considered to be the most effective hydrological model to explain the origin of pervasive early-formed dolomites [2]. Reflux of dolomitization brines is initiated by a phase of evaporation in shallow marine and/or lagoonal environments that produce a salinity gradient [1,4]. The fluids for seepage-reflux dolomitization are seawater or modified seawater, such as hypersaline seawater that results in the precipitation of evaporites (Adams and Rhodes, 1960), and mesosaline (44‰ ~ 85‰) to penesaline (72‰~199‰) seawater [10]. Because of the difference in density between fluids and the underlying pore water, aragonitic and magnesian calcite could be dolomitized by reflux in high-frequency cycles. If there is a large amount of Mg^{2+}, then the fluids would migrate laterally toward the basin interior, resulting in massive dolomitization. Numerical simulations show that even in a carbonate platform where geothermal convection is well developed, the seepage reflux model more easily forms large-scale dolomites than the convection model does [65].

In the late Ediacaran, the study area was dominated by the inner ramp of a restricted carbonate platform (Figure 1c). In this limited evaporation setting, the restricted platform environment led to higher salinity than normal seawater. However, no evaporitic minerals (e.g., gypsum and anhydrite) were found in the field or under a microscope, suggesting that the salinity of the seawater did not reach the critical value of gypsum precipitation. Therefore, we propose that seepage-reflux dolomitization by mesosaline to penesaline seawater is a more appropriate explanation of the massive early-formed dolomites in the Qigebrak Formation (Figure 11). During the depositional period, high salinity and consequently Mg^{2+} concentrations occurred because of the evaporation effect. At that time, MD-1 was seemingly formed when carbonate sediments got dolomitized before consolidation by brines with high Mg/Ca (Figure 11a). Thereafter, the downward seepage reflux of brines with high Mg^{2+} content caused further dolomitization (Figure 11b). The growth of microbial mats may lead to the precipitation of high-Mg calcite [79], of which the high-Mg calcite could release Mg^{2+} to the intergranular water, providing a large amount of Mg^{2+} for the dolomitization of the surrounding deposits. This may explain the large volume of MD-2. When the Qigebrak Formation entered the burial realm, the high temperature led to more extensive dolomite growth over MD-1 and MD-2, thus forming the coarser and more curved crystals of MD-3 [80] (Figure 11c).

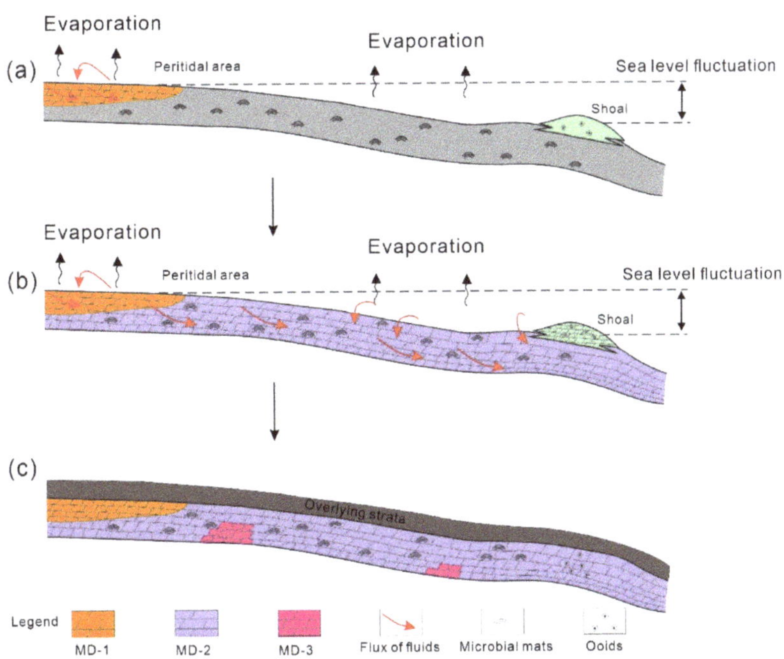

Figure 11. Schematic model for the large-scale dolomitization from the Qigebrak Formation, Aksu area. (**a**) The formation processes of MD-1 during the Qigebrak depositional period; (**b**) Large-scale MD-2 were formed by the seepage reflux of brines with mesosaline to penesaline seawater; (**c**) MD-3 were formed via recrystallization of MD-1/MD-2 in the burial realm.

5.4. Implications for Hydrocarbon Exploration

The potential for pervasive early dolomitization to significantly increase porosities and permeabilities of dolostones, as compared to the precursor limestone, has been shown in several studies [81–83]. MD-2, in the present study, has a stable lateral distribution with porosities between 2.5% and 6.0%, which can be regarded as potential reservoir rocks [84]. As presented above, MD-2 is dominated by planar-e(s) textures associated with very finely crystalline dolomite rhombs (5–20 μm), which are beneficial for the preservation of intercrystalline pores. Numerical simulations demonstrated that mesosaline reflux without evaporite precipitation could result in a porosity increase of up to 8% by "mole for mole" replacement [85,86]. Other pervasive matrix dolomites with similar reflux dolomitization models, such as the Lower Cambrian Longwangmiao Formation of the Sichuan Basin in China [12], and the Lower Carboniferous of the Mississippian Alida Bed of Williston Basin in Canada [80] have proven to be good dolomite reservoirs. This study indicates seepage-reflux dolomitization by mesosaline to penesaline seawater is possibly the fundamental controlling factor in forming high-quality dolomite reservoirs.

6. Conclusions

Coupling petrography with geochemistry analysis, the origin of early-formed matrix dolomites in the Qigebrak Formation of the northwestern Tarim Basin was investigated. The main conclusions include the following:

1. Three distinct dolomite types were recognized: (1) dolomicrite (MD-1); (2) fabric-preserving dolomite (MD-2); and (3) fabric-destructive dolomite (MD-3).
2. MD-1 represents the earliest dolomitization phase in syndepositional to near-surface settings. MD-2 was probably generated by the same brine reflux with slightly elevated temperature in shallow burial settings (600–1000 m).
3. Seepage-reflux dolomitization with mesosaline to penesaline seawater was probably responsible for the origin of MD-1 and MD-2. MD-3 likely resulted from the recrystallization of previously formed MD-1 and/or MD-2 dolomites and precipitated in greater depths (1000–2143 m).

Author Contributions: Conceptualization, P.T. and D.C.; methodology, P.T. and Y.W.; validation, P.T. and B.Y.; formal analysis, P.T. and D.C.; investigation, P.T. and S.L.; resources, P.T. and M.E.-S.; data curation, S.L. and B.Y.; writing—original draft preparation, P.T. and D.C.; writing—review and editing, P.T.; visualization, P.T. and S.L.; supervision, D.C.; project administration, M.E.-S. and B.Y.; funding acquisition, P.T. and D.C. All authors have read and agreed to the published version of the manuscript.

Funding: This research was funded by the National Natural Science Foundation of China (U19B6003).

Institutional Review Board Statement: Not applicable.

Informed Consent Statement: Not applicable.

Data Availability Statement: Data are available upon reasonable request. The data that support the findings of this study are available on request from the corresponding author. The data are not publicly available due to privacy or ethical restrictions.

Acknowledgments: We are especially grateful for the assistance in geochemical analysis made by Chaofeng Li and Hongwei Li (Institute of Geology and Geophysics, CAS).

Conflicts of Interest: The authors have no conflicts of interest to declare regarding the publication of this article.

References

1. Warren, J. Dolomite: Occurrence, evolution and economically important associations. *Earth Sci. Rev.* **2000**, *52*, 1–81. [CrossRef]
2. Machel, H.G. Concepts and models of dolomitization: A critical reappraisal. *Geol. Soc. Lond. Spec. Publ.* **2004**, *235*, 7–63. [CrossRef]
3. Friedman, G.M.; Sanders, J.E. Origin and occurrence of dolostones. In *Developments in Sedimentology*; Elsevier: Amsterdam, The Netherlands, 1967; Volume 9B, pp. 267–348.

4. Adams, J.E.; Rhodes, M.L. Dolomitization by seepage refluxion. *AAPG (Am. Assoc. Pet. Geol.) Bull.* **1960**, *44*, 1912–1920.
5. Davies, G.R.; Smith, L.B. Structurally controlled hydrothermal dolomite reservoir facies: An overview. *AAPG (Am. Assoc. Pet. Geol.) Bull.* **2006**, *90*, 1641–1690. [CrossRef]
6. Vahrenkamp, V.C.; Swart, P.K. *Late Cenozoic Dolomites of the Bahamas: Metastable Analogues for the Genesis of Ancient Platform Dolomites*; International Association of Sedimentologists Special Publication; Blackwell Science: Oxford, UK, 1994; Volume 21, pp. 133–153.
7. Shields, M.J.; Brady, P.V. Mass balance and fluid flow constraints on regional-scale dolomitization, Late Devonian, Western Canada Sedimentary Basin. *Bull. Can. Pet. Geol.* **1995**, *43*, 371–392.
8. Jiang, L.; Cai, C.; Worden, R.H.; Crowley, S.F.; Jia, L.; Zhang, K.; Duncan, I.J.; Hollis, C. Multiphase dolomitization of deeply buried Cambrian petroleum reservoirs, Tarim Basin, north-west China. *Sedimentology* **2016**, *63*, 2130–2157. [CrossRef]
9. Saller, A.H.; Henderson, N. Distribution of porosity and permeability in platform dolomites: Insight from the Permian of West Texas. *AAPG (Am. Assoc. Pet. Geol.) Bull.* **1998**, *82*, 1528–1550.
10. Qing, H.R.; Bosence, D.W.J.; Rose, E.P.F. Dolomitization by penesaline sea water in Early Jurassic peritidal platform carbonates, Gibraltar, western Mediterranean. *Sedimentology* **2001**, *48*, 153–163. [CrossRef]
11. Jiang, L.; Worden, R.H.; Cai, C. Thermochemical sulfate reduction and fluid evolution of the Lower Triassic Feixianguan Formation sour gas reservoirs, northeast Sichuan Basin, China. *AAPG (Am. Assoc. Pet. Geol.) Bull.* **2014**, *98*, 947–973. [CrossRef]
12. Liu, D.; Cai, C.; Hu, Y.; Peng, Y.; Jiang, L. Multistage dolomitization and formation of ultra-deep Lower Cambrian Longwangmiao Formation reservoir in central Sichuan Basin, China. *Mar. Pet. Geol.* **2021**, *123*, 104752. [CrossRef]
13. Yang, H.J.; Chen, Y.Q.; Tian, J.; Du, J.H.; Zhu, Y.F.; Li, H.H.; Pan, W.Q.; Yang, P.F.; Li, Y.; An, H.T. Great discovery and its significance of ultra-deep oil and gas exploration in well Luntan-1 of the Tarim Basin. *China Pet. Explor.* **2020**, *25*, 62–72.
14. Tang, P.; Chen, D.; Wang, Y.; Ding, Y.; El-Shafeiy, M.; Yang, B. Diagenesis of microbialite-dominated carbonates in the Upper Ediacaran Qigebrak Formation, NW Tarim Basin, China: Implications for reservoir development. *Mar. Pet. Geol.* **2022**, *136*, 105476. [CrossRef]
15. Bao, Z.D.; Ji, H.C.; Wang, Y.; Li, Z.F.; Liang, T.; Niu, B.; Wei, M.Y.; Lu, K.; Shi, Y.Q.; Zhang, H.; et al. The primary dolostone in the Meso-Neoproterozoic: Cases study on platforms in China. *J. Palaeogeogr.* **2022**, *11*, 151–172. [CrossRef]
16. Zhu, G.Y.; Li, X.; Li, T.T.; Zhou, L.; Wu, Y.X.; Shen, B.; Ning, M. Genesis mechanism and Mg isotope difference between the Ediacaran and Cambrian dolomites in Tarim Basin. *Sci. China Earth Sci.* **2023**, *66*, 334–357. [CrossRef]
17. Zheng, J.F.; Liu, Y.; Zhu, Y.J.; Liang, F. Geochemical features and its geological significances of the Upper Ediacaran Qigeblak Formation in Wushi area, Tarim Basin. *J. Palaeogeogr.* **2021**, *23*, 983–998.
18. Jia, C. Structural characteristics and oil/gas accumulative regularity in Tarim Basin. *Xinjiang Pet. Geol.* **1999**, *20*, 177–183.
19. Turner, S.A. Sedimentary record of late Neoproterozoic rifting in the NW Tarim Basin, China. *Precambrian Res.* **2010**, *181*, 85–96. [CrossRef]
20. Wu, L.; Guan, S.; Yang, H.; Ren, R.; Zhu, G.; Jin, J.; Zhang, C. The paleogeographic framework and hydrocarbon exporation potential of Neoproterozoic rift basin in northern Tarim Basin. *Acta Pet. Sin.* **2017**, *38*, 375–385.
21. Zheng, J.F.; Shen, A.J.; Yang, H.X.; Zhu, Y.J.; Liang, F. Geochemistry and geochronology characteristics and their geological significance of microbial dolomite in Upper Sinian, NW Tarim Basin. *Acta Pet. Sin.* **2021**, *37*, 2189–2202.
22. Lin, C.; Li, H.; Liu, J. Major unconformities, tectonostratigraphic frameword, and evolution of the superimposed Tarim basin, Northwest China. *J. Earth Sci.* **2012**, *23*, 395–407. [CrossRef]
23. He, J.; Qing, H.; Xu, B. The unconformity-related palaeokarst in the uppermost Ediacaran carbonate rocks in the northwestern Tarim Block, NW China: Implication for sedimentary evolution during the Ediacaran–Cambrian transition. *Int. Geol. Rev.* **2018**, *61*, 839–852. [CrossRef]
24. Hoefs, J. *Stable Isotope Geochemistry*; Springer: Berlin/Heidelberg, Germany, 2009; p. 285.
25. Du, Y.; Fan, T.; Machel, H.G.; Gao, Z. Genesis of Upper Cambrian-Lower Ordovician dolomites in the Tahe Oilfield, Tarim Basin, NW China: Several limitations from petrology, geochemistry, and fluid inclusions. *Mar. Pet. Geol.* **2018**, *91*, 43–70. [CrossRef]
26. Bau, M.; Dulski, P. Distribution of yttrium and rare-earth elements in the Penge and Kuruman iron-formations, Transvaal Supergroup, South Africa. *Precambrian Res.* **1996**, *79*, 37–55. [CrossRef]
27. Bolhar, R.; Kamber, B.S.; Moorbath, S.; Fedo, C.M.; Whitehouse, M.J. Characterisation of early Archaean chemical sediments by trace element signatures. *Earth Planet. Sci. Lett.* **2004**, *222*, 43–60. [CrossRef]
28. Wang, X.; Cui, L.L.; Li, Y.Y.; Huang, X.F.; Zhai, J.X.; Ding, Z.L. Determination of clumped isotopes in carbonate using isotope ratio mass spectrometry: Toward a systematic evaluation of a sample extraction method using a static Porapak™ Q absorbent trap. *Int. J. Mass Spectrom.* **2016**, *403*, 8–14. [CrossRef]
29. Dennis, K.J.; Affek, H.P.; Passy, B.H.; Schrag, D.P.; Eiler, J.M. Defining an absolute reference frame for 'clumped' isotope studies of CO_2. *Geochim. Cosmochim. Acta* **2011**, *75*, 7117–7131. [CrossRef]
30. Bernasconi, S.M.; Mueller, I.A.; Bergmann, K.D.; Breitenbach, S.F.M.; Fernandez, A.; Hodell, D.A.; Jaggi, M.; Meckler, A.N.; Millan, I.; Ziegler, M. Reducing Uncertainties in Carbonate Clumped Isotope Analysis Through Consistent Carbonate-Based Standardization. *Geochem. Geophys. Geosyst.* **2018**, *19*, 2895–2914. [CrossRef]
31. Gregg, J.M.; Sibley, D.F. Epigenetic dolomitization and the origin of xenotopic dolomite texture. *J. Sediment. Res.* **1984**, *54*, 908–931.

32. Dong, S.; Chen, D.; Qing, H.; Zhou, X.; Wang, D.; Guo, Z.; Jiang, M.; Qian, Y. Hydrothermal alteration of dolostones in the Lower Ordovician, Tarim Basin, NW China: Multiple constraints from petrology, isotope geochemistry and fluid inclusion microthermometry. *Mar. Pet. Geol.* **2013**, *46*, 270–286. [CrossRef]
33. Zhang, Y.; Yang, T.; Hohl, S.V.; Zhu, B.; He, T.; Pan, W.; Chen, Y.; Yao, X.; Jiang, S. Seawater carbon and strontium isotope variations through the late Ediacaran to late Cambrian in the Tarim Basin. *Precambrian Res.* **2020**, *345*, 105769. [CrossRef]
34. Shen, A.J.; Hu, A.P.; Zheng, J.F.; Liang, F.; Wang, Y.S. Reconstruction of tectonic-burial evolution based on the constraints of laser in situ U-Pb date and clumped isotopic temperature: A case study from Ediacaran Qigebrak Formation in Akesu area, Tarim Basin. *Mar. Orig. Pet. Geol.* **2021**, *26*, 200–210.
35. Qing, H.R.; Mountjoy, E.W. Rare earth element geochemistry of dolomites in the Middle Devonian Presqu'ile barrier, Western Canada Sedimentary Basin: Implications for fluid-rock ratios during dolomitization. *Sedimentology* **1994**, *41*, 787–804. [CrossRef]
36. Wu, S.Q.; Zhu, J.Q.; Hu, W.X.; Zhang, J.T.; Wang, X.L.; Su, Y.B. Rare earth element geochemistry characteristics of Cambrian-Ordovician dolostones in the Tarim Basin and their implications for the origin. *Geoscience* **2009**, *23*, 638–647.
37. Shields, G.A.; Webb, G.E. Has the REE composition of seawater changed over geological time? *Chem. Geol.* **2004**, *204*, 103–107. [CrossRef]
38. Miura, N.; Kawabe, I. Dolomitization of limestone with $MgCl_2$ solution at 150 °C: Preserved original signatures of rare earth elements and yttrium as marine limestone. *Geochem. J.* **2000**, *34*, 223–227. [CrossRef]
39. Chen, Q.; Hu, W.X.; Wang, X.L.; Li, Q.; Hu, G.; Zhu, J.Q.; Yao, S.P.; Cao, J. Characteristics and genesis of REE patterns of Changxing and Feixianguan dolomites in Panlongdong, northeastern Sichuan. *Pet. Geol. Exp.* **2011**, *33*, 624–633+638.
40. Xiang, P.; Ji, H.; Shi, Y.; Huang, Y.; Sun, Y.; Xu, X.; Zou, S. Petrographic, rare earth elements and isotope constraints on the dolomite origin of Ordovician Majiagou Formation (Jizhong Depression, North China). *Mar. Pet. Geol.* **2020**, *117*, 104374. [CrossRef]
41. Kawabe, I.; Toriumi, T.; Ohta, A.; Miura, N. Monoisotopic REE abundances in seawater and the origin of seawater tetrad effect. *Geochem. J.* **1998**, *32*, 213–229. [CrossRef]
42. Nothdurft, L.D.; Webb, G.E.; Kamber, B.S. Rare earth element geochemistry of Late Devonian reefal carbonates, Canning Basin, Western Australia: Confirmation of a seawater REE proxy in ancient limestones. Geochim. Cosmochim. *Acta* **2004**, *68*, 263–283. [CrossRef]
43. Bolhar, R.; Van Kranendonk, M.J. A non-marine depositional setting for the northern Fortescue Group, Pilbara Craton, inferred from trace element geochemistry of stromatolitic carbonates. *Precambrian Res.* **2007**, *155*, 229–250. [CrossRef]
44. Jiang, L.; Cai, C.; Worden, R.H.; Li, K.; Xiang, L.; Chu, X.; Shen, A.; Li, W. Rare earth element and yttrium (REY) geochemistry in carbonate reservoirs during deep burial diagenesis: Implications for REY mobility during thermochemical sulfate reduction. *Chem. Geol.* **2015**, *415*, 87–101. [CrossRef]
45. German, C.R.; Elderfield, H. Application of the Ce anomaly as a paleoredox indicator: The ground rules. *Paleoceanography* **1990**, *5*, 823–833. [CrossRef]
46. Alibo, D.S.; Nozaki, Y. Rare earth elements in seawater: Particle association, shale-normalization, and Ce oxidation. *Geochim. Cosmochim. Acta* **1999**, *63*, 363–372. [CrossRef]
47. Tostevin, R.; Shields, G.A.; Tarbuck, G.M.; He, T.; Clarkson, M.O.; Wood, R.A. Effective use of cerium anomalies as a redox proxy in carbonate-dominated marine settings. *Chem. Geol.* **2016**, *438*, 146–162. [CrossRef]
48. Wang, L.; Hu, W.; Wang, X.; Cao, J.; Chen, Q. Seawater normalized REE patterns of dolomites in Geshan and Panlongdong sections, China: Implications for tracing dolomitization and diagenetic fluids. *Mar. Pet. Geol.* **2014**, *56*, 63–73. [CrossRef]
49. Planavsky, N.; Bekker, A.; Rouxel, O.J.; Kamber, B.; Hofmann, A.; Knudsen, A.; Lyons, T.W. Rare earth element and yttrium compositions of Archean and Paleoproterozoic Fe formations revisited: New perspectives on the significance and mechanisms of deposition. *Geochim. Cosmochim. Acta* **2010**, *74*, 6387–6405. [CrossRef]
50. Bertram, C.J.; Elderfield, H. The geochemical balance of the rare earth elements and neodymium isotopes in the oceans. *Geochim. Cosmochim. Acta* **1993**, *57*, 1957–1986. [CrossRef]
51. Brookins, D.G. Aqueous geochemistry of rare earth elements. *Rev. Mineral. Geochem.* **1989**, *21*, 201–225.
52. Bau, M. Rare-earth element mobility during hydrothermal and metamorphic fluid-rock interaction and the significance of the oxidation state of europium. *Chem. Geol.* **1991**, *93*, 219–230. [CrossRef]
53. Michard, A.; Albarède, F. The REE content of some hydrothermal fluids. *Chem. Geol.* **1986**, *55*, 51–60. [CrossRef]
54. Bau, M.; Balan, S.; Schmidt, K.; Koschinsky, A. Rare earth elements in mussel shells of the Mytilidae family as tracers for hidden and fossil high-temperature hydrothermal systems. *Earth Planet. Sci. Lett.* **2010**, *299*, 310–316. [CrossRef]
55. Chen, H.L.; Yang, S.F.; Dong, C.W.; Zhu, G.Q.; Jia, C.Z.; Wei, G.Q.; Wang, Z.G. Study on geological thermal events in Tarim Basin. *Chin. Sci. Bull.* **1997**, *42*, 1096–1099.
56. Zhou, X.Q.; Chen, D.Z.; Zhang, L.Y.; Tang, D.J.; Guo, C. Silica-rich seawater in the early Cambrian: Sedimentological evidence from bedded cherts. *Terra Nova* **2021**, *33*, 494–501. [CrossRef]
57. Li, P.P.; Wang, C.; Zou, H.Y.; Yu, X.Y. Application of clumped isotopes to restoration of dolomitizing fluids and its limitations. *Oil Gas Geol.* **2021**, *42*, 738–746.
58. Dennis, K.J.; Schrag, D.P. Clumped isotope thermometry of carbonatites as an indicator of diagenetic alteration. Geochim. Cosmochim. *Acta* **2010**, *74*, 4110–4122.
59. Lloyd, M.K.; Ryb, U.; Eiler, J.M. Experimental calibration of clumped isotope reordering in dolomite. *Geochim. Cosmochim. Acta* **2018**, *242*, 1–20. [CrossRef]

60. Stolper, D.A.; Eiler, J.M. The kinetics of solid-state isotope-exchange reactions for clumped isotopes: A study of inorganic calcites and apatites from natural and experimental samples. *Am. J. Sci.* **2015**, *315*, 363. [CrossRef]
61. Hood, A.v.S.; Wallace, M.W. Neoproterozoic marine carbonates and their paleoceanographic significance. *Glob. Planet. Chang.* **2018**, *160*, 28–45. [CrossRef]
62. Macdonald, J.M.; John, C.M.; Girard, J.P. Testing clumped isotopes as a reservoir characterization tool: A comparison with fluid inclusions in a dolomitized sedimentary carbonate reservoir buried to 2–4 km. In *From Source to Seep: Geochemical Applications in Hydrocarbon Systems*; Lawson, M., Formolo, M.J., Eiler, J.M., Eds.; Geological Society Special Publication; Geological Society of London: London, UK, 2018; Volume 468, pp. 189–202.
63. Lukoczki, G.; Haas, J.; Gregg, J.M.; Machel, H.G.; Kele, S.; John, C.M. Early dolomitization and partial burial recrystallization: A case study of Middle Triassic peritidal dolomites in the Villany Hills (SW Hungary) using petrography, carbon, oxygen, strontium and clumped isotope data. *Int. J. Earth Sci.* **2020**, *109*, 1051–1070. [CrossRef]
64. Fu, Q.; Qing, H.; Bergman, K.M. Early dolomitization and recrystallization of carbonate in an evaporite basin: The Middle Devonian Ratner laminite in southern Saskatchewan, Canada. *J. Geol. Soc.* **2006**, *163*, 937–948. [CrossRef]
65. Al-Helal, A.B.; Whitaker, F.F.; Xiao, Y. Reactive transport modeling of brine reflux: Dolomitization, anhydrite precipitation, and porosity evolution. *J. Sediment. Res.* **2012**, *82*, 196–215. [CrossRef]
66. Chang, J.; Qiu, N.S.; Zuo, Y.H.; Li, C.C. The new evidence on tectonic uplift in Kepingtage area, Tarim Basin: Constraints from (U-Th)/He ages. *Chin. J. Geophys.* **2011**, *54*, 163–172.
67. Guo, C.; Chen, D.; Qing, H.; Zhou, X.; Ding, Y. Early dolomitization and recrystallization of the Lower-Middle Ordovician carbonates in western Tarim Basin (NW China). *Mar. Pet. Geol.* **2020**, *111*, 332–349. [CrossRef]
68. Land, L.S. The application of stable isotopes to studies of the origin of dolomite and to problems of diagenesis of clastic sediments. In *Stable Isotopes in Sedimentary Geology*; SEPM Short Course; SEPM Society for Sedimentary Geology: Tulsa, OK, USA, 1983; Volume 10, pp. 4.1–4.22.
69. Mattews, A.; Katz, A. Oxygen isotope fractionation during the dolomitization of calcium carbonate. *Geochim. Cosmochim. Acta* **1977**, *41*, 1431–1438. [CrossRef]
70. Galili, N.; Shemesh, A.; Yam, R.; Brailovsky, I.; Sela-Adler, M.; Schuster, E.M.; Collom, C.; Bekker, A.; Planavsky, N.; Macdonald, F.A.; et al. The geologic history of seawater oxygen isotopes from marine iron oxides. *Science* **2019**, *365*, 469–473. [CrossRef] [PubMed]
71. Chen, L.P.; Zhang, H.; Cai, Z.X.; Cong, F.Y.; Huang, S.; Tang, P. Characteristics and formation mechanisms of the unconformity-related paleokarst reservoirs in the Upper Sinian, Northwestern Tarim Basin, China. *Mar. Pet. Geol.* **2020**, *120*, 104559. [CrossRef]
72. Hunt, J.M. *Petroleum Geochemistry and Geology*, 2nd ed.; W.H. Freeman: New York, NY, USA, 1996.
73. Wang, J.; He, Z.; Zhu, D.; Liu, Q.; Ding, Q.; Li, S.; Zhang, D. Petrological and geochemical characteristics of the botryoidal dolomite of Dengying Formation in the Yangtze Craton, South China: Constraints on terminal Ediacaran "dolomite seas". *Sediment. Geol.* **2020**, *406*, 105722. [CrossRef]
74. Burns, S.J.; Mckenzie, J.A.; Vasconcelos, C. Dolomite formation and biogeochemical cycles in the Phanerozoic. *Sedimentology* **2000**, *47*, 49–61. [CrossRef]
75. You, X.; Sun, S.; Zhu, J.; Li, Q.; Hu, W.; Dong, H. Microbially mediated dolomite in Cambrian stromatolites from the Tarim Basin, north-west China: Implications for the role of organic substrate on dolomite precipitation. *Terra Nova* **2013**, *25*, 387–395. [CrossRef]
76. Kirkland, B.L.; Leo Lynch, F.; Rahnis, M.A.; Folk, R.L.; Molineux, I.J.; McLean, R.J.C. Alternative origins for nannobacteria-like objects in calcite. *Geology* **1999**, *27*, 347–350. [CrossRef]
77. Ruiz-Agudo, E.; Burgos-Cara, A.; Ruiz-Agudo, C.; Ibañez-Velasco, A.; Cölfen, H.; Rodriguez-Navarro, C. A non-classical view on calcium oxalate precipitation and the role of citrate. *Nat. Commun.* **2017**, *8*, 768. [CrossRef] [PubMed]
78. Lowenstam, H.A. Minerals formed by organisms. *Science* **1981**, *211*, 1126–1131. [CrossRef] [PubMed]
79. Pratt, B.R. Epiphyton and Renalcis; diagenetic microfossils from calcification of coccoid blue-green algae. *J. Sediment. Res.* **1984**, *54*, 948–971.
80. Rott, C.M.; Qing, H.R. Early dolomitization and recrystallization in shallow marine carbonates, Mississippian Alida beds, Williston Basin (Canada): Evidence from petrography and isotope geochemistry. *J. Sediment. Res.* **2013**, *83*, 928–941. [CrossRef]
81. Maliva, R.G.; Budd, D.A.; Clayton, E.A.; Missimer, T.M.; Dickson, J.A.D. Insights into the dolomitization process and porosity modification in sucrosic dolostones, Avon Park Formation (Middle Eocene), East-Central Florida, U.S.A. *J. Sediment. Res.* **2011**, *81*, 218–232. [CrossRef]
82. Jiang, L.; Xu, Z.; Shi, S.; Liu, W. Multiphase dolomitization of a microbialite-dominated gas reservoir, the middle Triassic Leikoupo Formation, Sichuan Basin, China. *J. Pet. Sci. Eng.* **2019**, *180*, 820–834. [CrossRef]
83. Tavakoli, V.; Jamalian, A. Porosity evolution in dolomitized Permian–Triassic strata of the Persian Gulf, insights into the porosity origin of dolomite reservoirs. *J. Pet. Sci. Eng.* **2019**, *181*, 106191. [CrossRef]
84. Li, P.W.; Luo, P.; Song, J.M.; Jin, T.F.; Wang, G.Q. Characteristics of Upper Sinian dolostone reservoirs in northwestern Margin of Tarim Basin. *Mar. Orig. Pet. Geol.* **2015**, *20*, 1–12.

85. Jones, G.D.; Xiao, Y. Dolomitization, anhydrite cementation, and porosity evolution in a reflux system: Insights from reactive transport models. *AAPG (Am. Assoc. Pet. Geol.) Bull.* **2005**, *89*, 577–601. [CrossRef]
86. Weyl, P.K. Porosity through dolomitization: Conservation of mass requirements. *J. Sediment. Res.* **1960**, *30*, 85–90.

minerals

MDPI

Article

Pore-Type Characterization and Reservoir Zonation of the Sarvak Formation in the Abadan Plain, Zagros Basin, Iran

Hamzeh Mehrabi [1,*], Fatemeh Karami [1], Nafiseh Fakhar-Shahreza [1] and Javad Honarmand [2]

1 School of Geology, College of Science, University of Tehran, Tehran 14398-14151, Iran; fateme.karami@ut.ac.ir (F.K.); nafisehfakhar94@gmail.com (N.F.-S.)
2 Research Institute of Petroleum Industry (RIPI), Tehran 14778-93855, Iran; honarmandj@gmail.com
* Correspondence: mehrabi.hamze@ut.ac.ir; Tel.: +98-9120762988

Abstract: Carbonate reservoirs are characterized by their complex depositional–diagenetic history and, consequently, intricate pore networks. In this study, pore types and reservoir zones of the Sarvak Formation, which is the second most important oil reservoir in Iran, are investigated by integrating core, thin-section, porosity–permeability, scanning electron microscopy (SEM), and mercury injection capillary pressure (MICP) data. Depositional, diagenetic, and fracture pores are identified in macroscopic to microscopic scales. Frequency analysis of pore types revealed that diagenetic pores, including vuggy, moldic, intercrystalline, and fracture pores, formed the majority of pore spaces of the Sarvak Formation. They are dominantly recorded in the regressive systems tracts of both the Cenomanian and Turonian sequences, just below the paleoexposure surfaces. Away from the disconformities, there are two general trends regarding the pore-types distribution: (a) grain-supported facies of shoal and talus settings with dominant primary (depositional) pores including interparticle, intra-skeletal, and growth-framework pores, which are concentrated in the middle part of the formation, and (b) microporous mud-dominated facies with various types of primary and secondary pore types in the TSTs of third-order sequences, especially in the lower part of the Sarvak Formation. Stratigraphic modified Lorenz plot differentiated reservoir, speed, barrier, and baffle zones in this formation. Depositional–diagenetic characters, petrophysical properties, and dominant pore types are defined for each zone in a sequence stratigraphic framework.

Keywords: Sarvak Formation; Cretaceous; pore-type; Lorenz plot; heterogeneity; microporosity; Abadan Plain; Zagros

Citation: Mehrabi, H.; Karami, F.; Fakhar-Shahreza, N.; Honarmand, J. Pore-Type Characterization and Reservoir Zonation of the Sarvak Formation in the Abadan Plain, Zagros Basin, Iran. *Minerals* **2023**, *13*, 1464. https://doi.org/10.3390/min13121464

Academic Editor: János Haas

Received: 23 September 2023
Revised: 12 November 2023
Accepted: 17 November 2023
Published: 22 November 2023

1. Introduction

Carbonate rocks are widely regarded as some of the most significant hydrocarbon reservoirs globally [1]. The reservoir quality of carbonate reservoirs is influenced by three interconnected factors: depositional facies, diagenetic processes, and tectonic events [2,3]. Due to the diverse depositional facies and the impact of various diagenetic processes, the pore system in carbonate reservoirs is more heterogeneous compared to the clastic rocks [4,5]. The pore system and its associated heterogeneities play a pivotal role in fluid flow, porosity–permeability values, water saturation, relative permeability, and capillary pressure in carbonate reservoirs [6–9]. Recognizing and interpreting variations in the pore system, along with its classification from a petrophysical standpoint, contribute to a better comprehension of reservoir heterogeneities [10–13].

The Upper Cretaceous (Cenomanian–Turonian) Sarvak Formation is the primary reservoir unit within the Bangestan Group and is considered the most significant carbonate reservoir rock in the Zagros Basin, following the Asmari Formation [14]. This formation holds immense importance for exploration and production across multiple oil fields in the Abadan Plain region [15–17]. Numerous studies have been conducted on the sedimentology, geochemistry, and reservoir quality of the Sarvak Formation in the southwest of Iran (e.g., [11,18–24]). These studies reveal a complex depositional–diagenetic history that has contributed to the creation and evolution of various pore types within the Sarvak Formation. However, a comprehensive study that thoroughly examines the Sarvak Formation's pore systems and establishes corresponding reservoir zones has not been undertaken yet.

This study represents the first research endeavor that integrates routine (RCAL) and special (SCAL) core data with petrographic evidence and scanning electron microscopy (SEM) to investigate the various types of pore systems in the Sarvak Formation. After identifying the controlling factors of these pore systems and their distribution, they are classified accordingly. This classification results in the definition of reservoir zones within the Sarvak Formation, and the impact of pore system diversity on the reservoir quality of these zones is assessed within the Abadan Plain. Acquiring a detailed understanding of the distribution of reservoir zones in this formation is crucial for a comprehensive understanding and modeling of this reservoir. On a wider scale, this study provides a standard methodology of pore-network characterization and pore-based reservoir zonation of carbonate reservoirs all around the world. The aims of this study include (1) defining the various pore types in macro-, meso-, and microscopic scales within the Sarvak Formation; (2) the genetical classification of pore types (i.e., depositional, diagenetic, and fracture pores); (3) reservoir zonation of the formation based on the stratigraphic modified Lorenz plot, (4) defining the controls of pore types on the development of reservoir and non-reservoir (barrier to baffle) zones, and (5) sequence stratigraphic framework of pore types and reservoir zones of the Sarvak Formation in the Abadan Plain.

2. Geological Setting and Stratigraphy

In the geological history of the Arabian Plate and the Zagros Basin, the Upper Cretaceous depositional successions contain multiple hydrocarbon reserves, which are economically significant [25,26]. The study area, Abadan Plain (Figure 1A), is situated in the northwestern part of the Persian Gulf and is recognized as one of the most prosperous hydrocarbon basins in the Zagros Basin [15].

In a broader context, the Upper Cretaceous is known as a warm period on a global scale, particularly within the Middle East [27]. At this time, the northeastern margin of the Arabian Plate, including the Zagros Basin of Iran, was situated close to the Equator, at approximately 0 to 5° in the northern hemisphere [28,29]. Consequently, the depositional setting of this area is expected to feature a warm and humid tropical paleoclimate with heavy precipitation [30]. Additionally, during the Upper Cretaceous, the Arabian Plate experienced high relative sea levels [28]. However, various instances of relative sea-level fluctuations occurred due to a combination of tectonic movements and eustatic changes [31].

The complex tectonic history of the Zagros Basin, following a huge change in tectonic regime from an extensional to a compressional regime during the closure of the Neo-Tethys Ocean, results in significant variations in the depositional, diagenesis, and reservoir characteristics of the Sarvak Formation and its equivalents across the Middle East [26,32,33].

During the Cenomanian–Turonian interval, both the Arabian Platform and the Zagros Basin underwent three notable emergence phases, leading to the formation of disconformities. The disconformities and their related diagenetic features significantly impacted the reservoir characteristics of the Sarvak Formation [14,34,35]. Meteoric dissolution (karstification) played a substantial role in diversifying the pore systems and enhancing porosity–permeability beneath the paleoexposure surfaces [36]. The Sarvak Formation, primarily composed of thick limestone and dolomitic limestone with minor clayey and argillaceous intervals, is located in the south and southwest of Iran (Figure 1B). It correlates with the Mishrif, Ahmadi, and Rumaila Formations in Saudi Arabia; the Natih Formation in Oman; the Derdere Formation in southeastern Turkey; and the Mishrif Formation in Iraq (Figure 1C) [25]. At its type section, the Sarvak Formation overlies the Kazhdumi Formation with a transitional contact, but its upper boundary with the shales and marls of the Gurpi Formation is sharp (Figure 1B,C) [37]. In our study area (Abadan Plain), a thin-bedded argillaceous–shaly limestone referred to as the Laffan Formation, with a total thickness of about 10–15 m, separates the Sarvak from the Ilam Formation (Figure 1B).

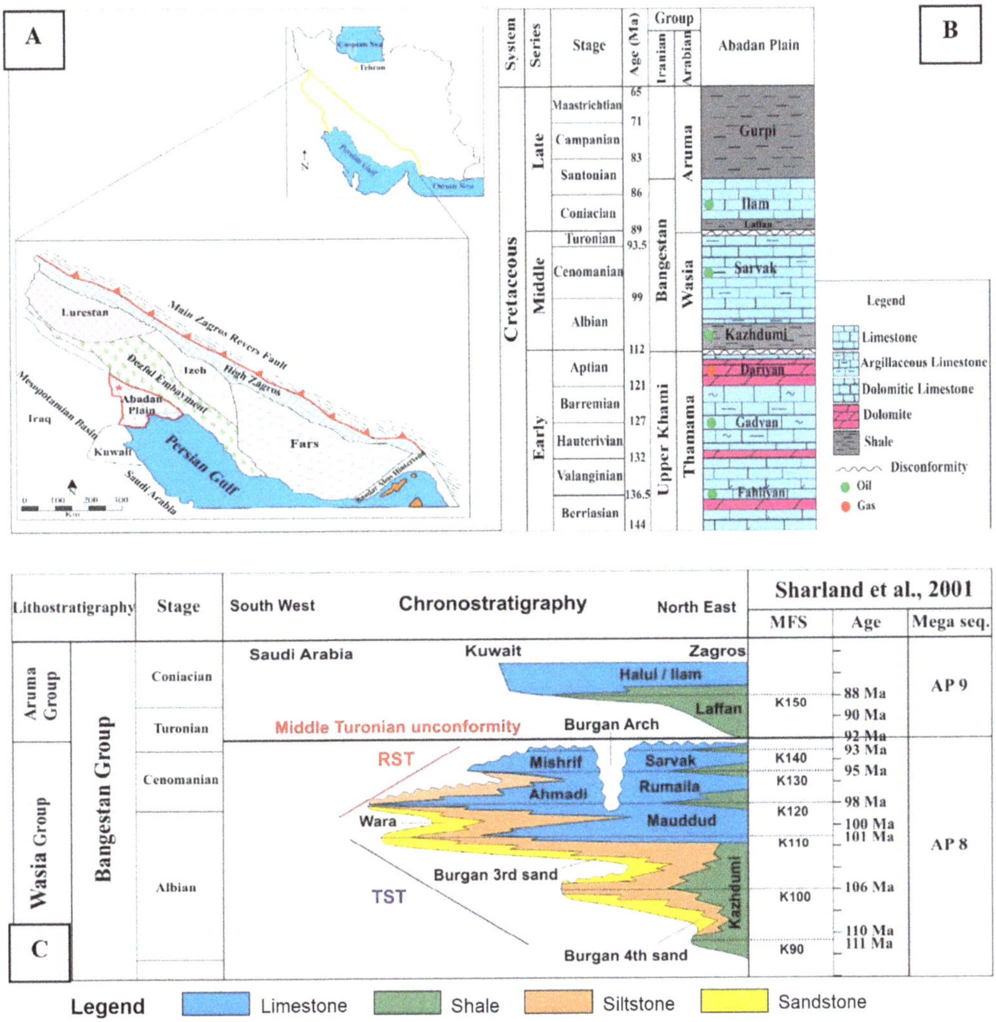

Figure 1. Location map of Iran and the Abadan Plain in the Zagros Basin (**A**). Stratigraphic charts of Cretaceous in the Abadan Plain of Iran (**B**) and the Middle East, modified after [28] (**C**). Figure adapted with some modifications from [38].

3. Materials and Methods

This study is based on a complete dataset of cores, core plugs, thin sections, scanning electron microscopy (SEM), mercury injection capillary pressure (MICP), porosity–permeability, and petrophysical log data from the Sarvak Formation in three wells located in the Abadan Plain (Figure 1A). In total, 430 m of drilled cores, 1300 thin sections, 985 core plugs, SEM analysis of 65 samples, MICP results of 25 samples, porosity–permeability data of 950 samples, and full-set logs (including gamma-ray, neutron, density, resistivity, and water saturation) were available for this study. Total of 50 thin-section samples were blue-dyed for better recognition of pore types, and 1250 thin sections were stained with Alizarin Red-S for mineral differentiation. MICP data are used in conjunction with R35 values (i.e., the pore throat size at 35% mercury saturation) calculated from Winland's equation to classify the various pore types in the Sarvak Fm. All core analyses (i.e., porosity–permeability and MICP) are implemented in the core analysis laboratory of National Iranian Oil Company (NIOC, Tehran, Iran). The SEM analyses were performed using the MIRA 3-XMU field emission scanning electron microscope at the Razi Metallurgical Research Center (RMRC), Tehran, Iran. Petrographic studies are performed by using the Leitz- ORTHOLUX II (Germany) polarizing microscope at the University of Tehran, Iran.

Facies nomenclature and classification was based on the Dunham [39] and Embry and Klovan [40]. Interpretation of depositional facies was based on the standard facies models presented by Flügel [41]. The pore-type classification was based on the schemes of Choquette and Pray [42], along with genetic pore classification of Ahr [43]. Two methods are used to define the reservoir zones:

(1) The Winland plot and calculated R35 values (Equation (1));

The empirical relationship between the radius of pore throat with porosity and permeability is revealed by this method [44]:

$$\mathrm{Log(R35)} = 0.732 + 0.588\,\mathrm{Log(K)} - 0.864\,\mathrm{Log}(\phi) \tag{1}$$

where the R35 is pore-radius corresponds to a level of 35% mercury saturation (μm), K is permeability (mD), and ϕ is porosity (%). Based on the calculated R35 values, reservoir-rock samples can be separated into five petrophysical categories [45]: mega-porous, R35 > 10 μm; macro-porous, 2.5 μm < R35 < 10 μm; mesoporous, 0.5 μm < R35 < 2.5 μm; micro-porous, 0.2 μm < R35 < 0.5 μm; nano-porous, R35 < 0.2 μm.

(2) The stratigraphic modified Lorenz plot (SMLP)

The SML plot provides an optimum number of zones in a reservoir formation [44]. It creates a continuous stratigraphic of core porosity (Ø) and permeability (K) with unique K/Ø ratios, known as reservoir process speed (RPS). Reservoir storage capacity (Øh) and flow capacity (Kh) are calculated using the following Equations (2) and (3) [46–48]:

$$Kh_{cum} = K_1(h_1 - h_0)/Kh_{total} + K_2\,(h_2 - h_1)/Kh_{total} + \ldots + K_n\,(h_n - h_{n-1})/Kh_{total} \tag{2}$$

$$\varnothing h_{cum} = \varnothing_1(h_1 - h_0)/\varnothing h_{total} + \varnothing_2(h_2 - h_1)/\varnothing h_{total} + \ldots + \varnothing_n\,(h_n - h_{n-1})/\varnothing h_{total} \tag{3}$$

where Ø is porosity, K is permeability (mD), and h is thickness (m) of sample interval.

On the SMLP, the slope of each segment reflects the reservoir's flow characteristics. Segments with steeper slopes indicate higher reservoir quality and storage capacity, while gentle slopes or horizontal segments represent zones of lower quality or barrier zones [46,49].

4. Results

4.1. Sedimentology

Petrographic facies analyses led to the recognition of eight microfacies types within the Sarvak Formation, labeled as SMF1 to SMF8. A summary of the facies characteristics of this formation is presented in Table 1. Representative photomicrographs of these facies are depicted in Figure 2.

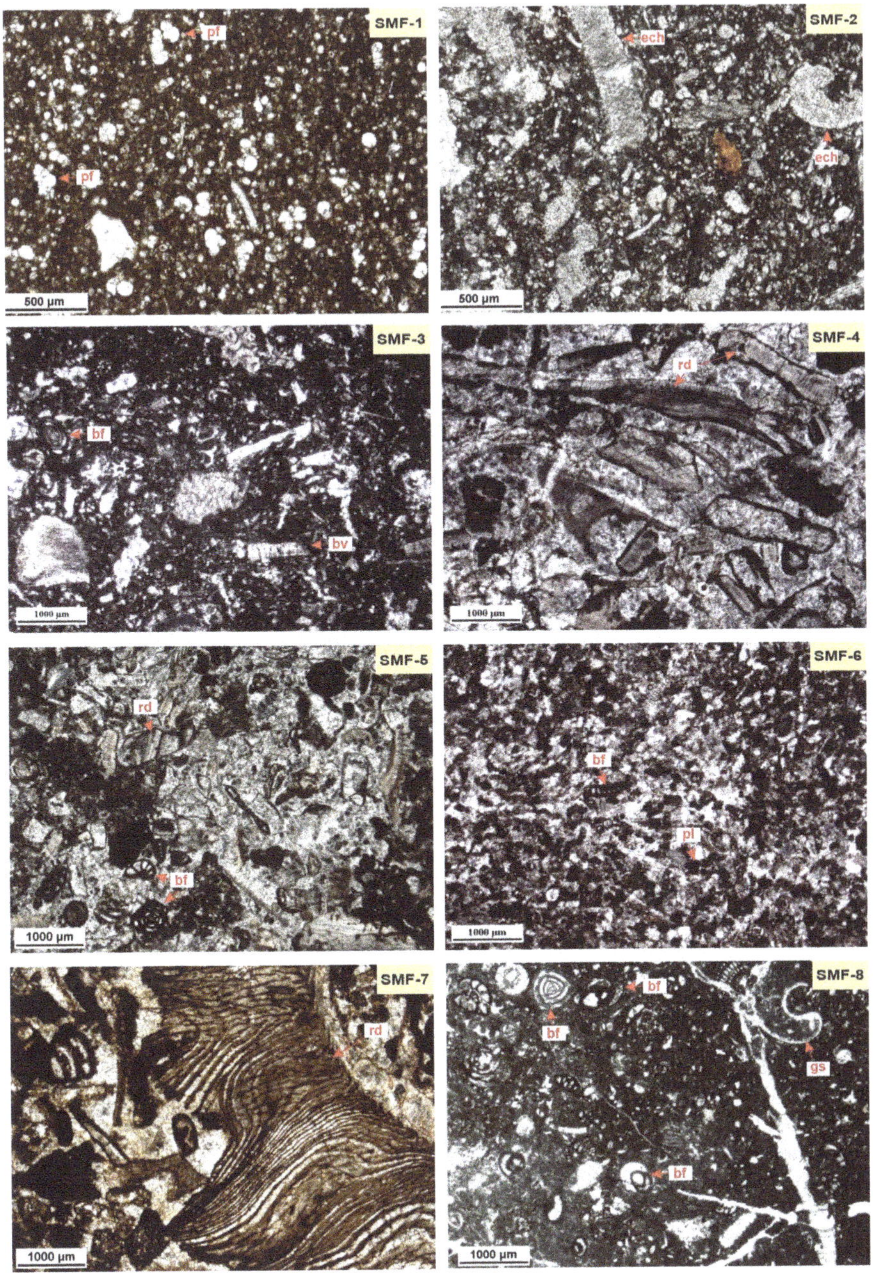

Figure 2. Photomicrographs of depositional facies of the Sarvak Formation. **SMF-1**: Planktonic Foraminifera microbioclast Wackestone, **SMF-2**: Foraminifera (B&P) bioclast Wackestone to Packstone, **SMF-3**: Benthic foraminifera bioclast (echinoids) Wackestone, **SMF-4**: Bioclastic Grainstone, **SMF-5**: Bioclast benthic foraminifera Packstone to Grainstone, **SMF-6**: Benthic foraminifera peloidal Packstone, **SMF-7**: Bioclast (rudist and coral) Rudstone/Floatstone, **SMF-8**: Bioturbated benthic foraminifera peloidal Mudstone to Wackestone. All photos are in PPL. Abbreviations: pf: planktic foraminifera, ech: echinoids, bv: bivalve, rd: rudist, bf: benthic foraminifera, pl: peloid, gs: gastropod.

Table 1. Summarized facies characteristics of the Sarvak Formation in the Abadan Plain.

Facies Code	Microfacies Name	Lithology	Allochems (Skeletal and Non-Skeletal)	Deposition Setting
SMF-1	Planktonic Foraminifera microbioclast Wackestone	(Dolomitic) Limestone	planktonic foraminifera, oligosteginids, echinoderm debris, fine peloids	Outer Ramp
SMF-2	Foraminifera (B&P) bioclast Wackestone to Packstone	(Dolomitic) Limestone	planktonic and small benthic foraminifera, echinoderms, bivalves, rudists, red algae, and bryozoan fragments, and oligosteginids, fine peloids	Middle Ramp
SMF-3	Benthic foraminifera bioclast (echinoids) Wackestone	(Dolomitic) Limestone	whole fossils of echinoderms, bivalves, red algae and bryozoans, foraminifera (both benthic and planktonic), peloids, intraclasts	Middle Ramp
SMF-4	Bioclastic Grainstone	Limestone	rudist debris, echinoderms, and bivalves	Inner Ramp—Shoal (Seaward)
SMF-5	Bioclast benthic foraminifera Packstone to Grainstone	Limestone	rudist debris (well sorted and rounded), benthic foraminifera, green algae intraclasts	Inner Ramp—Shoal (Central)
SMF-6	Benthic foraminifera peloidal Packstone	Limestone	peloids and benthic foraminifers, shell fragments (largely rudists and bivalve debris), and intraclasts	Inner Ramp—Shoal (Leeward)
SMF-7	Bioclast (rudist and coral) Rudstone/Floatstone	Limestone	rudist debris and green to red algae fragments, corals, echinoderms, and benthic foraminifera	Inner Ramp—Reef talus
SMF-8	Bioturbated benthic foraminifera peloidal Mudstone to Wackestone	(Dolomitic) Limestone	benthic foraminifera and peloids, rudist and green algae fragments	Inner Ramp—Lagoon

Detailed facies description and interpretation were not the aims of this study. By analyzing the sample's textures and sedimentary structures as well as fossil content and utilizing standard facies models, it can be inferred that they were deposited in a spectrum of depositional settings, ranging from shallow to deep settings of a ramp-like carbonate platform model.

The inner ramp setting includes mud-dominated lagoonal facies (SMF8), rudist patch reefs (SMF7), and shoal facies (SMF4 to 6). Middle ramp facies include wackestones to packstones containing echinoderms and a mixture of benthic and planktic fauna (SMF2 and 3). The outer ramp setting is represented by mud-dominated facies containing planktic foraminifera and clacispheres (SMF1) (Table 1 and Figure 2).

Diagenetic processes of the Sarvak Formation include micritization, isopachous cementation, recrystallization, dissolution, calcite (equant, drusy, bladed, and blocky) cementation, brecciation, silicification, pressure dissolution (stylolites and solution seams), dolomitization, pyritization, and fracturing (Figure 3). The paragenesis of these features in petrographic studies indicates a complex diagenesis history for this formation in marine, meteoric, and shallow-to-deep burial realms. Meteoric diagenesis was the main phase of diagenesis that altered the Sarvak Formation carbonates through intensive dissolution (karstification), cementation, brecciation, and development of paleosols (Figure 3).

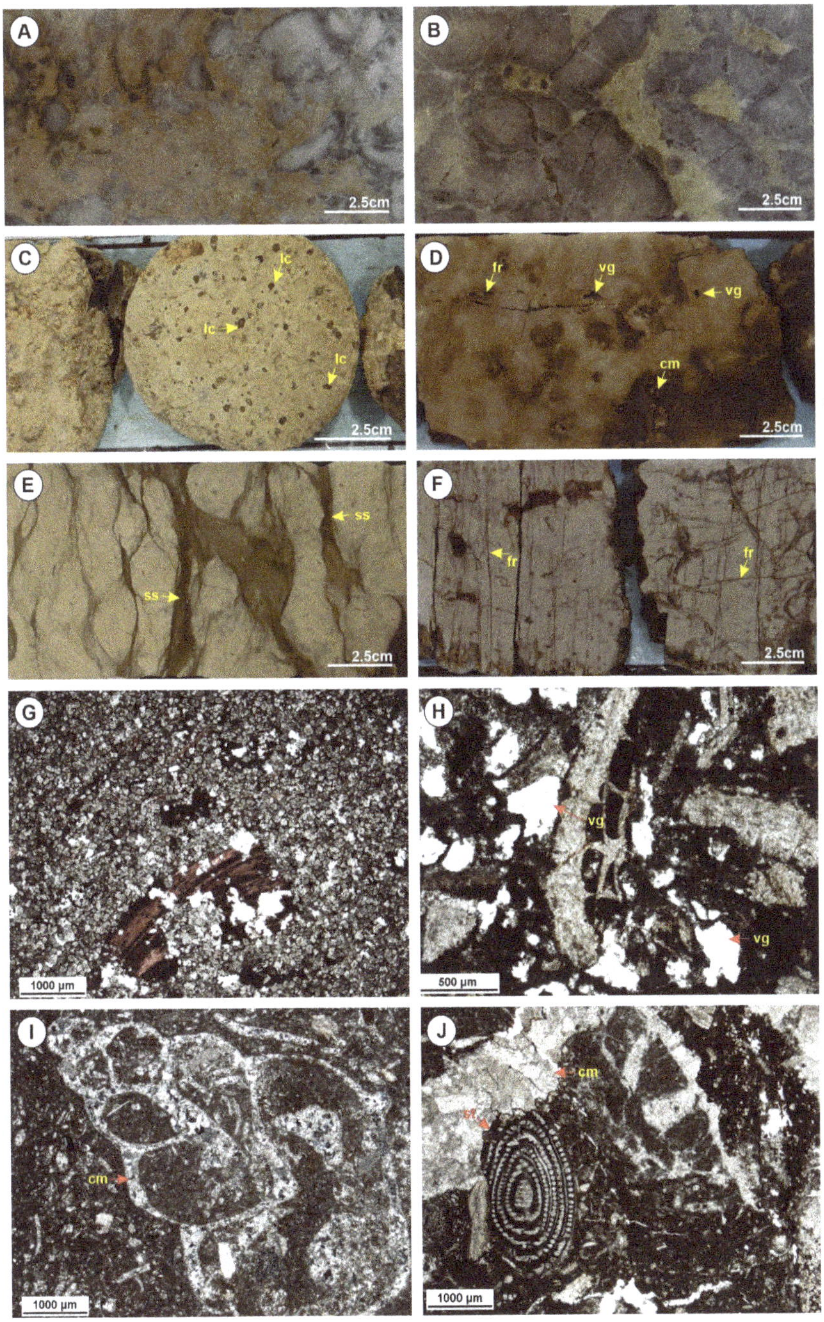

Figure 3. Some of the important diagenetic alterations of the Sarvak Formation in the Abadan Plain. They include brecciation (**A**,**B**); development of paleosol (**C**); meteoric dissolution (**D**,**H**); chemical compaction (**E**); fracturing (**F**); dolomitization (**G**); cementation (**I**,**J**). Abbreviations—cm: cementation, st: stylolite, vg: vug, fr: fracture, ss: solution seam, Ic: iron-rich concretion.

Cenomanian–Turonian strata of the Sarvak Formation comprise two third-order depositional sequences coded as SV_Cen (Cenomanian sequence) and SV_Tur (Turonian sequence) (Figures 4 and 5).

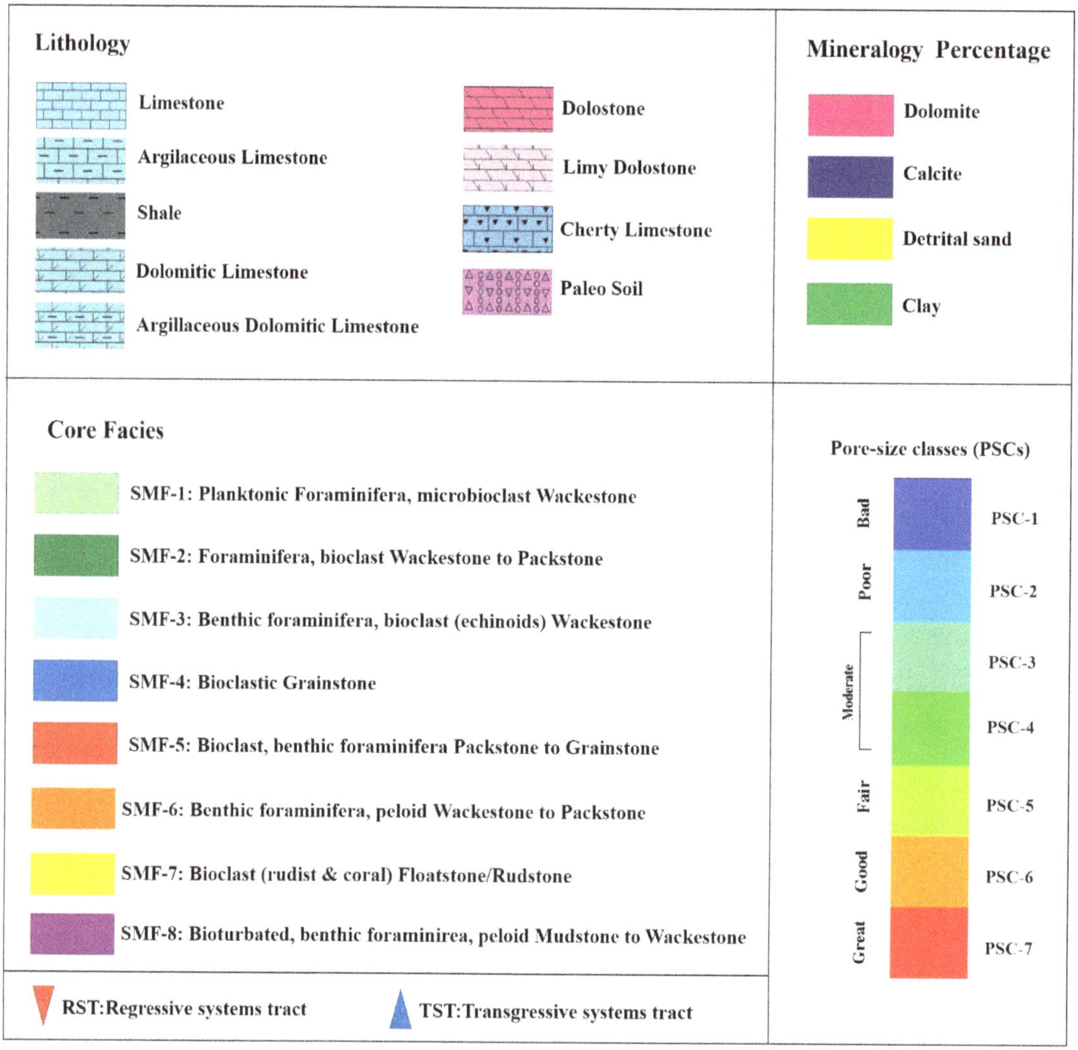

Figure 4. Legend for Figures 5, 13, and 15.

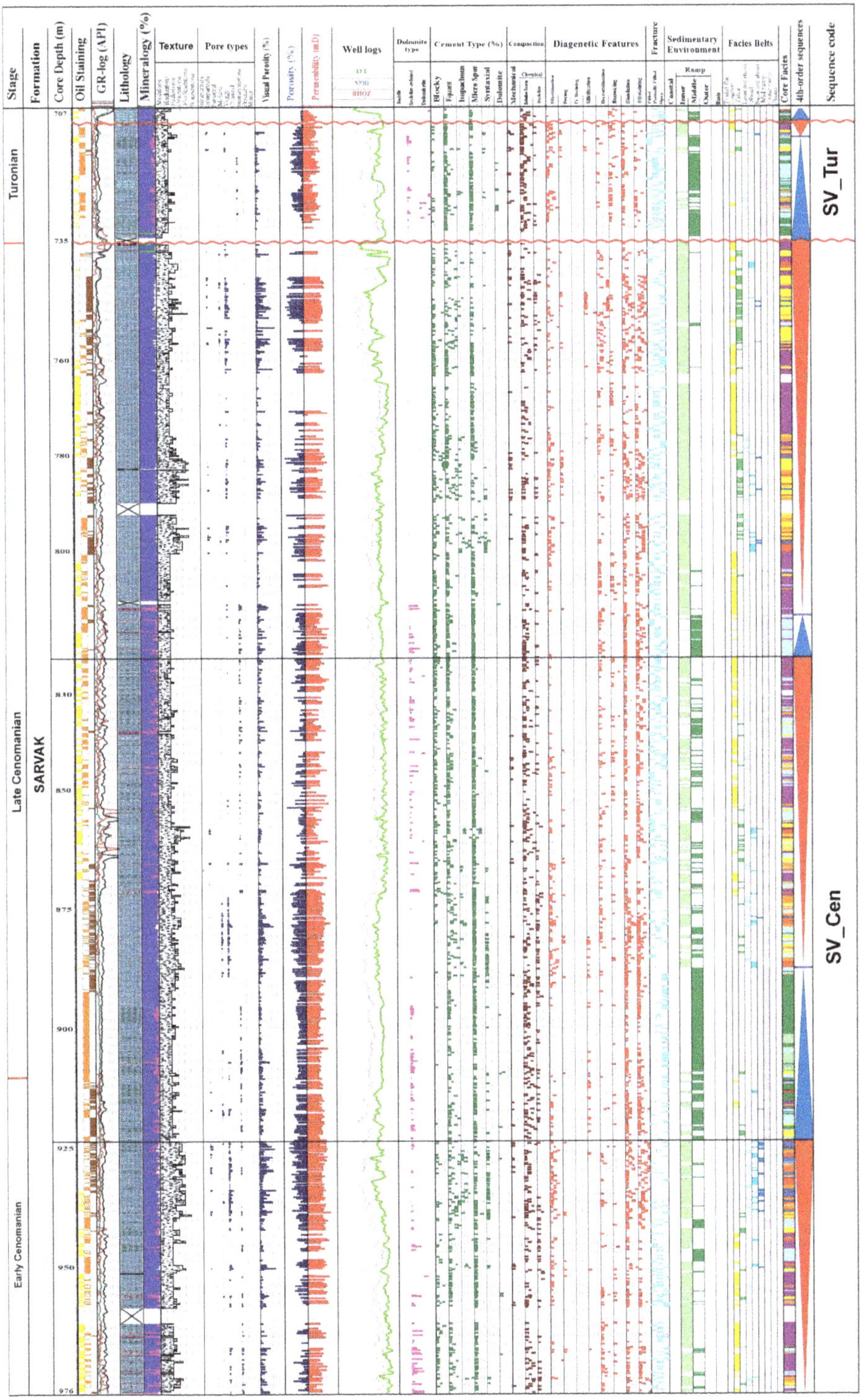

Figure 5. Sedimentological log of the Sarvak Formation in the Abadan Plain. See Figure 4 for legend.

Both sequences are bounded by two regional disconformable surfaces at the Cenomanian-Turonian boundary (CT-ES) and middle Turonian (mT-ES). Middle and outer ramp facies formed the transgressive systems tracts (TST), and inner ramp facies shaped the regressive systems tracts (RST) of these sequences. The RSTs are extensively affected by meteoric diagenetic processes.

4.2. Petrographic Pore Types

According to Choquette and Pray [42], various pore types are recorded in the Sarvak Formation, including the primary and secondary pores. These pores are evident at both macroscopic (core samples) and microscopic (thin sections) scales.

4.2.1. Primary Pore Types

This group includes interparticle, intra-skeletal, and growth-framework pores (Figure 6). Interparticle pores are predominantly observed between the allochems (skeletal and non-skeletal) in grain-supported facies with a packstone to grainstone texture, which was deposited in high-energy shoal settings (i.e., SMF4, 5, and 6; Table 1). Intra-skeletal or intra-particle and growth-framework pores are commonly recorded within the unfilled chambers of rudists, corals, and algae in reef-talus facies (i.e., SMF7; Table 1).

4.2.2. Secondary Pore Types

Secondary pore types encompass vuggy, moldic, intercrystalline, and fracture pores (Figure 6). Vuggy pores are commonly found in dissolved shoal and reef-talus facies (Figure 5). The moldic pores are observed in grain-dominated facies, especially in rudist facies, where only the molds of dissolved rudist fragments remain. Both vuggy and moldic pores were formed as a result of intense meteoric dissolution in rudist-dominated facies below the paleoexposure surfaces (i.e., CT–ES and mT–ES in Figure 5).

Intercrystalline pores between the dolomite crystals are sporadically observed, primarily within the lower parts of the Sarvak Formation (Figure 6). This occurs in mud-dominated facies (mudstone and wackestone) influenced by stylolite-related (burial) dolomitization. Fractures are frequently recorded across various parts of the Sarvak Formation, with a higher prevalence in the upper parts, particularly below the disconformities, where karstification has occurred.

The frequency of various pore types in the Sarvak Formation is measured based on the microscopic image analysis of this formation in the studied field. Results are presented as a frequency diagram of pore types in Figure 6. According to this plot, primary (depositional) pores form 44% of the total pore volume of the formation in the Abadan Plain. A majority (56%) of pore types in this formation belong to the secondary (diagenetic) pores, such as the vuggy, moldic, intercrystalline, and fractures (Figure 7).

Among the primary pores, interparticle pore types are predominant, with a frequency of 24%. They are common pore types within the grain-supported facies. On the other hand, vuggy pores are the predominant secondary pore type of the Sarvak Formation, with a frequency of 29% (Figure 7). They are predominantly recorded from the meteorically dissolved, rudist-dominated facies.

4.3. Scanning Electron Microscopy (SEM)

The average diameter size of 10 μm is considered as the pore size threshold separating the macro- and micropores [50,51]. These micropores, as classified by Choquette and Pray [42], encompass primary and secondary pore types.

Scanning electron microscopy (SEM) images can be utilized for the characterization of pore networks on a microscopic scale. The term 'micropores' refers to pores that are only discernible in SEM images and are not visible even at the highest magnifications in thin-section microscopic scale.

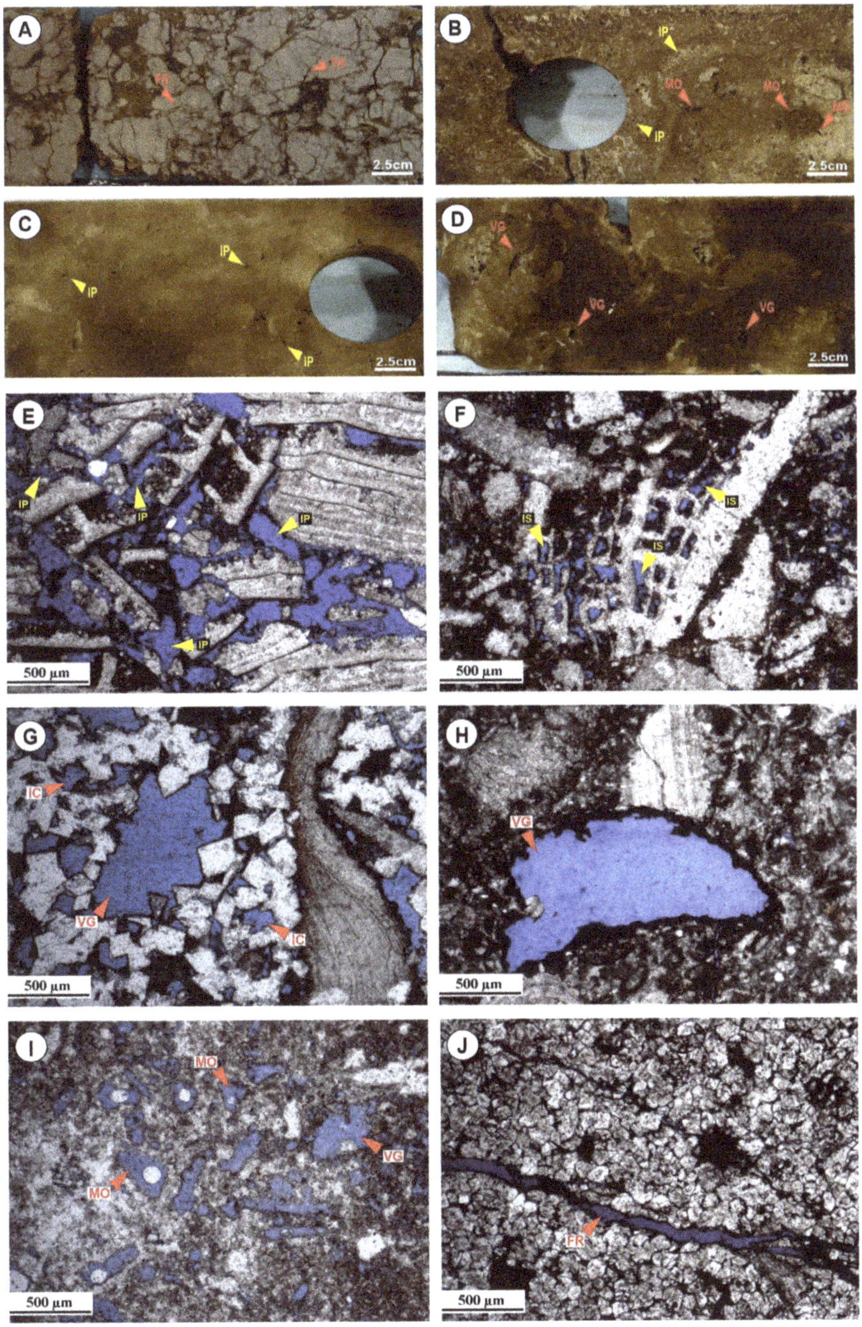

Figure 6. Core photos and photomicrographs of dominant pore types of the Sarvak Formation in the Abadan Plain. They include vuggy (**D**,**I**); moldic (**B**,**H**,**I**); interparticle (**B**,**C**,**E**); intra-skeletal (**F**); intercrystalline (**G**,**J**); and fracture (**A**,**J**). Abbreviations—IP: interparticle, IS: intraskeletal, MO: moldic, VG: vuggy, FR: fracture. Primary pores are marked by yellow arrows, and secondary pores are marked by red arrows.

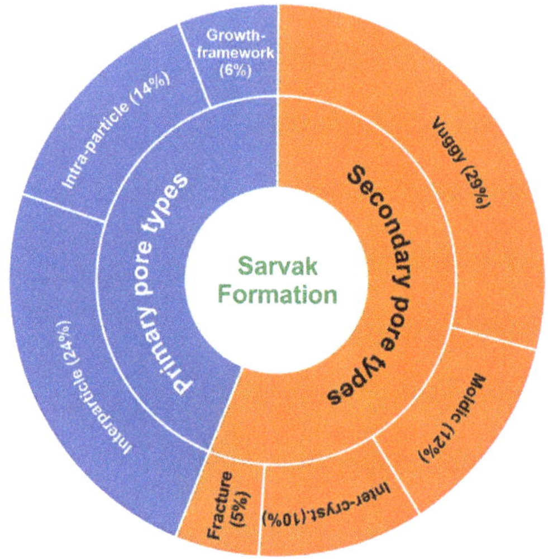

Figure 7. Frequency diagram of various pore types of the Sarvak Formation in the Abadan Plain.

4.3.1. Primary Micropores

Primary micropores comprise interparticle pores (between micrite particles) and intra-skeletal pores within the chambers of foraminifers and in the central parts of clacispheres (Figure 8). Various micrite textures can be observed in SEM images from different facies of the Sarvak Formation. Common micrite textural classes include rounded, sub-rounded, angular, semi-compact, and compact textures. Distinct pore spaces are evident between micrite particles in each of these textures. Notably, significant values of intra-particle or intra-skeletal micropores can be observed in the mud-dominated, open-marine facies of this formation, including the outer ramp and middle ramp facies (i.e., SMF1, 2, and 3; Table 1).

4.3.2. Secondary Micropores

Secondary micropores include intercrystalline, fracture, dissolution vugs, and moldic pores. Dissolution of micrite particles at the microscopic scale by meteoric or burial fluids leads to the formation of vugs that may be touching or separated, as well as dissolution molds within the Sarvak Formation (Figure 8). These pores are frequently recorded in mud-dominated facies. Pore spaces between dolomites and crystals of calcite cement are distinguishable in SEM images and categorized as intercrystalline pores (Figure 8).

4.4. Pore Size Classes

Porosity–permeability data of the Sarvak Formation are plotted on Winland's standard diagram, and seven pore types, named pore-size classes (PSC), are differentiated according to the iso-pore-throat lines and calculated R_{35} values (Figure 9). Summarized characteristics of pore-size classes are presented in Table 2.

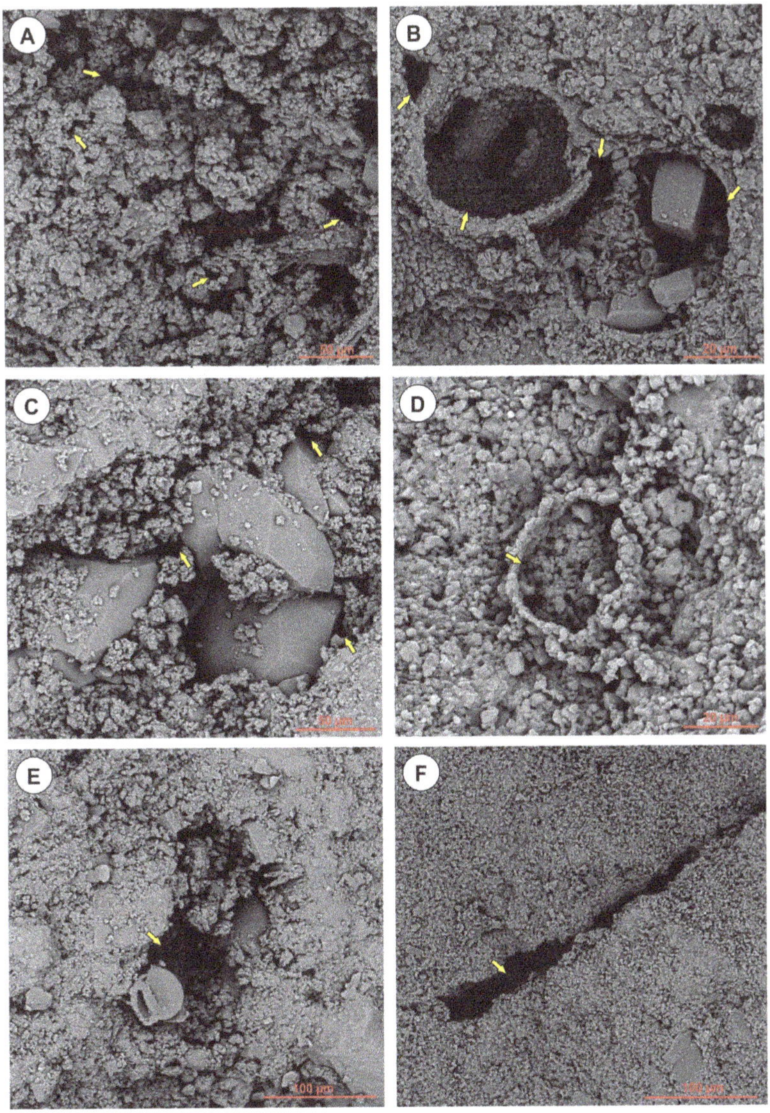

Figure 8. Scanning electron microscopy (SEM) images of dominant micropore types of the Sarvak Formation in the Abadan Plain. They include interparticle (**A**); intraparticle (**B**); intercrystalline (**C**); micro-moldic (**D**); micro-vuggy (**E**); and fracture (**F**). Arrows showing the mentioned pore types.

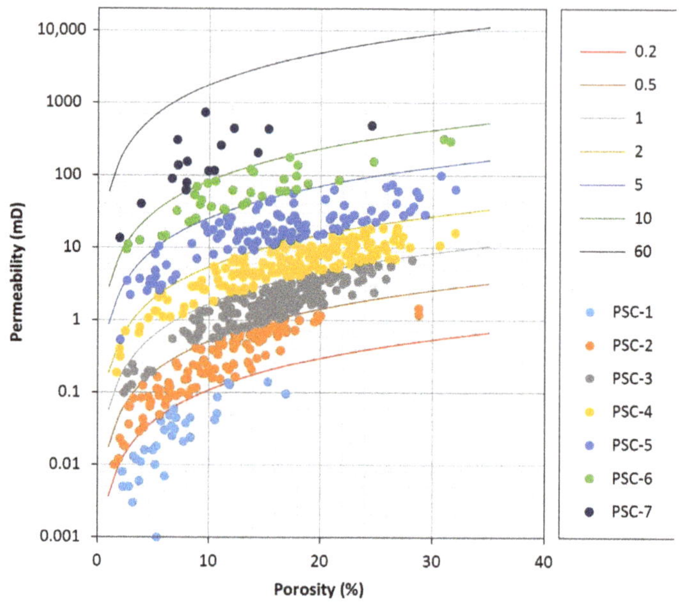

Figure 9. Porosity–permeability cross plot showing seven pore size classes (PSCs) identified in the Sarvak Formation, Abadan Plain.

Table 2. Summarized characteristics of pore-size classes of the Sarvak Formation in the studied field, Abadan Plain.

Pore-Size Class (PSC)	R_{35} Range (μm)	Porosity Range (%)	Permeability Range (mD)	Dominant Facies Types	Dominant Pore Types
PSC-1	<0.2	2.2 to 16.9	0.001 to 0.14	SMF1 to 3	Micropores
PSC-2	0.2 to 0.5	1.5 to 28.8	0.01 to 1.48	SMF1 to 3, SMF8	Micropores, Fractures
PSC-3	0.5 to 1	2.3 to 28.2	0.1 to 6.76	SMF3 to 6	Interparticle, Moldic
PSC-4	1 to 2	1.7 to 32	0.18 to 19.2	SMF4 to 7	Interparticle, Intraparticle, Moldic, Vuggy
PSC-5	2 to 5	21.6 to 32	28.7 to 101	SMF5, 7, 8	Vuggy, Moldic, Interparticle
PSC-6	5 to 10	2.5 to 31.5	9.5 to 321	SMF5, 7, 8	Fractures, Vuggy, Moldic, Interparticle
PSC-7	>10	1.3 to 30	0.18 to 19.2	SMF7, 8	Vuggy, Moldic

According to this table, PSCs 1 and 2 encompass open marine, mud-dominated facies (SMFs 1 to 3) in which micropores are dominant pore types. They are marked with a wide range of porosity (<2% to >28%) and low permeability (<1 mD) values. Fracturing enhanced the permeability of these samples to some extent, especially in PSC-2.

Pore-size classes 3 and 4 include mud- to grain-supported facies of shoal (SMFs 4 to 6) and reef-talus (SMF-7) settings in which interparticle and intraparticle pores are predominant pore types. Consequently, they show higher permeability values in comparison to PSCs 1 and 2.

Pore-size classes 5 to 7 correspond to intensively dissolved lagoonal (SMF-8) and reef-talus facies (SMF-7), in which vuggy, moldic, and fracture pores are the dominant pore types. The highest values of porosity (up to 32%) and permeability (up to 321 mD) are measured from these PSCs.

4.5. Mercury Injection Capillary Pressure (MICP)

Results of mercury injection capillary pressure (MICP) tests of the Sarvak Formation are plotted on Hg saturation vs. pressure and pore-radius vs. Hg saturation cross plots in Figure 10. As shown, most of the tested samples were selected from the high-quality reservoir intervals of the formation, and consequently, units of poor reservoir properties were not sampled for MICP analysis. Therefore, it was not possible in this study to perform a comprehensive pore-facies analysis according to these biased data. However, all tested samples can be categorized into three classes (Figure 11):

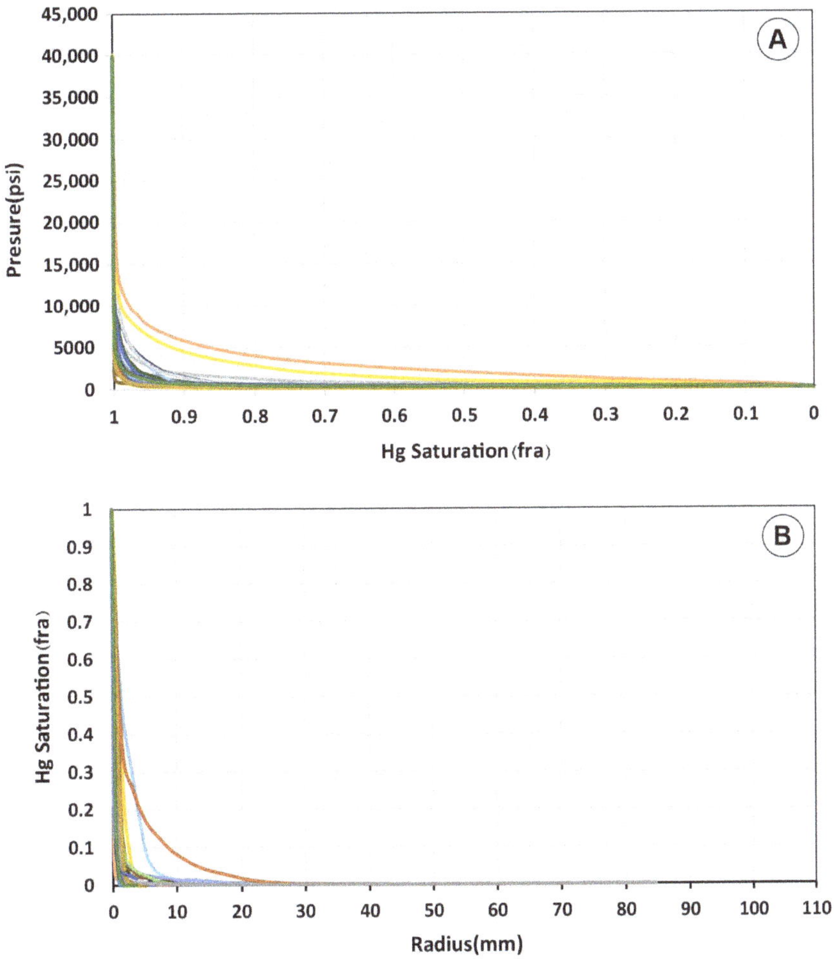

Figure 10. Cross plots of Hg saturation vs. pressure (**A**) and pore radius vs. Hg saturation (**B**) for analyzed samples in the Sarvak Formation, Abadan Plain. Lines with different colors belong to the different pore types.

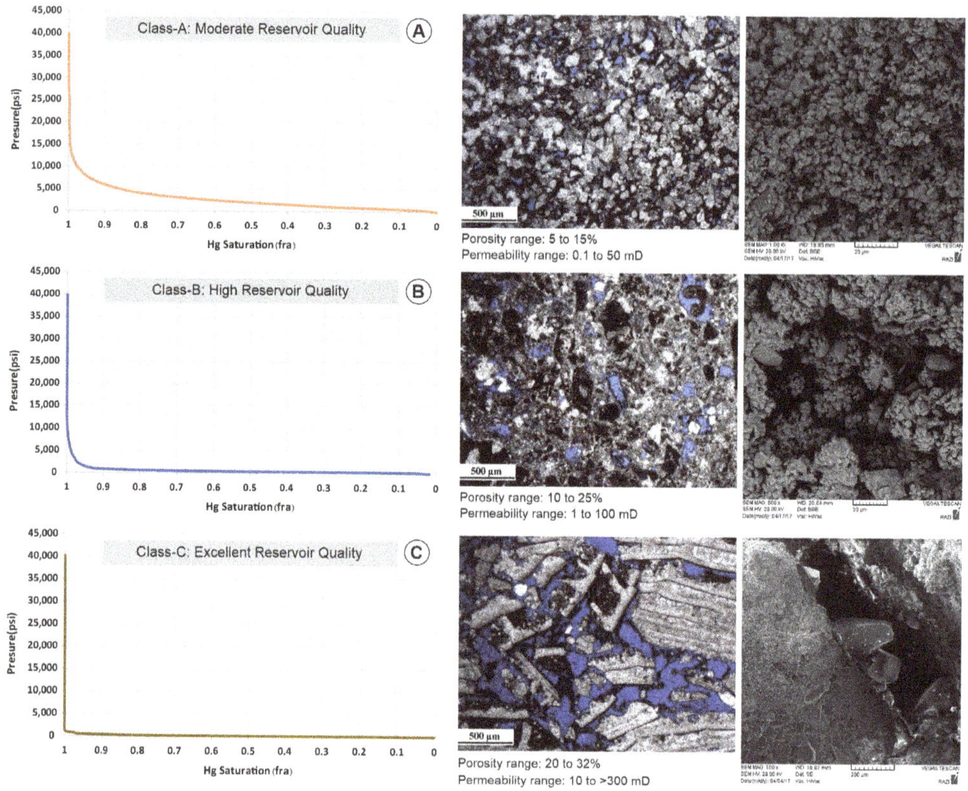

Figure 11. Cross plots of Hg saturation vs. pressure (**A–C**) and representative photomicrographs and SEM images for analyzed samples in the Sarvak Formation, Abadan Plain.

Class-A are samples of moderate reservoir quality that encompass recrystallized and dolomitized mud-dominated facies of the Sarvak Formation in which porosity ranges from 5% to 15%, and permeability varies between 0.1 mD to >50 mD. MICP curves of these samples shifted towards the right in the Hg saturation (fraction) versus pressure (psi) cross plot (Figure 11A).

Class-B are samples of high reservoir quality that include grain-dominated facies of middle ramp and shoal settings in which interparticle pores are the major pore type. High porosity (10% to 25%) and permeability (1 mD to >100 mD) values are measured from these samples. MICP curves of this class shifted toward the left in the Hg saturation (fraction) versus pressure (psi) cross plot (Figure 11B).

Class-C are samples of excellent reservoir quality, including the intensively dissolved reef-talus and shoal facies in which vuggy and moldic pores are the dominant pore type. The highest values of porosity (20% to 32%) and permeability (10 mD to >300 mD) and perpendicular MICP curves indicate excellent reservoir quality (Figure 11C).

4.6. Reservoir Zonation

A stratigraphic modified Lorenz plot (SMLP) is generated for the Sarvak Formation using the cumulative flow capacity (Kh_{cum}) and cumulative storage capacity ($Øh_{cum}$) (Figure 12). Distinct segments of the graph are differentiated based on the major inflection points, which correspond to different flow units. As a result, 13 reservoir and non-reservoir (i.e., barrier and baffle) zones are identified in the studied well (Figure 12). Using this plot, the studied interval can be categorized into four types of zones (Figure 13):

1. Reservoir zones, which are characterized by high flow and storage capacities (zones 4, 6, 7, and 8; Figure 13);
2. Barrier zones, which encompass zones with very low storage and flow capacities (zones 1, 3, 5, 9, and 13; Figure 13);
3. Speed zones, which are exemplified by zone 12 and exhibit a higher percentage of flow capacity relative to storage capacity (Figure 13);
4. Baffle zones, represented by zones 2 and 10, which induce turbulence in fluid movement and possess significantly greater storage capacity compared to flow capacity (Figure 13).

Each identified zone possesses a distinct pore-type system, thereby displaying specific flow behaviors [3]. Depending on the available data and the study's scale, it becomes possible to determine the type, size, and distribution of pore types, along with their corresponding pore throat diameters. Summarized statistical parameters and the sedimentological and reservoir characteristics of the Lorenz zones are presented in Table 3.

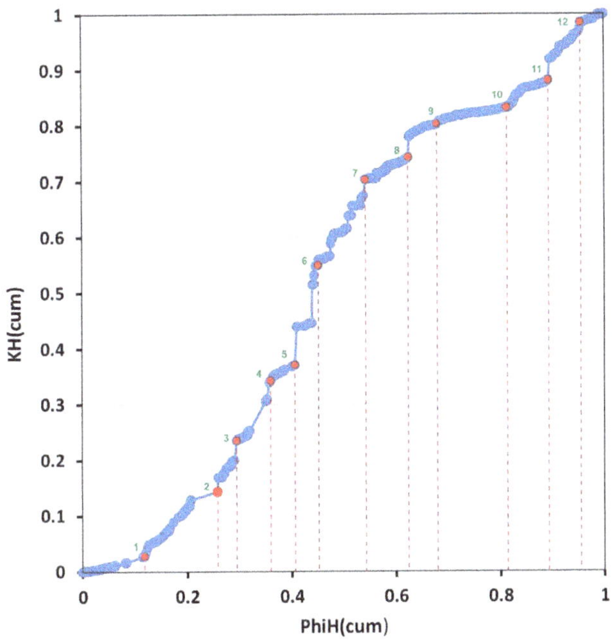

Figure 12. Cross plot of storage capacity (PhiH) vs. flow capacity (KH) and defined zones in the Sarvak Formation, Abadan Plain. According to this plot, 12 inflection points (numbered on the plot) and 13 zones have been identified in the reservoir.

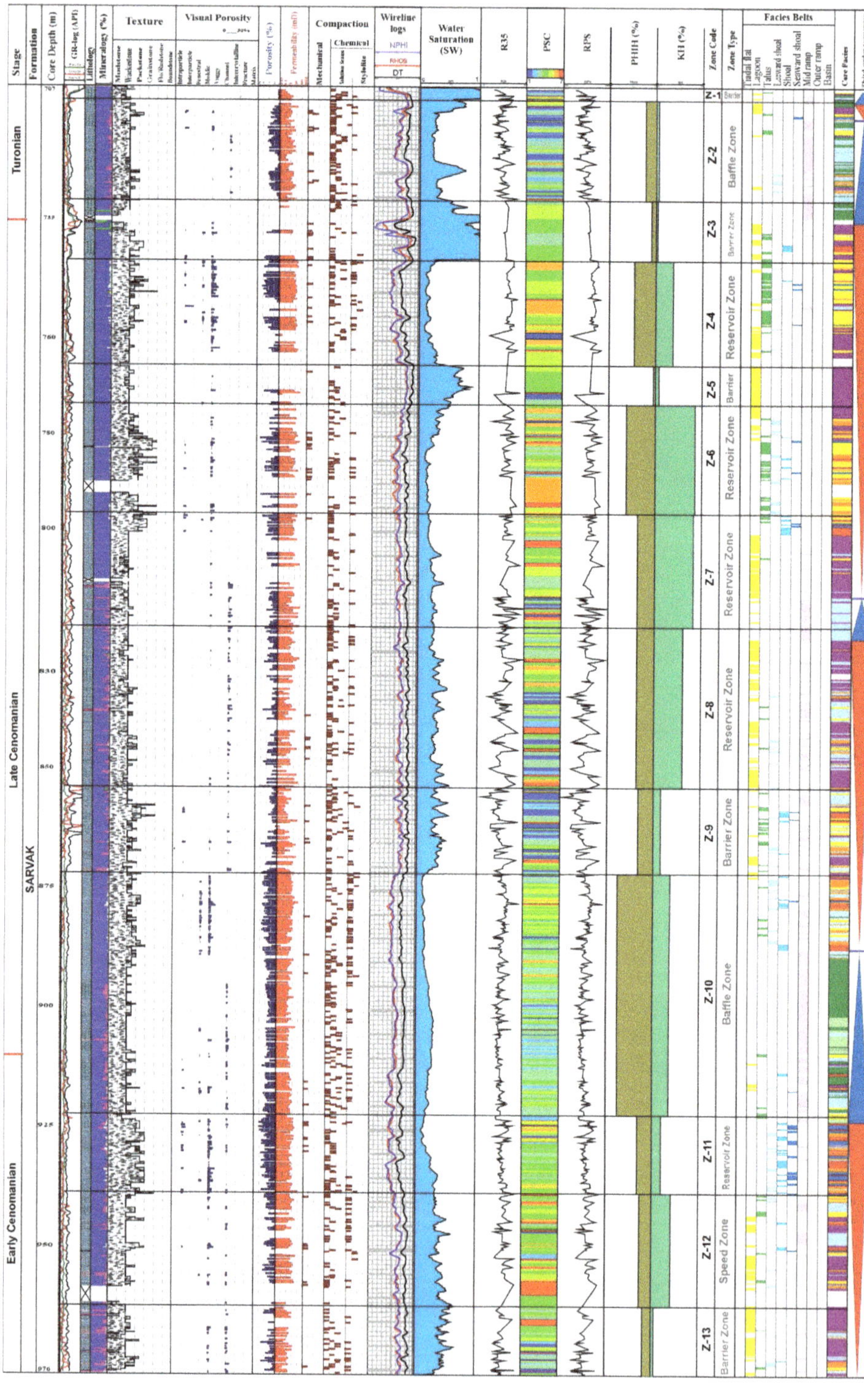

Figure 13. Reservoir quality log of the Sarvak Formation in the Abadan Plain. See Figure 4 for legend.

Table 3. Summarized characteristics of reservoir zones of the Sarvak Formation in the studied field, Abadan Plain.

Zone	Thickness (m)	KH (%)	PhiH (%)	RPS (Avg)	Poro (Avg)	Perm (Avg)	R35 (Avg)	Pore Size Class	PSC	SMF	Pore Type	Diagenesis	C
Z-1	3.38	0.03	0.29	22.3	3.7	0.71	0.08	Nano	1	SMF2	Microporosity	Compaction, Cementation	**Barrier**
Z-2	17.5	1.01	5.6	21.3	12.4	2.62	0.21	Micro	1, 3	SMF2,3	Vuggy Inter-crystalline	Compaction, Cementation, Neomor-phism	**Baffle**
Z-3	8.24	0.6	2.39	46	8.64	2.67	0.32	Micro	4	SMF2,8	Microporosity	Compaction, Cementation	**Barrier**
Z-4	16.4	9.38	11.4	127	20.8	24	0.39	Micro	5	SMF7	Vuggy, Moldic Intraskeletal Interparticle	Dissolution, Fracture	**Reservoir**
Z-5	1.97	1.86	0.91	271	12.6	33	0.44	Micro	4	SMF8	Microporosity	Compaction, Cementation	**Barrier**
Z-6	24.3	21.3	15.3	214	16.1	37.4	0.49	Micro	6	SMF7	Interparticle Intraparticle Vuggy, Moldic	Dissolution, Fracture	**Reservoir**
Z-7	23.2	20.7	9.19	318	14.6	33.6	0.59	Meso	4, 5	SMF7,8	Intercrystalline Interparticle Microporos-ity	Dissolution, Fracture	**Reservoir**
Z-8	32.7	15.3	9.19	178	11.3	18.1	0.74	Meso	3	SMF7,8	Microporosity Intercrys-talline	Dissolution, Fracture	**Reservoir**
Z-9	23.8	4.01	8.06	61.1	12.5	7.22	0.97	Meso	1, 3	SMF3	Microporosity Intercrys-talline	Compaction, Cementation	**Barrier**
Z-10	43.9	8.86	18.8	53.9	16.7	8.16	1.58	Meso	3	SMF2	Microporosity Intercrys-talline	Compaction, Cementation, Neomor-phism	**Baffle**
Z-11	16.0	4.94	7.82	77.2	19.3	17.1	2.72	Macro	4	SMF4–6	Interparticle	Dissolution, Fracture	**Reservoir**
Z-12	19.3	10.3	6.36	210	13.1	30.3	4.73	Macro	4	SMF8	Intercrystalline Microporos-ity Interparticle	Dolomitization and Fracturing	**Speed**
Z-13	14.2	1.55	4.57	64.7	11.1	4.89	12.3	Mega	3	SMF8	Intercrystalline Microporos-ity	Compaction, Cementation	**Barrier**

5. Discussion

5.1. Pore Type's Paragenesis

In this study, pore types of the Sarvak Formation are investigated in various scales. Accordingly, both primary and secondary pores are defined with strong effects on the petrophysical properties (i.e., porosity and permeability) of rocks. Understanding the relative time of occurrence and subsequent evolution of the pore network is essential for characterizing the final reservoir quality of studied sequences. Therefore, in this section, the paragenetic sequence of the Sarvak Formation is discussed with a special emphasis on its pore types and their evolutionary path, starting from deposition and passing through marine, meteoric, and shallow-to-deep burial environments (Figure 14).

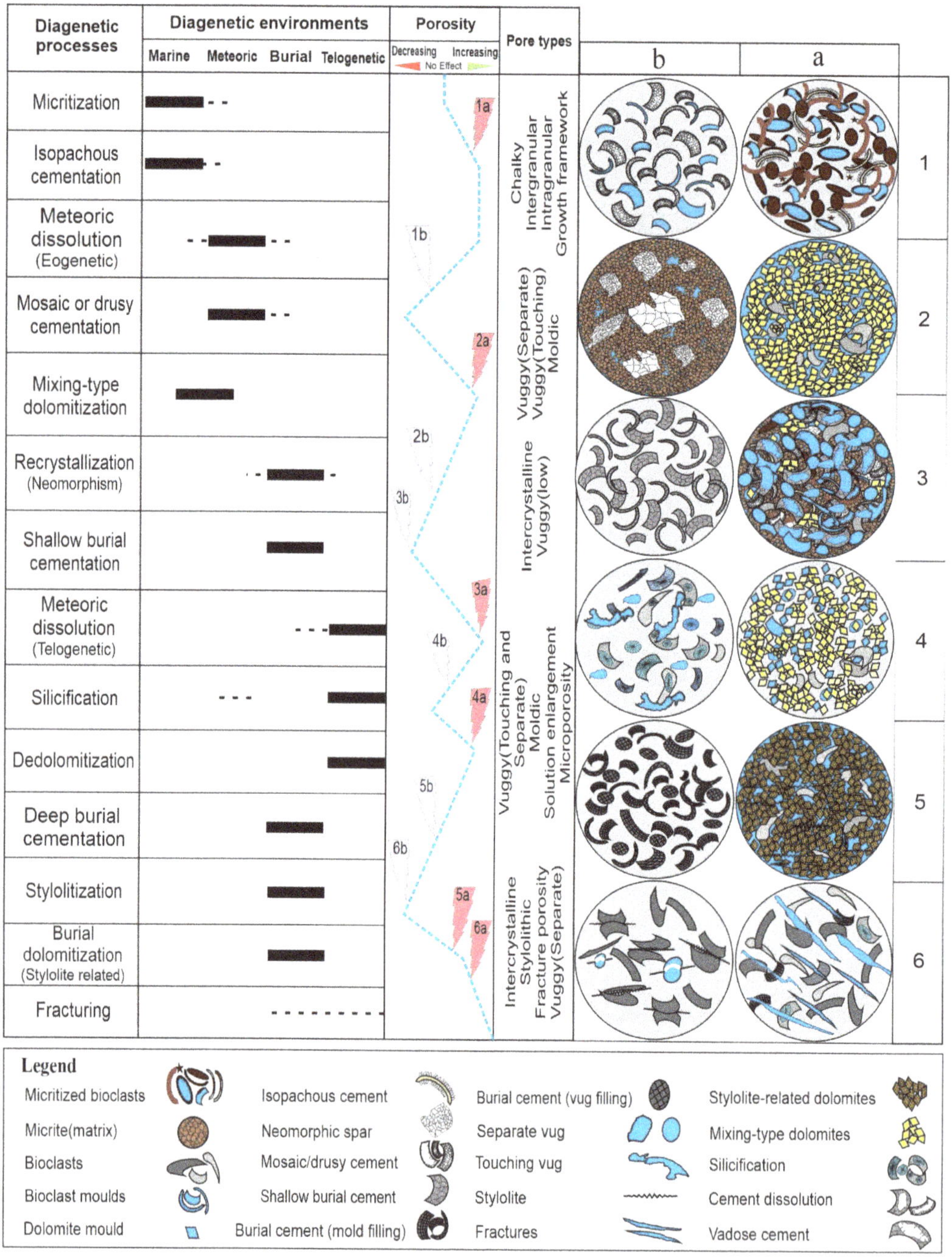

Figure 14. Paragenetic sequence of diagenetic processes and pore types of the Sarvak Formation in the Abadan Plain. Lables in porosity column indicate the representative cartoons on the right.

- Depositional pore-types

The primary (depositional) pore types, including interparticle (intergranular), intraparticle, and growth framework pores, form a major part of the pore spaces in the Sarvak Formation. The frequency analysis of pore types (Figure 6) showed that they form 44% of the total pore volume of the reservoir. Interparticle pores are the main primary pore types, which are usually recorded within the grain-supported facies of shoal and talus settings, which commonly correspond to the RSTs of depositional sequences.

The deposition of carbonate particles under the high-energy condition of these settings has resulted in texturally mature sediments with high levels of grain sorting, roundness, the scarcity of micrite, and, consequently, high volumes of primary pore spaces between the particles [52]. The lack of pore-occluding cements, such as drusy, blocky, and equant calcite, along with minor effects of compaction, have resulted in the preservation of pore spaces between the allochems and provide some high-quality (reservoir) units in the Sarvak Formation (e.g., zones 4, 6, 7, and 8 in Figure 15). Such facies are represented by Class-B of MICP curves and pore-size classes 3 and 4 of Winland's method.

Intra-particle and growth-framework pores are the second most important depositional pore types, which formed a total of 20% of pores in the Sarvak Formation. They are distinguished from coral-, algal-, and rudist-bearing facies (SMF7) in which intra-skeletal pore spaces are preserved. Such pore types are dominantly recorded from reservoir units of the Sarvak Formation (e.g., zones 4 and 6 in Figure 15). They are represented by Class-C of MICP curves and pore-size classes 3 and 4 of Winland's method.

Rudist–algal–coral facies are frequent facies types in the Upper Cretaceous carbonates of the Tethyan realm, including the Zagros Basin [53,54]. They form some important hydrocarbon reservoirs in the Middle East, and most of them are concentrated within the Sarvak Formation and its stratigraphic counterparts (e.g., Mishrif Formation) [55]. A major portion of the pore network in these facies includes primary (intra-skeletal and growth-framework) pores within the chambers of the mentioned skeletal grains.

- Diagenetic pore-types

Secondary or diagenetic pore types, including the moldic, vuggy, intercrystalline, and fracture pores, form the major portion of pore volume (56%) in the Sarvak Formation. Dissolution molds and vugs form 41% of the total pores within this reservoir (Figure 6). They are limited to the dissolved lagoonal (SMF8), reef-talus (SMF7), and shoal facies (SMFs 4 to 6) that are stratigraphically located below the paleoexposure surfaces (sequence boundaries) at the Cenomanian–Turonian boundary (CT-ES) and middle Turonian (mT-ES) (Figure 15).

The best productive zones of the Sarvak Formation were formed within these dissolved units with high amounts (15% to 32%) of porosity, which substantially includes vuggy and moldic pore types (e.g., zones 4, 6, 7, and 8 in Figure 15). Such pore facies are represented by Class-C of the MICP curves and pore-size classes 5 to 7 of Winland's method. From a sequence stratigraphic point of view, these pore types are concentrated within the late RSTs of depositional sequences.

Subaerial exposure of carbonate platforms during the relative sea-level falls, under the warm and humid (tropical) climatic condition governing the Zagros Basin during the Late Cretaceous, has resulted in the development of extensively dissolved (namely karstified) carbonate successions with a great potential for hydrocarbon accumulation [14]. Meteoric dissolution (karstification) was the most important diagenetic process in the Sarvak Formation that provided vuggy and moldic pores, especially below the paleoexposure surfaces [5,19,21,23,24,33]. In some cases, long-term exposures have resulted in the development of over-matured (brecciated) karstified intervals with poor reservoir properties (barrier zone). This happens at the topmost parts of the Cenomanian and Turonian sequences (see photos C and D in Figure 15).

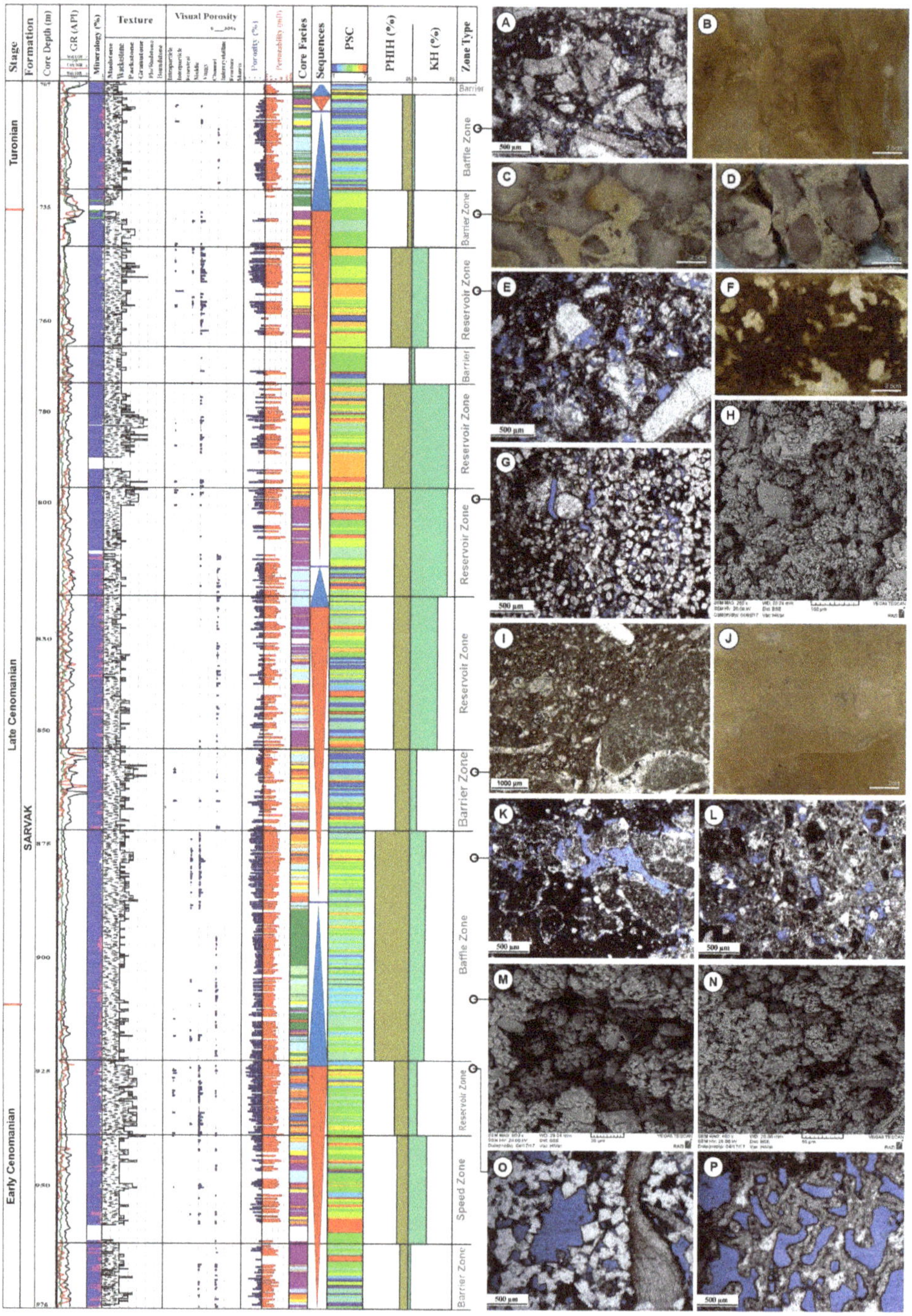

Figure 15. Stratigraphic occurrence of reservoir zones and their dominant pore types in the Sarvak Formation in the Abadan Plain. Depositional textures, petrographic pore types, porosity–permeability profiles, and depositional sequences are also shown. Photos (**A**–**P**) represent the main zones. See Figure 4 for legend.

Intercrystalline pores form 10% of the total pore spaces of the Sarvak Formation in the studied field. They are dominant pore types in the dolomitized intervals, especially in the lower half part of the formation (e.g., zone 7 in Figure 15). As mentioned in many previous studies, dolomitization routinely improves the permeability of carbonate rocks by providing an interconnected network of intercrystalline pores [56–58]. Dolomites of the Sarvak Formation were mainly formed in the deep burial realm, along with the chemical compaction features, including the solution seams and stylolites [34].

Fractures are important features in the Sarvak Formation. They are recorded on the core samples and in thin sections with variable lengths, widths, and filling percentages. Most of the fractures are open, and some of them are filled, partially to completely, by calcite cements. It seems that most of the fractures within the karstified units of the Sarvak Formation were diagenetic in origin. Far from the karstified intervals, tectonically induced fractures are recorded in this formation. A detailed fracture analysis of this formation in the Abadan Plain was previously presented by Ezati et al. [59].

5.2. Micropores

Microporous mud-dominated carbonate sequences, known as tight carbonates, are among the important unconventional reservoirs in the Middle East [60]. In Iran, they are commonly concentrated within the Cretaceous rock record in the Zagros Basin [61]. Microporous intervals in the lower half parts of the Sarvak and Ilam formations have been previously reported from the Abadan Plain [11].

In this study, various types of micropores were detected from middle to outer ramp facies (SMFs 1 to 3) and lagoonal facies (SMF8) in the Sarvak Formation. Stratigraphically, microporous open marine facies are located in the lower parts of the formation, in the TST of the Cenomanian sequence (zone 9 in Figure 15). As shown, this zone is detected as a baffle zone with high storage capacity and low flow capacity, which is an expected trend in microporous reservoirs (Figure 15, zone 9).

A major part of micro-pore types in mud-dominated facies includes interparticle pores between the micrite particles, which are well-documented by SEM image analysis (Figure 7). In open marine facies, intra-skeletal pores are recognizable within the unfilled chambers of planktic foraminifera and in the central parts of clacispheres. Moreover, microscopic dissolution provided some micro-vug and micro-mold pore types within the mud-dominated facies (lower parts of zone 10 in Figure 15). Microscopic dissolution of micrite was the subject of some previous studies in the Zagros Basin [11] and other places in the world [62].

6. Conclusions

The Pore network of the Sarvak Formation, the second important oil reservoir of Iran, is investigated using the petrographic and petrophysical data in the Abadan Plain. Petrographic studies of whole cores, core plugs, thin sections, and scanning electron microscopy (SEM) images resulted in the recognition of primary (depositional), secondary (diagenetic), and fracture pores in macroscopic, mesoscopic, and microscopic scales.

Porosity–permeability and mercury injection capillary pressure (MICP) data are used to discriminate the pore-size classes of the Sarvak Formation. MICP curves showed three distinct classes of excellent-to-moderate reservoir quality. Micropore types of the formation are differentiated using the SEM image analysis. The Sarvak Formation is subdivided into 13 reservoir, speed, barrier, and baffle zones according to the stratigraphic modified Lorenz plot (SMLP). The sedimentological and petrophysical properties and dominant pore types of these zones were discussed.

The frequency analysis of the pores indicated that the majority of pore types in the Sarvak Formation are diagenetic in origin. Vuggy, moldic, intercrystalline, and fracture pores formed 56% of the total pore spaces in this formation. Among these, vuggy and moldic pores are predominant pore types within the meteorically dissolved intervals located below the paleoexposure surfaces at the upper half part of the Sarvak Formation.

Away from the paleoexposure surfaces, in the lower half part of the formation, depositional pores become predominant. They form 44% of the total pore spaces of the Sarvak Formation in forms of interparticle, intra-particle (intra-skeletal), and growth-framework pore types within the grain-supported facies of shoal complexes and reefal facies.

Microporous mud-dominated facies provide some baffle zones in the lowermost parts of the Sarvak Formation. In these units, both depositional and diagenetic pore types are frequently recognizable in the SEM images. Interparticle pores between the micrite particles, intra-skeletal pores within the chambers of foraminifera and in the central parts of clacispheres, intercrystalline pores in the dolomitized intervals, and fractures are predominant pore types within these microporous units. They are majorly formed within the open marine (middle to outer ramp) facies and, with lesser importance, in lagoonal facies with mudstone-to-wackestone textures.

Author Contributions: Conceptualization, H.M., F.K., N.F.-S. and J.H.; methodology, H.M., F.K. and N.F.-S.; software, H.M., F.K. and N.F.-S.; validation, H.M.; formal analysis, H.M., F.K., N.F.-S. and J.H.; investigation, F.K. and N.F.-S.; resources, H.M., F.K., N.F.-S. and J.H.; data curation, F.K. and N.F.-S.; writing—original draft preparation, F.K. and N.F.-S.; writing—review and editing, H.M.; visualization, H.M., F.K., N.F.-S. and J.H.; supervision, H.M.; project administration, H.M. All authors have read and agreed to the published version of the manuscript.

Funding: This research received no external funding.

Data Availability Statement: The data presented in this study are available upon request from the corresponding author.

Acknowledgments: The University of Tehran and the Research Institute of Petroleum Industry (RIPI) are acknowledged for their data and research facility support.

Conflicts of Interest: The authors declare no conflict of interest. We confirm that the manuscript has been read and approved by all named authors and that there are no other persons who satisfied the criteria for authorship but are not listed. We further confirm that the order of authors listed in the manuscript has been approved by all of us. Finally, we confirm this manuscript is not being submitted to any journal.

References

1. Lucia, F.J. *Carbonate Reservoir Characterization: An Integrated Approach*; Springer: Berlin/Heidelberg, Germany, 2007; ISBN 978-3-540-72742-2.
2. Moore, C.H.; Wade, W.J. Carbonate Reservoirs. In *Porosity and Diagenesis in a Sequence Stratigraphic Framework*, 2nd ed.; Elsevier: Amsterdam, The Netherlands, 2013.
3. Mehrabi, H.; Bagherpour, B. Scale, origin, and predictability of reservoir heterogeneities in shallow-marine carbonate sequences: A case from Cretaceous of Zagros, Iran. *J. Pet. Sci. Eng.* **2022**, *214*, 110571. [CrossRef]
4. Tavoosi Iraj, P.; Mehrabi, H.; Rahimpour-Bonab, H.; Ranjbar-Karami, R. Quantitative analysis of geological attributes for reservoir heterogeneity assessment in carbonate sequences; A case from Permian–Triassic reservoirs of the Persian Gulf. *J. Pet. Sci. Eng.* **2021**, *200*, 108356. [CrossRef]
5. Foroshani, J.S.; Mehrabi, H.; Rahimpour-Bonab, H. Reservoir heterogeneity of Upper Cretaceous Sarvak Formation in the Dezful Embayment, SW Iran: Implications of flow unit distribution, electrofacies analysis and geological-based reservoir zonation. *J. Afr. Earth Sci.* **2023**, *200*, 104882. [CrossRef]
6. Chehrazi, A.; Rezaee, R.; Rahimpour, H. Pore-facies as a tool for incorporation of small-scale dynamic information in integrated reservoir studies. *J. Geophys. Eng.* **2011**, *8*, 202–224. [CrossRef]
7. Hollis, C.; Lawrence, D.A.; Deville de Perière, M.; Al Darmaki, F. Controls on porosity preservation within a Jurassic oolitic reservoir complex, UAE. *Mar. Pet. Geol.* **2017**, *88*, 888–906. [CrossRef]
8. Assadi, A.; Honarmand, J.; Moallemi, S.A.; Abdollahie-Fard, I. An integrated approach for identification and characterization of Palaeo-exposure surfaces in the Upper Sarvak Formation of Abadan Plain, SW Iran. *J. Afr. Earth Sci.* **2018**, *94*, 1432–1450. [CrossRef]
9. Mahaman Salifou, I.A.; Zhang, H.; Boukari, I.O.; Harouna, M.; Cai, Z. New vuggy porosity models-based interpretation methodology for reliable pore system characterization, Ordovician carbonate reservoirs in Tahe oilfield, North Tarim Basin. *J. Pet. Sci. Eng.* **2021**, *196*, 1077–1100. [CrossRef]
10. Tavakoli, V. Microscopic heterogeneity. In *Carbonate Reservoir Heterogeneity*; Springer: Berlin/Heidelberg, Germany, 2020; pp. 17–51.

11. Mehrabi, H.; Bagherpour, B.; Honarmand, J. Reservoir quality and micrite textures of microporous intervals in the Upper Cretaceous successions in the Zagros area, SW Iran. *J. Pet. Sci. Eng.* **2020**, *192*, 107292. [CrossRef]
12. Mehrabi, H.; Bahrehvar, M.; Rahimpour-Bonab, H. Porosity evolution in sequence stratigraphic framework: A case from cretaceous carbonate reservoir in the Persian Gulf, southern Iran. *J. Pet. Sci. Eng.* **2021**, *196*, 107699. [CrossRef]
13. Ranjbar-Karami, R.; Tavoosi Iraj, P.; Mehrabi, H. Integrated rock typing and pore facies analyses in a heterogeneous carbonate for saturation height modelling, A case study from Fahliyan Formation, the Persian Gulf. *J. Pet. Explor. Prod.* **2021**, *11*, 1577–1595. [CrossRef]
14. Mehrabi, H. Deposition, diagenesis, and geochemistry of Upper Cretaceous carbonates (Sarvak Formation) in the Zagros Basin and the Persian Gulf, Iran. *Minerals* **2023**, *13*, 1078. [CrossRef]
15. Abdollahie-Fard, I.; Braathen, A.; Mokhtari, M.; Alavi, S.A. Interaction of the Zagros Fold-Thrust Belt and the Arabian-Type, Deep-Seated Folds in the Abadan Plain and the Dezful Embayment, SW Iran. *Pet. Geosci.* **2006**, *12*, 347–362. [CrossRef]
16. Abdollahie-Fard, I.; Sherkati, S.; McClay, K.; Haq, B.U. Tectono-sedimentary evolution of the Iranian Zagros in a global context and its impact on petroleum habitats. In *Developments in Structural Geology and Tectonics*; Elsevier: Amsterdam, The Netherlands, 2019; Volume 3, pp. 17–28.
17. Assadi, A.; Honarmand, J.; Moallemi, S.A.; Abdollahie-Fard, I. Depositional environments and sequence stratigraphy of the Sarvak Formation in an oil field in the Abadan Plain, SW Iran. *Facies* **2016**, *62*, 1–22. [CrossRef]
18. Razin, P.; Taati, F.; van Buchem, F.S.P. Sequence stratigraphy of Cenomanian-Turonian carbonate platform margins (Sarvak Formation) in the High Zagros, SW Iran: An outcrop reference model for the Arabian Plate. *Geol. Soc. Spec. Publ.* **2010**, *329*, 187–218. [CrossRef]
19. Hajikazemi, E.; Al-Aasm, I.S.; Coniglio, M. Subaerial exposure and meteoric diagenesis of the Cenomanian-Turonian Upper Sarvak Formation, southwestern Iran. *Geol. Soc. Spec. Publ.* **2010**, *330*, 253–272. [CrossRef]
20. Hajikazemi, E.; Al-Aasm, I.S.; Coniglio, M. Chemostratigraphy of Cenomanian-Turonian Carbonates of the Sarvak Formation, Southern Iran. *J. Pet. Geol.* **2012**, *35*, 187–205. [CrossRef]
21. Navidtalab, A.; Rahimpour-Bonab, H.; Huck, S.; Heimhofer, U. Elemental geochemistry and strontium-isotope stratigraphy of Cenomanian to Santonian neritic carbonates in the Zagros Basin, Iran. *Sediment. Geol.* **2016**, *346*, 35–48. [CrossRef]
22. Navidtalab, A.; Heimhofer, U.; Huck, S.; Omidvar, M.; Rahimpour-Bonab, H.; Aharipour, R.; Shakeri, A. Biochemostratigraphy of an Upper Albian–Turonian Succession from the Southeastern Neo-Tethys Margin, SW Iran. *Palaeogeogr. Palaeoclimatol. Palaeoecol.* **2019**, *533*, 109255. [CrossRef]
23. Mehrabi, H.; Navidtalab, A.; Enayati, A.; Bagherpour, B. Age, duration, and geochemical signatures of paleo-exposure events in Cenomanian–Santonian Sequences (Sarvak and Ilam Formations) in SW Iran: Insights from carbon and strontium isotopes chemostratigraphy. *Sediment. Geol.* **2022**, *434*, 106136. [CrossRef]
24. Mehrabi, H.; Navidtalab, A.; Rahimpour-Bonab, H.; Heimhofer, U. Geochemical expression of sequence stratigraphic surfaces: A case from Upper Cretaceous shallow-water carbonates of southeastern Neo-Tethys Margin, SW Iran. *Cretac. Res.* **2022**, *140*, 105329. [CrossRef]
25. Alsharhan, A.S. Petroleum systems in the Middle East. In *Tectonic Evolution of the Oman Mountains*; Rollinson, H.R., Searle, M.P., Abbasi, A.I., Al-Lazki, A.I., Al Kindi, M.H., Eds.; Geological Society, London, Special Publications: London, UK, 2014; Volume 392, pp. 361–408.
26. Hollis, C. Diagenetic controls on reservoir properties of carbonate successions within the Albian-Turonian of the Arabian Plate. *Pet. Geosci.* **2011**, *17*, 223–241. [CrossRef]
27. Huber, B.T.; Norris, R.D.; MacLeod, K.G. Deep-sea paleotemperature record of extreme warmth during the Cretaceous. *Geology* **2002**, *30*, 123–126. [CrossRef]
28. Sharland, P.R.; Archer, D.M.; Casey, R.B.; Davies, S.H.; Hall, A.P.; Heward, A.D.; Horbury, A.D.; Simmons, M.D. *Arabian Plate Sequence Stratigraphy*; GeoArabia Special Publication 2; Gulf Petrolink: Manama, Bahrain, 2001; Volume 2, p. 371.
29. Scotese, C.R. An atlas of phanerozoic paleogeographic maps: The seas come in and the seas go out. *Annu. Rev. Earth Planet. Sci.* **2021**, *49*, 679–728. [CrossRef]
30. Keller, G.; Adatte, T.; Berner, Z.; Chellai, E.H.; Stueben, D. Oceanic events and biotic effects of the Cenomanian-Turonian anoxic event, Tarfaya Basin, Morocco. *Cretac. Res.* **2008**, *29*, 976–994. [CrossRef]
31. Immenhauser, A.; Creusen, A.; Esteban, M.; Vonhof, H.B. Recognition and interpretation of polygenic discontinuity surfaces in the middle Cretaceous Shu'aiba, Nahr Umr, and Natih Formations of Northern Oman. *GeoArabia* **2000**, *5*, 299–322. [CrossRef]
32. Mehrabi, H.; Yahyaei, E.; Navidtalab, A.; Rahimpour-Bonab, H.; Abbasi, R.; Omidvar, M.; Assadi, A.; Honarmand, J. Depositional and diagenetic controls on reservoir properties along the shallow-marine carbonates of the Sarvak Formation, Zagros Basin: Petrographic, petrophysical, and geochemical evidence. *Sediment. Geol.* **2023**, *454*, 106457. [CrossRef]
33. Mohseni, H.; Zeybaram Javanmard, R. New data on sequence stratigraphy of the Sarvak Formation in Malekshahi city, (Ilam province) Zagros basin, Iran. *Mar. Pet. Geol.* **2020**, *112*, 104035. [CrossRef]
34. Taghavi, A.A.; Mørk, A.; Emadi, M.A. Sequence stratigraphically controlled diagenesis governs reservoir quality in the carbonate Dehluran Field, southwest Iran. *Pet. Geosci.* **2006**, *12*, 115–126. [CrossRef]

35. Hajikazemi, E.; Al-Aasm, I.S.; Coniglio, M. Diagenetic history and reservoir properties of the Cenomanian-Turonian carbonates in southwestern Iran and the Persian Gulf. *Mar. Pet. Geol.* **2017**, *88*, 845–857. [CrossRef]
36. van Buchem, F.S.P.; Simmons, M.D.; Droste, H.J.; Davies, R.B. Late Aptian to Turonian stratigraphy of the eastern Arabian Plate—Depositional sequences and lithostratigraphic nomenclature. *Pet. Geosci.* **2011**, *17*, 211–222. [CrossRef]
37. Motiei, H. *Stratigraphy of Zagros*; Geological Survey of Iran Publication: Tehran, Iran, 1993; p. 536.
38. Kiani, A.; Saberi, M.H.; ZareNezhad, B.; Mehmandosti, E.A. Reservoir zonation in the framework of sequence stratigraphy: A case study from Sarvak Formation, Abadan Plain, SW Iran. *J. Pet. Sci. Eng.* **2022**, *208*, 109560. [CrossRef]
39. Dunham, R.J. Classification of carbonate rocks according to depositional textures. In *Classification of Carbonate Rocks*; Ham, W.E., Ed.; AAPG: Tulsa, OK, USA, 1962; pp. 108–121.
40. Embry, A.F.; Klovan, J.E. A Late Devonian Reef Tract on Northeastern Banks Island, NWT. *Bull. Can. Pet. Geol.* **1971**, *19*, 730–781.
41. Flügel, E. *Microfacies of Carbonate Rocks, Analysis, Interpretation and Application*; Springer: Berlin/Heidelberg, Germany, 2010; 976p.
42. Choquette, P.W.; Pray, L.C. Geologic nomenclature and classification of proposity in sedimentary carbonates. *Am. Assoc. Pet. Geol. Bull.* **1970**, *54*, 207–250.
43. Ahr, W.M. *Geology of Carbonate Reservoirs: The Identification, Description, and Characterization of Hydrocarbon Reservoirs in Carbonate Rocks*; John Wiley & Sons: Hoboken, NJ, USA, 2008; ISBN 9780470164914.
44. Kolodzie, S. Analysis of pore throat size and use of the waxman-smits equation to determine ooip in spindle field, colorado. In Proceedings of the SPE Annual Technical Conference and Exhibition, Dallas, TX, USA, 21 September 1980. [CrossRef]
45. Porras, J.C.; Campos, O. Rock Typing: A Key Approach for Petrophysical Characterization and Definition of Flow Units, Santa Barbara Field, Eastern Venezuela Basin. In Proceedings of the SPE Latin American and Caribbean Petroleum Engineering Conference, Buenos Aires, Argentina, 25–28 March 2001; OnePetro: Richardson, TX, USA, 2001.
46. Gunter, G.W.; Finneran, J.M.; Hartmann, D.J.; Miller, J.D. Early Determination of Reservoir Flow Units Using an Integrated Petrophysical Method. In Proceedings of the SPE Annual Technical Conference and Exhibition, San Antonio, TX, USA, 5–8 October 1997. [CrossRef]
47. Skalinski, M.; Kenter, J.A.M. Carbonate petrophysical rock typing: Integrating geological attributes and petrophysical properties while linking with dynamic behaviour. *Geol. Soc. Spec. Publ.* **2015**, *406*, 229–259. [CrossRef]
48. Mehrabi, H.; Ranjbar-Karami, R.; Roshani-Nejad, M. Reservoir rock typing and zonation in sequence stratigraphic framework of the Cretaceous Dariyan Formation, Persian Gulf. *Carbonates Evaporites* **2019**, *34*, 1833–1853. [CrossRef]
49. Gomes, J.S.; Ribeiro, M.T.; Strohmenger, C.J.; Negahban, S.; Kalam, M.Z. Carbonate Reservoir Rock Typing—The Link between Geology and SCAL. Society of Petroleum Engineers 118284, manuscript. In Proceedings of the Abu Dhabi International Petroleum Exhibition and Conference, Abu Dhabi, United Arab Emirates, 3–6 November 2008.
50. Coalson, E.B.; Hartmann, D.J. Rock Types, Pore Types, and Hydrocarbon Exploration: ABSTRACT. *Am. Assoc. Pet. Geol. Bull.* **1985**, *69*, 845. [CrossRef]
51. Asquith, G.B. Microporosity in the O'Hara Oolite Zone of the Mississippian Ste. Genevieve Limestone, Hopkins County, Kentucky, and Its Implications for Formation Evaluation. *Carbonates Evaporites* **1986**, *1*, 7–12. [CrossRef]
52. Tucker, M.E.; Wright, V.P.; Dickson, J.A.D. *Carbonate Sedimentology*; Blackwell Science Ltd.: Hoboken, NJ, USA, 2009; ISBN 9780632014729.
53. Steuber, T.; Löser, H. Species richness and abundance patterns of Tethyan Cretaceous rudist bivalves (mollusca: Hippuritacea) in the central-eastern Mediterranean and Middle East, analysed from a palaeontological database. *Palaeogeogr. Palaeoclimatol. Palaeoecol.* **2000**, *162*, 75–104. [CrossRef]
54. Esrafili-Dizaji, B.; Rahimpour-Bonab, H.; Mehrabi, H.; Afshin, S.; Kiani Harchegani, F.; Shahverdi, N. Characterization of rudist-dominated units as potential reservoirs in the Middle Cretaceous Sarvak Formation, SW Iran. *Facies* **2015**, *61*, 14. [CrossRef]
55. Yu, Y.; Sun, L.; Song, X.; Guo, R.; Gao, X.; Lin, M.; Yi, L.; Han, H.; Li, F.; Liu, H. Sedimentary diagenesis of rudist shoal and its control on reservoirs: A case study of Cretaceous Mishrif Formation, H Oilfield, Iraq. *Pet. Explor. Dev.* **2018**, *45*, 1075–1087. [CrossRef]
56. Tavakoli, V. Permeability's response to dolomitization, clues from Permian–Triassic reservoirs of the central Persian Gulf. *Mar. Pet. Geol.* **2021**, *123*, 104723. [CrossRef]
57. Omidpour, A.; Mahboubi, A.; Moussavi-Harami, R.; Rahimpour-Bonab, H. Effects of dolomitization on porosity—Permeability distribution in depositional sequences and its effects on reservoir quality, A case from Asmari Formation, SW Iran. *J. Pet. Sci. Eng.* **2022**, *208*, 109348. [CrossRef]
58. Zhang, J.; Wang, J.; Zheng, B.; Sheng, Q.; Wei, H.; Shen, L.; Xiong, S.; Mansour, A. Age assignment of dolomite in palaeo-reservoirsof the Qiangtang Basin: New evidence from palaeontology and carbonate in situ U-Pb dating. *Mar. Pet. Geol.* **2023**, *158*, 106545. [CrossRef]
59. Ezati, M.; Azizzadeh, M.; Riahi, M.A.; Fattahpour, V.; Honarmand, J. Characterization of micro-fractures in carbonate Sarvak reservoir, using petrophysical and geological data, SW Iran. *J. Pet. Sci. Eng.* **2018**, *170*, 675–695. [CrossRef]
60. Volery, C.; Davaud, E.; Foubert, A.; Caline, B. Shallow-marine microporous carbonate reservoir rocks in the Middle East: Relationship with seawater Mg/Ca ratio and eustatic sea level. *J. Pet. Geol.* **2009**, *32*, 313–326. [CrossRef]

61. Tavakoli, V.; Jamalian, A. Microporosity evolution in Iranian reservoirs, Dalan and Dariyan Formations, the central Persian Gulf. *J. Nat. Gas Sci. Eng.* **2018**, *52*, 155–165. [CrossRef]
62. Lambert, L.; Durlet, C.; Loreau, J.P.; Marnier, G. Burial dissolution of micrite in Middle East carbonate reservoirs (Jurassic-Cretaceous): Keys for recognition and timing. *Mar. Pet. Geol.* **2006**, *23*, 79–92. [CrossRef]

minerals

Article

Diagenetic Evolution of Upper Cretaceous Kawagarh Carbonates from Attock Hazara Fold and Thrust Belt, Pakistan

Saif Ur Rehman [1,*], Muhammad Jawad Munawar [1], Mumtaz Muhammad Shah [2], Naveed Ahsan [1], Muhammad Kashif [3], Hammad Tariq Janjuhah [4,*], Vasiliki Lianou [5] and George Kontakiotis [5]

1 Institute of Geology, University of the Punjab, Lahore 54590, Pakistan; jawad.geo@pu.edu.pk (M.J.M.); naveedahsan.geol@gmail.com (N.A.)
2 Department of Earth Sciences, Quaid-e-Azam University, Islamabad 44000, Pakistan; mshah@qau.edu.pk
3 Department of Earth Sciences, University of Sargodha, Sargodha 40100, Pakistan; kashifyaqub@yahoo.com
4 Department of Geology, Shaheed Benazir Bhutto University, Sheringal 18050, Pakistan
5 Department of Historical Geology-Paleontology, Faculty of Geology and Geoenvironment, School of Earth Sciences, National and Kapodistrian University of Athens, 15784 Athens, Greece; vlianou@geol.uoa.gr (V.L.); gkontak@geol.uoa.gr (G.K.)
* Correspondence: saif.geo@pu.edu.pk (S.U.R.); hammad@sbbu.edu.pk (H.T.J.)

Abstract: A recent hydrocarbons discovery in 2021 in the Kawagarh Formation has brought attention to the significance of sedimentology and specifically diagenesis for understanding and characterizing the reservoir properties. The diagenetic history and multiscale processes that contributed to diagenesis were vaguely known. This study aimed to reconstruct various diagenetic phases, paragenetic sequences, and the interrelationship of these phases in the Kawagarh Formation. The diagenetic processes were identified and characterized through an integrated methodology utilizing the outcrop, petrographic, and geochemical analyses. Early calcite cementation was found to occur in the early stages of marine burial diagenesis involving pore fluid originating from the dissolution of aragonite in interlayer marl/mudstone beds and reprecipitating as microspar in adjacent limestone beds. The absence of mechanical compaction in wackstone and mudstone facies and the presence of late compaction in lithified packstones clearly imply that early calcite cementation occurred prior to compaction. Dolomitization with stylolites coupled with significant negative oxygen ($\delta^{18}O$) isotope values implies a fault-related hydrothermal dolomitization model. Uplift introduced the fractures and low Mg fresh fluids to the system which caused calcitisation in shallow burial settings. The depleted $\delta^{13}C$ and negative $\delta^{18}O$ values indicate the mixing of surface-derived waters with hot burial fluids during the calcitization. This study offers valuable insights into several aspects related to the formation and the basin itself, including burial depths, fluid influx, and geochemical gradients. It also sheds light on the evolution of reservoir properties such as porosity and permeability in dolomitization fronts. Such insights can be used to gain a deeper understanding about the burial history, basin evaluation, and reservoir characterization for hydrocarbon exploration.

Keywords: Kawagarh Formation; cretacrous; carbonate diagenesis; shallow marine depositional environment; early calcite cementation; dolomitization; reservoir characterization

Citation: Rehman, S.U.; Munawar, M.J.; Shah, M.M.; Ahsan, N.; Kashif, M.; Janjuhah, H.T.; Lianou, V.; Kontakiotis, G. Diagenetic Evolution of Upper Cretaceous Kawagarh Carbonates from Attock Hazara Fold and Thrust Belt, Pakistan. *Minerals* **2023**, *13*, 1438. https://doi.org/10.3390/min13111438

Academic Editors: Michael S. Zhdanov and José Javier Alvaro

Received: 1 September 2023
Revised: 23 October 2023
Accepted: 7 November 2023
Published: 13 November 2023

1. Introduction

Shallow water carbonates settings are sites where marine life flourishes; thus, they preserve and present excellent depositional and diagenetic records in oceanic environments [1]. Despite the simple mineralogy, carbonate rock poses extremely complex fabric and pore systems developed under composite diagenetic process [2]. These diagenetic processes are typically governed by the pore fluids which can significantly alter original depositional fabric and pore systems of the rock by dissolution, cementation, or mineral replacement. The chemical composition of pore fluids, the rate of fluid flux, and the temperature and pressure conditions under which rock-water interactions occur regulate the diagenetic processes manipulating texture and porosity of carbonates [2–5]. Such processes mainly involve dissolution of extant mineral phases and/or precipitation of new mineral phases. Globally, reservoirs characterization has significantly stressed upon the diagenetic studies and porosity evolution in carbonate rocks [2,6–8]. Information such as primary rock fabric, porosity types, and timing of porosity development plays a critical role in geological reservoir models. These models are very important for the prediction of the distribution of favorable reservoir rock during the hydrocarbon migration [9].

There are three principal carbonate diagenetic environments (meteoric, marine, and burial) within which porosity development, evolution, and their distribution on the earth's surface or in the subsurface can be discussed [4,10–13]. The marine and meteoric environments commonly associated with near surface and surface environments are also called eogenetic and telogenetic zones [14]. The marine environments are the main source of carbonate deposition where the diagenesis is predominantly influenced by the supersaturated marine pore fluids. Ideal conditions are present in marine environments for the precipitation of calcite cement during the fluid percolation through the pore system [12,14,15]. The precipitation of new mineral phase and/or replacement of existing mineral phase by enriched fluids can bring unprecedent changes and can produce highly complex carbonate fabric and pore systems. The meteoric diagenetic environment is generally associated with the subaerial shallow marine carbonate depositional sequence [12,16]. The meteoric environments are represented by the zone of dilute waters which are moving at a strong rate from the vadose zone to the phreatic zone. The vadose zone commonly constitutes under-saturated states of carbonate mineral species due to the presence of CO_2 in meteoric water [10]. The dissolution of $CaCO_3$ is more active in the vadose zone and it produces a more depleted zone of $CaCO_3$ which leads to the formation of vugs and cavernous structures in limestone deposits. Among the subsurface conditions, burial diagenesis is the most significant that is encountered upon the deep burial of sedimentary deposits within the earth. It is active beyond near surface environments and marks the mesogenetic porosity zone [14]. The major diagenetic controlling agent in burial zone environments is the pressure, temperature, and formation water [12,17,18].

The diagenetic process naturally distributes the various mineral and pore phases spatially and temporally in such naturally complex patterns making it very difficult to understand and reconstruct the diagenetic phase. The Late Turonian Kawagarh Formation exhibits a unique library possessing spectacular marine, sedimentation, diagenetic, climatic, and tectonic records. Depositional environments and tectonic history have already been established by [19] and [20] in the Hazara and Kalachitta Range. However, diagenetic history and meso to micro scale processes drive multiscale diagenetic events, especially calcitization of lime mud in thick marls, strata bound dolomitization, and dedolomitization which were vaguely known. The first objective of this study is to find out and discuss the source and mechanism of early calcite cementation in the Kawagarh Formation, and particularly in its marls. The second objective is to investigate the stylolite and its relationship with dolomitization. This work is accomplished by utilizing a highly integrated diagenetic workflow based on a multiproxy geochemical approach. Each carbonate diagenetic phase is recognized and characterized through geochemical analyses: optical petrography, atomic absorption spectroscopy (AAS), and carbon and oxygen isotope. The present study provides invaluable context for the reconstruction of the history of the shallow ramp carbonates of the Kawagarh Formation and can serve as a reference for studying and documenting equivalent carbonate systems.

2. Geological Setting

Attock Hazara Fold and Thrust Belt (AHFTB), which is itself part of the Northwest Lesser Himalayas, presents a narrow EW trending strip comprised of thick sedimentary package with low grade metamorphic rocks along the hanging wall of the Nathiagali Thrust (NGT) in Northern Pakistan [21]. Geologically, it is surrounded by the Panjal/Kherabad Fault in the north and bordered by the Main Boundary Fault (MBT) in the south (Figure 1) [19,22–24].

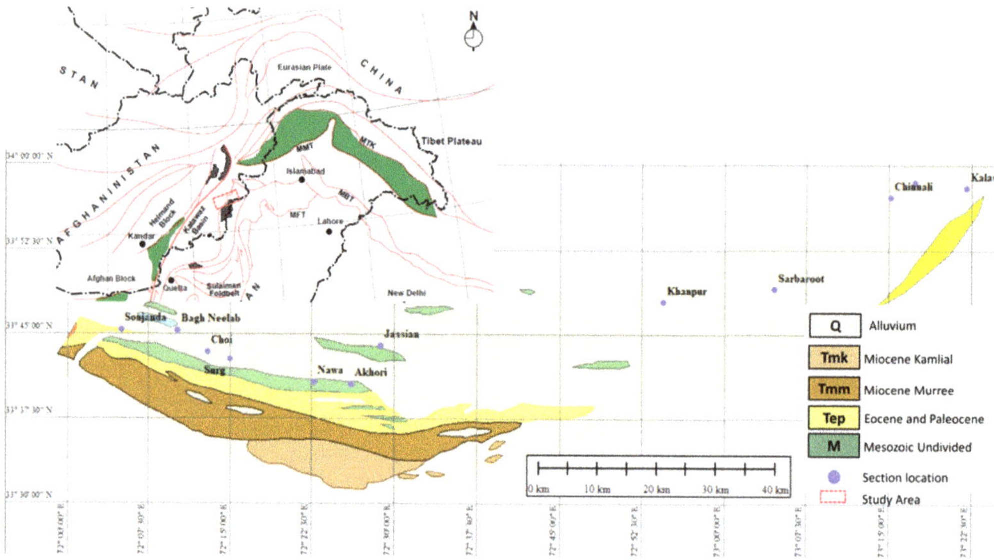

Figure 1. Geological map of the Hazar and Kala Chitta Basin showing the location of the studied sections.

Northern Pakistan is comprised of three convergence tectonic elements: the Indian Plate, the Kohistan-Ladakh Arc (KLA), and the Asian Plate from south to north [23,25–30]. This collision led to the development of the Himalaya Mountain range along the northern margin of the Indian Plate. The movement of the Indian Plate northwards caused older rocks to be thrust over younger strata, resulting in the formation of small fault-bounded blocks including the Trans Himalaya, Higher Himalaya, Lesser Himalaya, and Sub-Himalaya [31–33]. In the western Himalayan Basin (HB), the Lesser Himalayas of Pakistan are bounded by the Hissartang Fault in the north and the Indus River in the west, constituting the southernmost extension of the Himalayas [34]. The sedimentary succession in the region, from Triassic to Miocene-Pliocene, is tightly folded and cut by numerous thrust faults [35]. The Jurassic sedimentation consists of the Datta Formation, Shinawari Formation, and Samana Suk Formation (Figure 2). The Cretaceous succession includes the Chichali Shale, Lumshiwal Formation, and Kawagarh Formation. The Chichali Shale contains belemnite and pyrite and was deposited in anoxic restricted environments when the Indian Plate was separating from Gondwana [36]. The Kawagarh Formation is mainly comprised of limestone and marl with subordinate dolomite [20,21,24,37,38] that was deposited over a carbonate ramp formed due to the subsidence created by the northward drift of the Indian Plate [39].

Paleomagnetic data, zircon dating, and fossils in volcanic sediments suggest Early Cretaceous rifting and volcanism because of separation and drifting of the microcontinent and the opening of a new basin (Neo Tethys) at the northern margin of India [40,41]. Conferring to the faunal content and ecological attributes, the Kawagarh Formation is deposited in Neo Tethys under shallow open marine and well oxygenated conditions, suggesting a ramp setting [19,20,38,42]. The deposition initiated with transgressive depositional cycle at the base, with inner-ramp microfacies overlying the lateritised top of the underlying Lumshiwal Formation and overlain by mid-ramp and outer ramp microfacies [20,24]. The lateritisation suggests shallow water conditions and intense weathering induced by regression after the deposition of the Lumshiwal Formation [24]. This regression also corresponds to the early Turonian global sea level fall of [43], which partially exposed the Lumshiwal Formation sediments. The regression during the late Turonian was followed by a transgression characterized by minor sea level fluctuations until the end of the Cretaceous with a general deepening upward sequence. The deposition of the Kawagarh Formation ended with the uplift of carbonate sediments during the Maastrichtian as shown in paleontological data and the presence of residual deposits of the K-Pc boundary at the top of the Kawagarh Formation [37,44]. The regional distribution of unconformity-related residual clays over the top of the Kawagarh Formation in the Attock Hazara Fold and Thrust Belt (AHFTB) and the complete absence of Upper Cretaceous sediments in the Sub-Himalayas in the south of the AHFTB suggest a regional event that uplifted these areas at the end of the Cretaceous [23,27,29,45–50]. The Indian Plate first collided with the KLA at the end of the Cretaceous or the Cretaceous-Paleocene boundary, terminating the depositional trend of the Kawagarh Formation due to the regional tectonic settings [23,27,29,45–50]. Generalized stratigraphic sequences of this study area are provided in Figure 2.

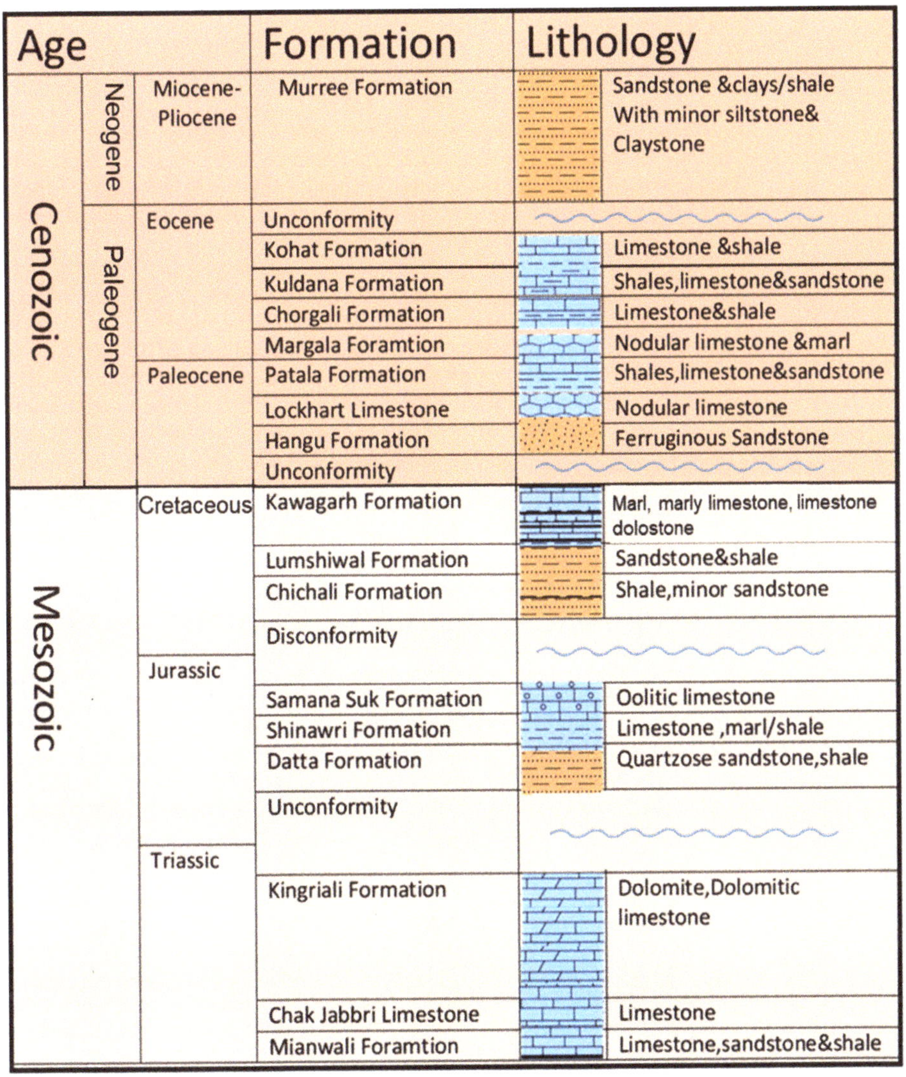

Figure 2. Generalized stratigraphy and formation lithology of the Hazara and Kala Chitta Basin.

The Kawagarh Formation is mainly composed of grey to light grey limestone. The limestone is generally fine grained, thin to thick bedded, and occasionally very thick bedded. It is dolomitic at places and shows a sandy texture at the basal part. Dolomite is yellowish brown to rusty brown and pale yellow and at some places exhibits a sandy texture.

3. Materials and Methods

In the current study, we employed an integrated petrographic and geochemical approach to unravel the complex diagenetic history of the Cretaceous Kawagarh Formation in the Hazara and Kalachitta regions. Field observations and thin section microscopy were integrated to delineate the spatial and temporal diagenetic framework. Element chemistry (major, minor) and stable carbon and oxygen isotopes ($\delta^{13}C$, $\delta^{18}O$) were utilized to delineate processes which governed these diagenetic changes. A total of 256 samples from limestone, dolomite, and marl were analyzed to determine major (Ca, Mg) elements and trace elements (Fe, Sr, Mn, Na) chemistry. A same number of samples were used to determine bulk rock $\delta^{13}C$ and $\delta^{18}O$ isotope values.

3.1. Fieldwork and Sampling

We investigated twelve sections of the Cretaceous carbonate platform Kawagarh Formation, along an E-W transect from the Kalachitta and Hazara ranges of northern Pakistan (Figure 1). Four sections (Chinali, Kalas, Khanpur, Sarbaroot) from the Hazara Basin and seven sections (Akhori, Nawa, Sugdara, Bagh Neelab, Sojhanda Dam, Surg, and Jassian) from the Kalachitta ranges were studied. Microfacies and depositional environments of the Kalachitta sections have already been reported by [20].

Bed to bed sampling was carried out for limestone and dolomite while the marl samples were collected from embedded horizons between limestone or dolomite beds. The sections were measured considering that each rock unit should have complete exposure and maximum features with assured lateral extension and minimum deformation, while top and bottom should be well preserved, and deformation should be removed in case of faulting or folding. Detailed graphical sedimentary logs of the stratigraphic sections were constructed along with the recording of all necessary information including color, lithology, fossil content, and sedimentary structures.

3.2. Thin Section Microscopy

The samples were sliced perpendicular to the bedding plane with a diamond cutter to prepare the rock chips. These chips were ground to flatten the surface and polished with 100 μm particle sized carborundum abrasive powder. The polished chips were mounted on glass slides with a mixture of epoxy resin and hardener and ground on the other side to attain about 30 μm thickness. After that, the rock was polished with corundum powder (up to 0.8 μm particle size) to remove the residual surface abrasion and to maximize the textural resolution. Thin sections were stained in a solution of Alizarian Red S and Potassium ferricyanide solution for the differentiation of calcite from dolomite and to distinguish the ferron or non-ferron character of calcite [51,52]. Thin sections were examined under a polarizing microscope to record the microscopic data to establish paragenetic sequence, based on the superposition, crosscutting relationship, or overlap of diagenetic features. Field observations were also integrated with the thin sections study.

3.3. Major and Trace Element Stable Isotope Geochemistry

Samples were analyzed on an Atomic Absorption Spectrometer (AAS, AA 6300, Shimadzo) at High Tech Laboratory, University of Sargodha, Sargodha to measure the major (Ca, Mg) and trace elements (Fe, Sr, Mn, Na) chemistry. Samples were prepared following the method of [53]. Selected rock samples were crushed into fine powder and the powder was placed in an oven for 12 h at 100 °C. One gram sample of dried powder was dissolved in 50 mL of solution of 1N HCl for two hours. The dissolved solution was filtered using filter papers. A 0.5% KCl solution was added to the aforesaid solution to avoid further reactions. Next, 100X solutions were prepared by adding 100mL of deionized water for the determination of major elements (Ca and Mg). Dilutions were prepared to the detectable limits of AAS.

Carbon ($\delta^{13}C$ (and oxygen ($\delta^{18}O$) isotope ratios from bulk rocks were analyzed at the Nanjing Institute of Geology and Paleontology, Nanjing, China. The samples were washed, cleaned, and powdered to 75 μm grain size. This powder was treated with phosphoric acid in an automated process. The $\delta^{13}C$ and $\delta^{18}O$ ratios of bulk rock were reported in standard delta permil (‰) notation relative to Vienna Pee Dee Belemnite (V-PDB) calibrated with the international standards NBS-18 ($\delta^{13}C$ = −5.01‰ VPDB; $\delta^{18}O$ of −23.1‰ VPDB) and NBS-19 ($\delta^{13}C$ of +1.95‰ VPDB; $\delta^{18}O$ of −2.20‰ VPDB).

4. Results

4.1. Petrography

Through extensive examination and delicate observation, we have documented the intricate details of diagenetic processes and their consequential effects on the fabric and mineralogical composition of the Kawagarh Formation. The petrography section presents an in-depth mineralogical, text, and fabric analysis, providing valuable insights into the complex diagenetic history of the Formation. We have undertaken conventional thin section microscopic analysis to unravel and elucidate five major diagenetic events that have significantly influenced the Kawagarh Formation. The identified diagenetic events encompass early calcite cementation (ECC) and neomorphism, compaction, and replacive dolomitization (RD), as well as the placement of fracture calcite (FC) and calcitization. The details of the diagenetic processes and their resultant fabrics are discussed in the following sections.

4.1.1. Early Calcite Cementation (ECC)

ECC is a well-studied and evolving aspect of carbonate diagenesis. Substantial literature is available on the morphology, growth, distribution, and types of cements [11,54–62]. Substantial features of early calcite cement such as microspar precipitation and neomorphism have been observed in the Kawagarh Formation. ECC is characterized by subhedral to anhedral crystals that can also appear as fibrous crystals. Three types of ECC are identified in the Kawagarh Formation.

In the Kawagarh Formation, microspar is found in both interparticle and intraparticle pore spaces. It is present in the peripheral regions of skeletal particles such as planktonics, calcispheres, ostracods, pelecypods, inoceramids, echinoids, crinoids, trigonia, and textularia, as well as in the chambers of skeletal particles (Figure 3a,b). It becomes more prominent along the peripheries of skeletal particles in mudstones and wackstones (Figure 3c). In wackstones and packstones microfacies where grains are in contact, bladed crystals of Low Magnesium Calcite (LMC) cement envelope particles as isopachous cement (Figure 3e). The needle-like appearance of the LMC is due to the relict geometry of aragonite. In intraparticle pores, it occurs as the chamber fills in the form of interlocking mosaics (Figure 3b). Interlocking mosaics (sparite) are mainly comprised of anhedral to subhedral coarse crystals with fused edges. The interlocking crystals show extinction and exhibit light color.

4.1.2. Neomorphism

Neomorphism refers to transformations of mineral crystals caused by dissolution and reprecipitation in such a way that bulk chemistry remains intact [54,60,63]. Neomorphism includes the various types of transformations occurring in carbonate sediments like replacement, inversion, and recrystallization. Neomorphism is prominent in Kawagarh where the skeletal particles are recrystallized into coarse calcite crystals and micritic lime mud groundmass and already developed microspar affected by neomorphism and generally show the replacement of finer crystals with coarsely crystalline calcite. In marl or mudstone facies, the microspar tiny patches occur indicating neomorphism of micrite or lime mud (Figure 3f), and similarly skeletal particles show less effects of neomorphism. In wackstone and packstone microfacies, micrite and skeletal grains are significantly altered by neomorphism (Figures 3c,d and 4a,b). The skeletal particles of calcispheres and planktons are extensively replaced by coarse calcite (Figure 4a,b) in wackstone and packstone facies. Planktons (Figure 4b), trigonia, textularia (Figure 4f), echinoid (Figure 4c), and calcispheres (Figure 4a) are largely affected by neomorphism. However, filaments, oysters (Figure 4d), inoceramids (Figure 4f), and ostracods are less affected.

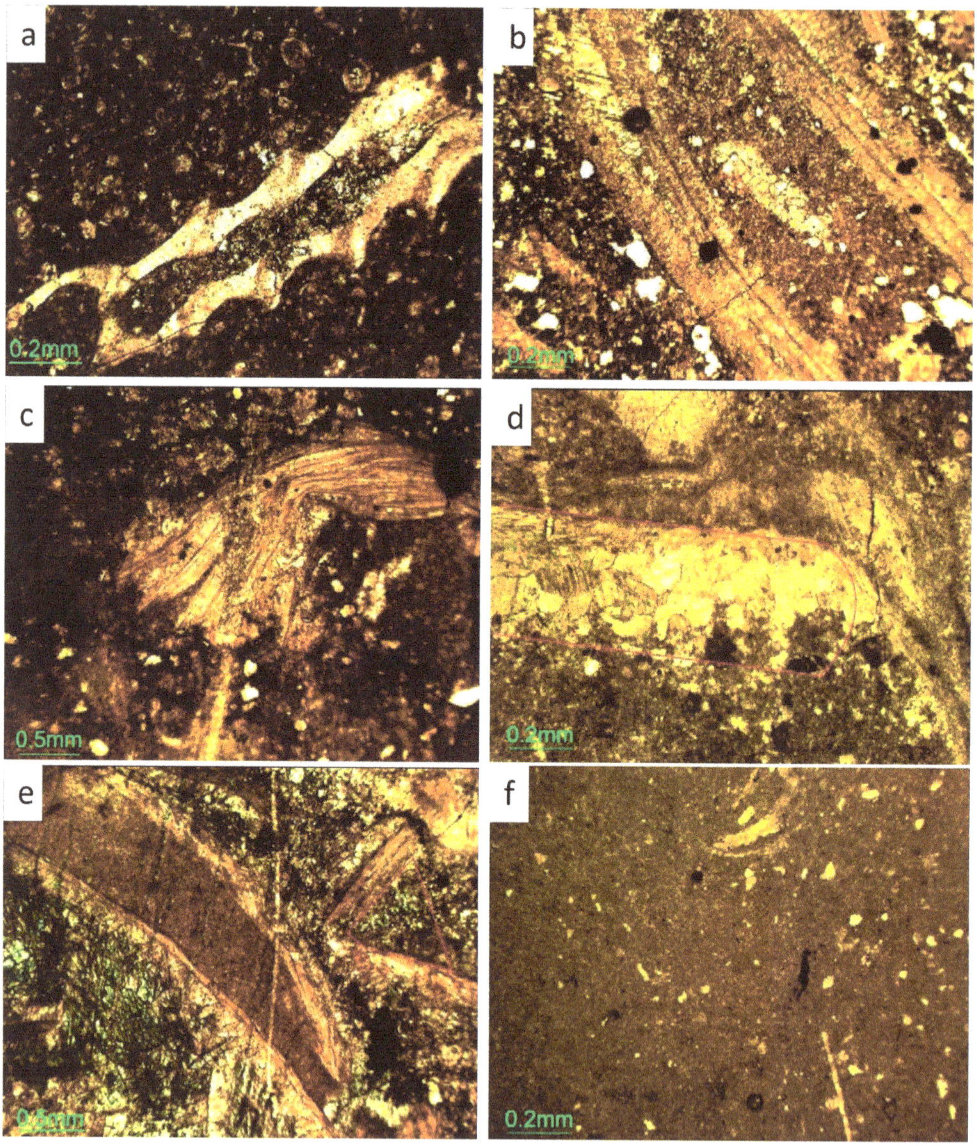

Figure 3. (**a**) Microspar filling in the bioclast, Sarbaroot Section. (**b**) Microspar filling in the oyster shell, Surg section. (**c**) Microspar along the lower periphery of oyster shell, Nawa section (**d**) Dripstone cement (indicated by red line) beneath the oyster shell, Jassian section. (**e**) Meniscus cement between the oyster shells (indicated by red line) and dripstone cement beneath the large oyster shell, Kalas Section. (**f**) Very minute effect of neomorphism.

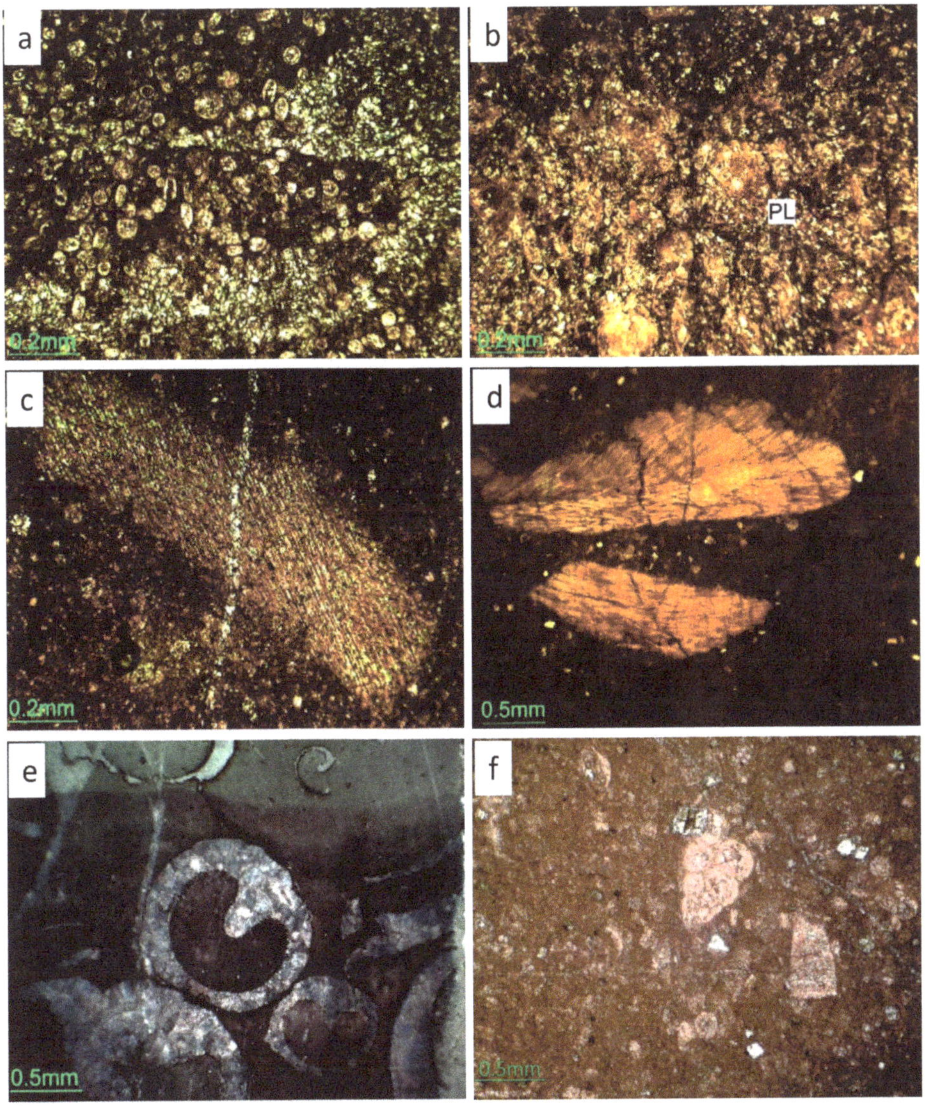

Figure 4. (**a**) Neomorphism in calcispheres packstone facies, Khanpur Section. (**b**) Neomorphism in planktone (PL) packstone facies, Bhag Neelab section. (**c**) Neomorphism in echinoid plate, Khanpur Section. (**d**) Unaltered oyster shells marking the absence of neomorphism in packstone, Akhori section. (**e**) Trigonia with neomorphosed periphery, Surg Section. (**f**) Textularia with neomorphosed chambers and blocky fragment of unaltered inoceramid, Thoba Section.

4.1.3. Compaction

Compaction is observed using common burial depth indicators such as orientation of burrows, particle interaction and deformation, porosity type, dissolution seams and their orientation, and overall texture of the rock.

- Mechanical Compaction

In carbonates, the mechanical compaction follows the distributed porosity and consequently from highly variable layer thickness with depth [64]. Since the recognition of mechanical compaction in fine-grained limestones is difficult, it is still evident largely in packstone microfacies (Figure 5a,b) by broken shells of skeletal particles, compressed particles, and close packing of skeletal particles. Two types of compactions are observed; the first type includes mechanical rearrangement of skeletal particles, which possibly occurred shortly after the aragonite dissolution or early calcite cementation; the second type of deformation brought ductile or brittle deformation in individual particles depending upon the particle mineralogy and shape. Breakage and deformation of skeletal particles are more distinct in plankton and pelecypods. Mechanical breakage is almost absent in calcispheres wherein instead these show close packing and concavo-convex contacts (Figure 4a). The breakage of skeletal particles in burrows is also extant (Figure 5a).

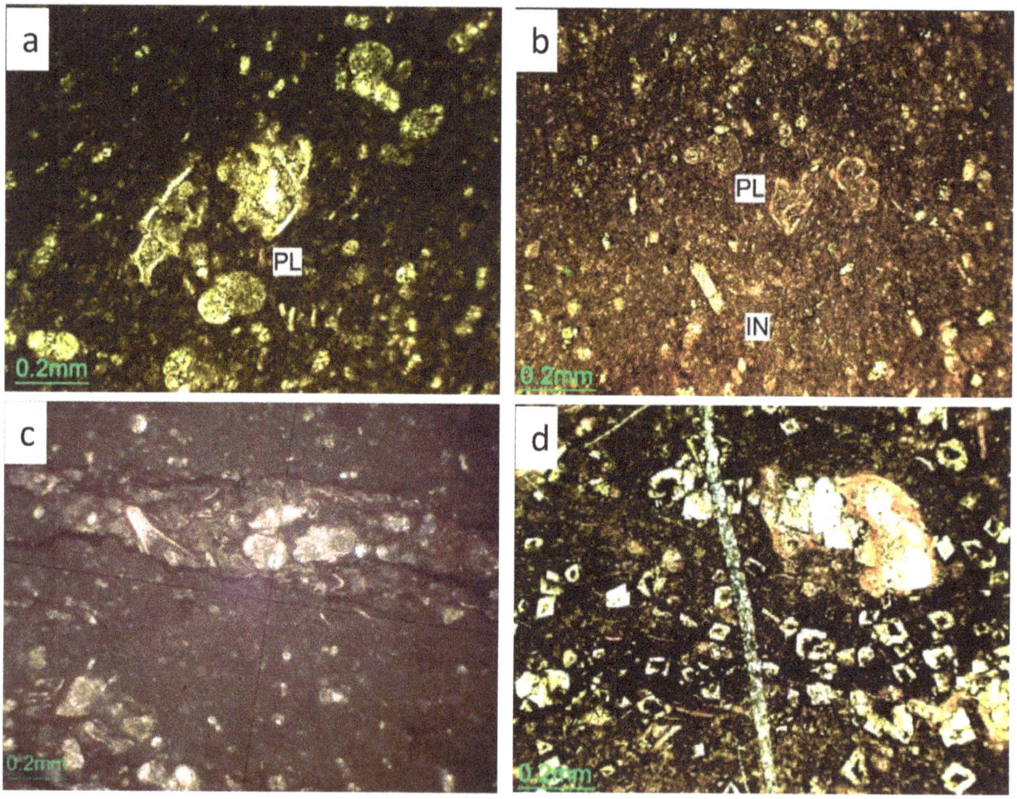

Figure 5. (**a**) Broken plankton (PL) in packstone, Kalas Section. (**b**) Broken and dissolved plankton (PL) and inoceramid (IN) in packstone, Chinali section. (**c**) Tightly packed broken plankton in burrow, Chinali Section. (**d**) Dolomitization along black stylolite structure, Thoba section.

- Chemical Compaction

Chemical compaction here mainly implies the removal of material via dissolution along pressure seams or stylolite, and subsequently transport of this material into the pores of the same system or elsewhere, causing overall compaction. Stylolites are irregular dissolution seams caused by the pressure solutions present in the intergranular space [65–67] which usually accommodate insoluble materials such as clay minerals, oxides, and organic matter transported by the pressure solution [68–70]. Initial chemical compaction starts with pressure solutioning of grains and grain contacts. Dissolved tests of skeletal particles have also been recorded, sometimes with broken shells. (Figure 5a,b). These dissolved shells are dominantly planktonics showing sutured contacts (microstylolites) at places (Figure 5a,b).

- Stylolites

Outcrop studies revealed that stylolites define the sharp contacts and boundaries of dolomitized fronts and layers. In stratabound dolostone, stylolites are bedding-parallels, marking the contact between dolostones and limestones and frequently develop dolomitization fronts (Figure 6a). Stylolites show a variety of amplitudes with the highest at the Chinali outcrop associated with dolostones. In wackestone and packstone microfacies, stylolites exhibit abundant suture and sharp peak (Figure 6b) morphology, while mudstone facies dominate in wavy morphologies (Figure 6c).

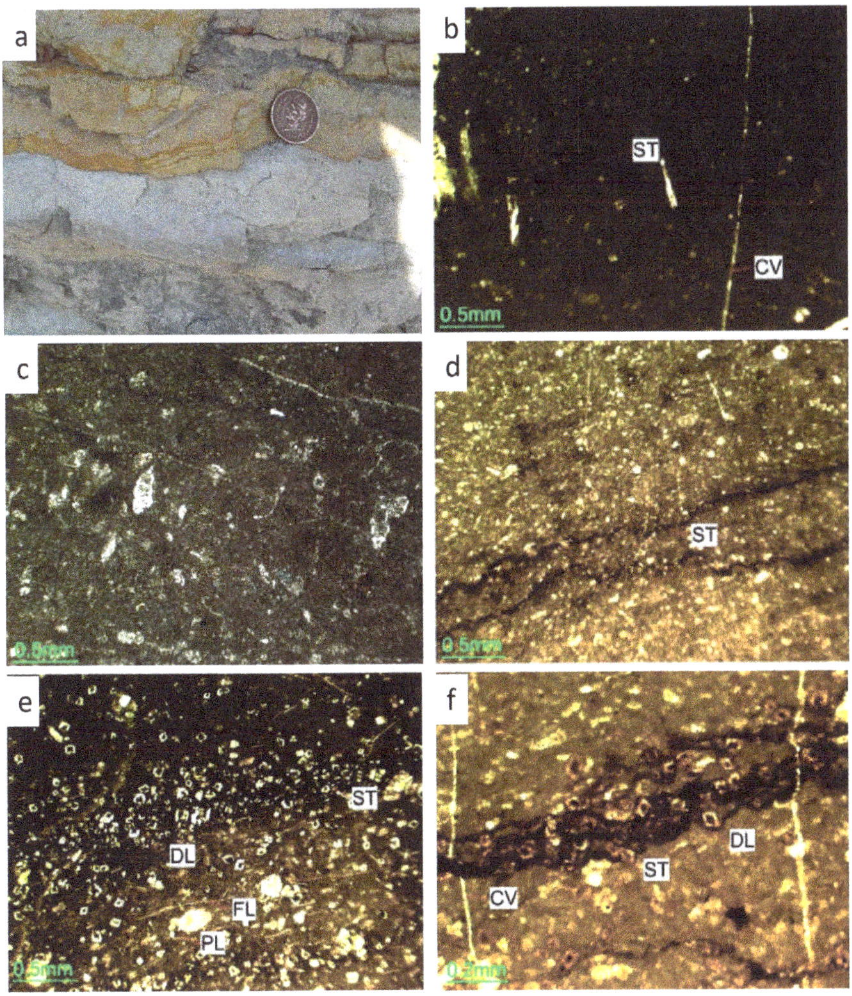

Figure 6. (**a**) Light grey limestone and yellowish-brown dolostone separated by stylolite (indicated by white arrows), Jassian section. (**b**) Sharpe peak stylolite, Chinali section. (**c**) Wavy stylolite in planktonic mudstone facies, Surg section. (**d**) Set of two wavy stylolites, Khanpur section. (**e**) Preferably developed rhombic dolomite (DL) crystals along black color stylolite (ST), Chinali section. (**f**) Rhombic dolomite (DL) with dark cores associated with yellowish and black colored stylolite (ST), calcite vein (CV) cutting through the stylolite.

In thin sections, the suture-and-sharp-peak (Figure 6b) and wave-like stylolites (Figure 6d), along with the associated reaction fronts, are not sharp but have a width of up to a few tens of microns. The single reaction front is bounded by a series of stylolites (Figures 6a and 7a,c). The seams of stylolites contain insoluble residues of iron oxide and clay material that can be recognized by the black and yellow colors, respectively (Figure 6e,f). In SEM, stylolites can easily be identified by the alignment of iron oxide between the rhombic dolomitic crystals (Figure 8a).

Figure 7. (**a**) Yellow to rusty brown, medium bedded dolostone, Jassian section. (**b**) Yellowish brown dolostone, Jassian section. (**c**) Dolomitization front (DF) and light yellow to yellowish grey dolomitic patches (DP) in limestone beds, Kalas section. (**d**) Mosaic of interlocking crystals of dolomite, Chinali Section. (**e**) Dolomitization along peripheries of trigonia indicated by arrow, Sarbaroot section. (**f**) Rhombic dolomite developed over the skeletal grain.

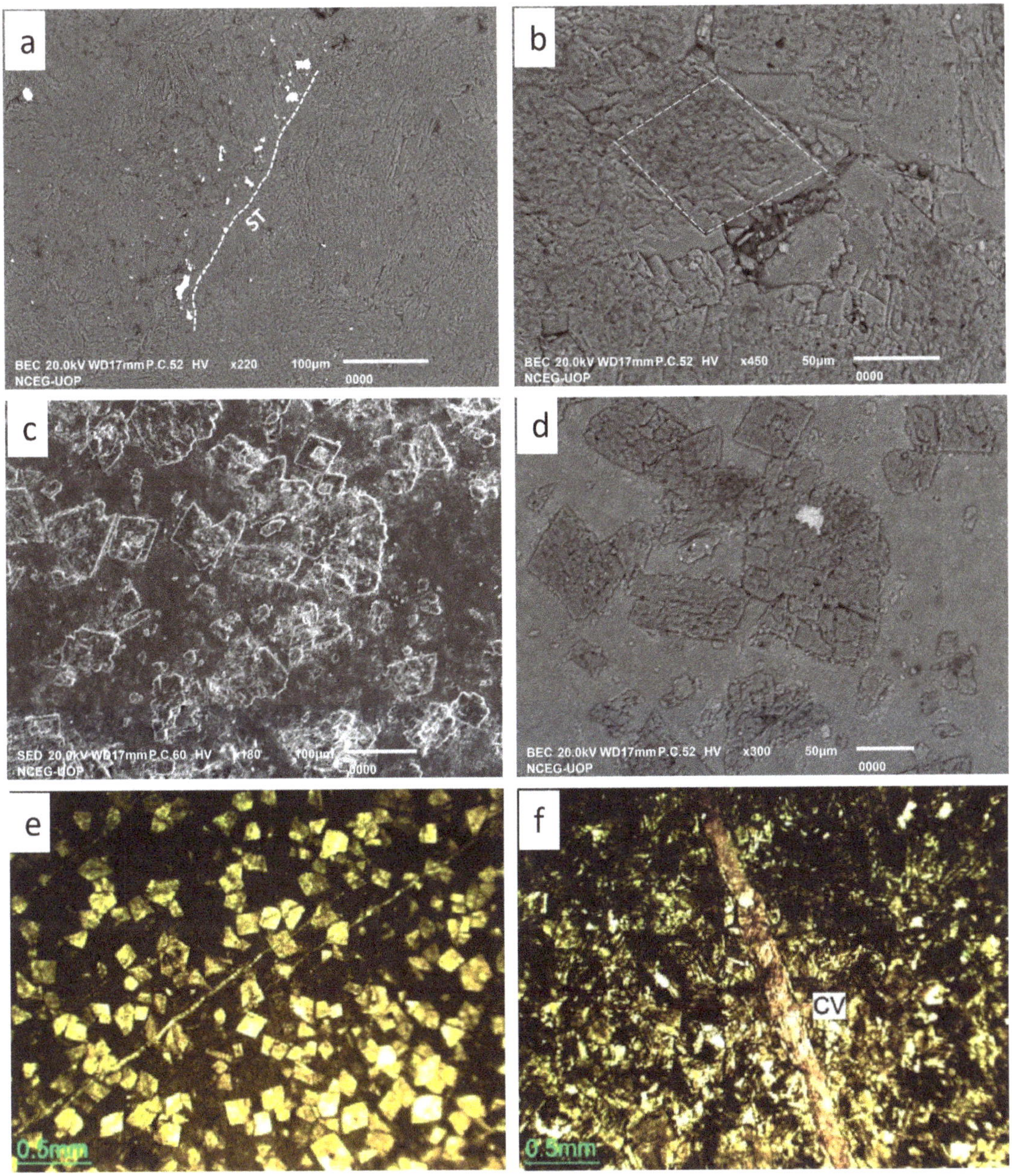

Figure 8. (**a**) BSE_SEM image showing stylolite identifiable by the alignment of iron oxide (white color), Chinali section. (**b**) A zoomed view "a" showing rhombic dolomite (DL). (**c**) SE_SEM image showing morphology of rhombic dolomite (DL) crystals, Jassian section. (**d**) BSE_SEM zoomed view of "b" showing rhombic polycrystalline dolomite (DL). (**e**) Individual rhombic dolomite crystals cross-cutted by calcite veins. (**f**) Calcite vein cutting through mosaic dolomite whereby engulfed dolomite crystals in the vein are very clear.

4.1.4. Dolomitization

In the Kawagarh Formation, dolomites are found in two forms: 1—dolostone in stratabound dolomites (Figure 7a,b) and 2—dolomitic patches in partially dolomitized limestones (Figure 7c). The dolostone bodies are developed in the form of dolomite fronts bounded by bedding surfaces where whole beds are dolomitized. Dolomitic patches formed as a result of the partial dolomitization of limestone. Outcrop observations revealed that dolomitic patches regularly coincide with stylolites with sharp boundaries often bounded by stylolites (Figures 6a and 7c).

In the Hazara Basin, the maximum dolostone contents are observed at the Chinali section which accounts for 12% of the total samples, while the minimum dolostone contents (3% of the total samples) are recorded at the Sarbaroot section. In Kalachitta, dolostone is observed at the Surg and Jassian sections only, whereas the Akhori, Nawa, Sugdara, Bagh Neelab, and Sojhanda Bata sections show an absence of dolostone. Dolomitic patches appear in variable amounts at various levels of the Formation, however, dolostone bodies are confined to the lower part of the formation near the lower contact with the Lumshiwal Formation. Dolomitic limestone and dolostone are generally characterized by yellowish brown, rusty brown, and often light grey color with sandy texture.

Microscopic studies revealed that dolomite appears as individual rhombic crystals (Figures 6e,f and 8b–d) and mosaics of crystals (Figure 7d). Induvial rhombic dolomite is referred to as replacive dolomite (RD1), while mosaic dolomite is referred as replacive dolomite (RD2) here. RD1 appears in both dolomitic limestones and dolostones, while RD2 is exclusively present in dolostones. The dolomite crystals are generally characterized by the brown or yellowish-brown color (Figure 7d) in RD2, while yellowish brown to off-white with black spotted or rhombic cores in RD1 (Figures 5d and 7e,f). sRD1 are mostly euhedral (planar-p) dolomites with sharp edges, while RD2 are euhedral to subhedral (planar-s) dolomites (Figures 6f and 8b–d). The crystal size of RD1 is relatively finer ranging from 20 μm to 250 μm, while RD2 is coarser with crystal size ranging from 50 μm to 500 μm. RD2 extensively replaced skeletal and non-skeletal particles (Figure 7e,f), calcite, and mircospar. The individual crystals of RD1 exhibit darker/black cores (Figures 5d and 7e,f). RD1 preferably populated along the stylolites (Figure 7e,f) where the dolomite crystals systematically superimpose over stylolites. Both destroy the fabric of the rock, however, RD1 exhibits intercrystallite porosity, while all the porosity is destroyed in RD2. RD1 and RD2 can easily be differentiated by distinct texture and color sometimes separated by stylolites.

4.1.5. Calcite Veins and Fractures

Veins filled with bladed and/or coarse equant blocky calcite (FC) crystals occur as replacement and engulfment of components in the host limestones/dolostones. Euhedral dolomite crystals RD1 in the host dolostone that are engulfed by coarse-crystalline blocky calcite are unaltered or display evidence of partial to extensive calcitization (Figure 8e,f). Some of the coarse calcite crystals display undulous extinction and/or the presence of numerous sub-crystals. Calcite veins are the only features which crosscut the skeletal particles (Figure 4d), stylolites (Figure 6b,f), and dolomite crystals (Figure 8e). These veins show perpendicular to sub-perpendicular orientation to the bedding. Dedolomitization of dolomite and formation of calcite is obvious along these features (Figure 8f). These veins are thick where they cut skeletal particles and thin out in the distal areas.

4.2. *Trace and Major Element Chemistry*

The major elements, calcium and magnesium, show covariance across the sections, with mean values of 20.6% and 18.8%, respectively, with Ca ranging from 14.8% in Nawa to 23.9% in Sojhanda, and Mg ranging from 14.0% in Jassian to 23.6% in Bagh Neelab. The Khanpur, Sojhanda, and Bagh Neelab sections exhibit higher Mg content in the upper parts of Kawagarh, whereas in Jassian the Mg contents are high from the basal to the middle parts, and in Kalas the basal and upper parts have higher Mg content. All other sections exhibit higher Mg occurrences only in the basal parts of Kawagarh (Figure 9).

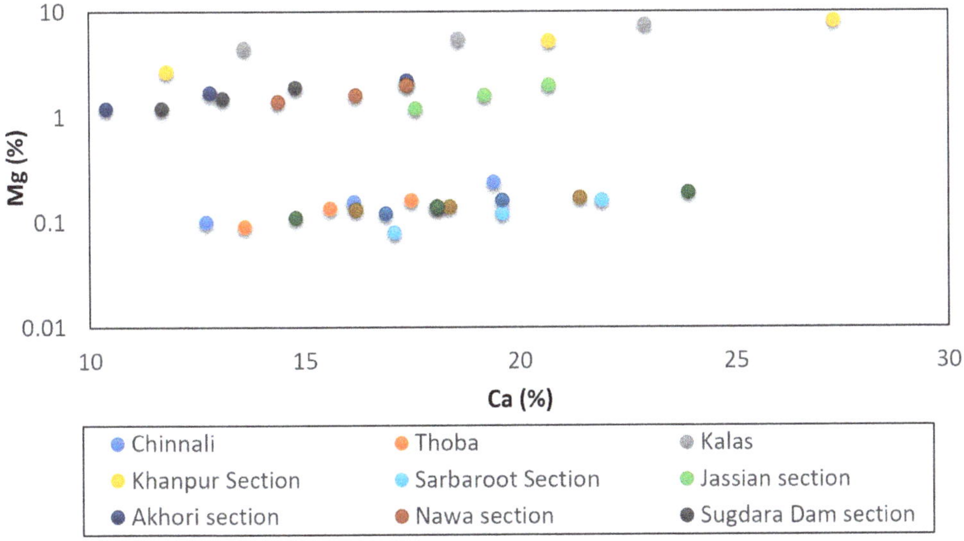

Figure 9. Calcium magnesium cross plot showing the variation of magnesium in study sections.

The trace elements, manganese, strontium, iron, and sodium, show much larger variations across the sections. For manganese, the maximum value of 358 ppm is found in Bagh Neelab, while the minimum value of 105 ppm is found in Khanpur. Strontium has the highest values in Bagh Neelab, with a maximum value of 1080 ppm, while the lowest value is found in Jassian with only 148 ppm. Iron has a minimum value of 187 ppm in Khanpur and a maximum value of 8163 ppm in Sojhanda, while sodium has a minimum value of 1657 ppm in Sojhanda and a maximum value of 6210 ppm in Thoba. In the Hazara sections, Fe and Mn correspond to the Mg rise, whereas the Sr depletes. On the other hand, it is not the same case in the Kalachitta sections, where Fe and Mn deplete, and Sr corresponds positively with the Mg rise or sometimes shows no particular relationship.

4.3. Carbon ($\delta^{13}C$) and Oxygen ($\delta^{18}O$) Isotopes

The stable isotopes, δ^{13}C and δ^{18}O, also show significant variation between the sections. The δ^{13}C values range from −6.069‰ PDB in Thoba to 1.142 in Sojhanda, while the δ^{18}O values range from −7.664‰ PDB in Sojhanda to −1.346‰ PDB in Surg (Figure 10). The mean value of δ^{13}C in the Jassian section (−5.09‰ PDB) is lower than the overall mean, while the mean value of δ^{18}O in the Surg section (−7.57‰ PDB) is noticeably lower than the overall mean.

δ^{13}C shows maximum negative value −5.14‰ PDB in marls, −3.91‰ PDB in limestone, and −2.67‰ PDB in dolostone. It marks positive excursion up to 3.30‰ PDB in limestone, 2.80‰ PDB in marl, and 2.69‰ PDB in dolostone. The studied samples of the Hazara Basin are more depleted in δ^{18}O as compared to samples of Kalachitta. δ^{18}O marks maximum negative excursion −10.73‰ PDB in dolostone, −10.48‰ PDB in marl, and −8.79‰ PDB in limestone. It shows minimum negative excursion in −2.70‰ PDB in marl, 2.99‰ PDB in limestone, and 3.40‰ PDB in dolostone. Higher δ^{18}O values are encountered in the Hazara Basin (−3.915‰ PDB to 3.304‰ PDB) than in Kalachitta (−3.467‰ PDB to 2.693‰ PDB).

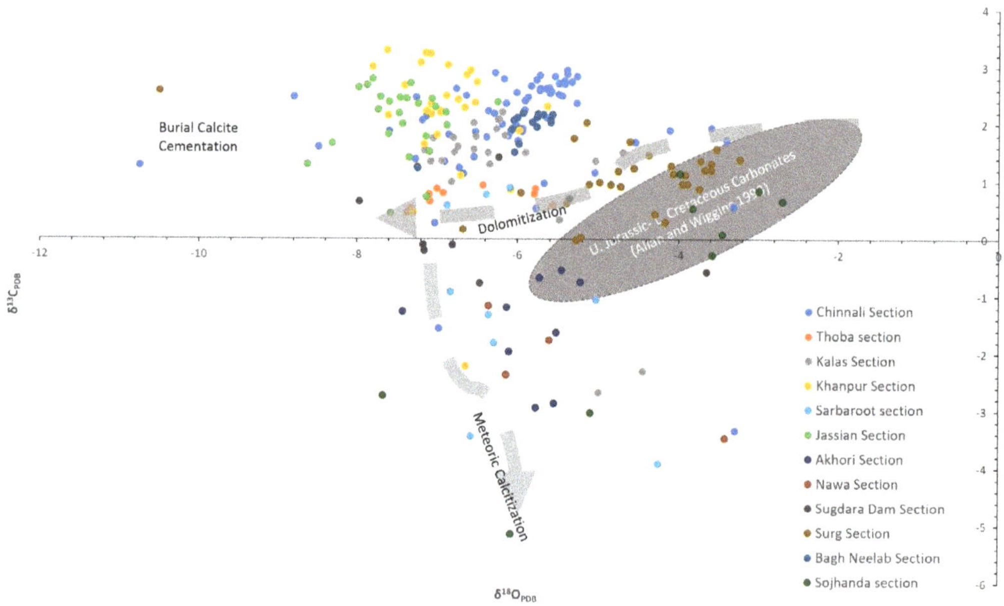

Figure 10. Carbon oxygen stable isotope cross plot showing the covariance in studied sections.

5. Discussion

5.1. Early Calcite Cementation

Microspar is a mosaic-like texture of inorganic calcite crystals ranging in grain size between 4 and 30 μm formed by secondary recrystallization of previously lithified micrite (<4 μm) after removal of Mg ions [71]. [72] discovered microspar cement in aragonite-dominated carbonate muds formed by meteoric diagenesis. The wider consensus has been established that microspar occurs from recrystallization of previously lithified micrite. However, there are examples [73,74] that chalks, although made of micrite, do not recrystallize to microspar even if subjected to fresh water meteoric diagenetic conditions. Many ancient carbonate systems do not experience meteoric conditions [75], which also raises the question about the origin of the microspar in them. [76] reported that microspar is recrystallized from lime mud as a primary cement in the early stages of marine burial diagenesis in Pliocene Bahamas carbonates and Silurian Gotland limestones. Despite the insufficient pressure and temperature condition, Munnecke suggested that strong geochemical gradients can be achieved by microbial decomposition of organic matter [77] in shallow burial environments which may dissolve and reprecipitate carbonates. According to [15,76,78], the pore fluid originated from the interlayer marls beds by the dissolution of aragonite, and then it was transported to the adjacent limestone bed where it was reprecipitated as LMC (microspar). Lately, a series of studies [78–81] led to a limestone–marl layer alternations model which is based on this diagenesis postulate that selective dissolution of aragonite in marl/limemud/shale beds and reprecipitation of calcite cement in limestone beds can result in a bed differentiation which can be confusing with the depositional cyclicity. The formation of microspar in the Kawagarh Formation is in accordance with the above model, evident by the alteration of marls with limestone, with no freshwater substantiations in the lower Kawagarh at early diagenetic stages. However, freshwater flux due to the subaerial exposure during the uplift events on the northern margin of the Indian plate is obvious owing to the India–Asia collision in the Upper Cretaceous (later stage of diagenesis). This cannot contribute to shallow burial early calcite cementation as the rock had already experienced the intermediate to deep burial.

The transformation of aragonite and high magnesium calcite (HMC) into microspar represents shallow marine burial diagenetic conditions. Furthermore, the isopachous cement is evidence of the marine Phreatic conditions.

5.2. Compaction

Mechanical compaction reflects in particle deformation, dewatering, and reduction in thickness and porosity [64,82–84]. Initially, mechanical compaction started with a few meters of burial depth characterized by dehydration and reduction in the thickness of sediment layers which led to close packing and a reduction in porosity. The dewatering of sediments resulted in plastic deformation which is characterized by the rearrangement of particles [85] and the deformation and breakage of skeletal particles [83]. Compaction-related porosity reduction is controlled by textures and dominant mineralogies. Other diagnostic features of mechanical compaction include deformed burrows, thinned and wispy laminations, broken and welded shells, broken micrite envelops, and rotation and particle flattening [85–89]. In Kawagarh, mechanical compaction indicated by the skeletal particle rearrangement, breakage, deformation, and close contacts suggests burial depths of up to 100m. Mechanical compaction always shows greater signals in granular facies as the bigger grains take the first impact of the compaction, whereas the absence of preferred orientations and the neomorphism of intact grains supports how cementation took place before compaction [15]. Sutured skeletal particles contact and stylolite is clear evidence of pressure solutioning (chemical compaction) on depths >300 m and is indicative of the onset of burial diagenesis.

5.3. Dolomitization

Dolomitization refers to the replacement of limestone ($CaCO_3$) by dolomite (CaMg $(CO_3)^2$), which can occur in a variety of diagenetic settings. It is critical to establish key geochemical and environmental controls for various dolomitization models [3,90–92]. The fault-related hydrothermal dolomitization (HTD) model claims to explain the source of Mg by the convection of seawater along a rift-related open fault system and basal clastic aquifers [93–96]. Some studies also suggest thermodynamic ultramafic carbonation by the interaction of water and mafic/ultramafic rocks as the source of magnesium [97–99]. These views have modernized the thinking about the sources for water and magnesium in HTD. The same seems appropriate for the Late Cretaceous rifted northern margin of India [100] where the Kawagarh Formation deposited. A similar kind of digenetic and dolomitization model has also been reported in Early Cretaceous Benicassim carbonates, Maestrat Basin, E. Spain [2,101,102]. Both the carbonates deposited in a rift basin share diagenetic similarities such as early calcite cementation, strata bound dolostones, stylolites, burial, and uplift. [6] has also reported similar Cambrian HTD dolomite bodies formed at shallow burial depths through hot brines, which were generated by high heat flows through faults in Cambrian rifting in dolomite from the Western Canadian Sedimentary Basin. These models justify the dolomitization in the Kawagargarh Formation supported by stylolization paired with significant negative $\delta^{18}O$ values suggesting that the dolomitization occurred at intermediate to deep burial settings at higher temperatures negating the meteoric diagenesis. These HTD dolomite bodies often involve multiple stages of dolomitization [94,103,104] where a paragenetic sequence can be established by observing the cross-cutting relationship of different dolomite fabrics. In the Kawagrh Formation, the overlay of dolomitic crystals over stylolites suggests that stylolites were formed first and fed Mg for dolomitization. RD1 developed firstly and subsequently replaced by RD2 indicates multistage dolomitization [7]. In Figure 11, a schematic illustration shows how the evolving of dolomitization fronts can initiate RD1 and with time how RD1 can be replaced with RD2 euhedral and RD2 subhedral. Along the length of such fronts as it is recrystallizing the host, the depletion of Mg occurs ahead of the front levels only, forming individual crystals of RD1 with intercrystallite porosity.

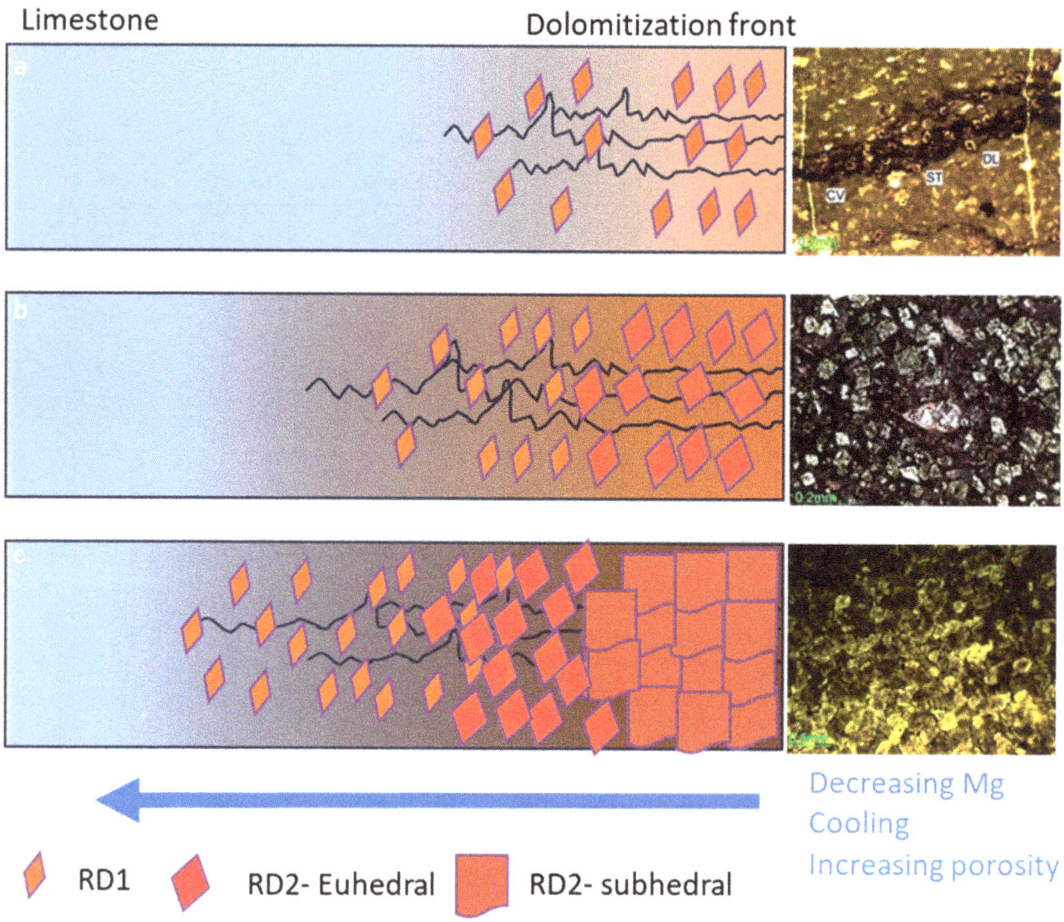

Figure 11. Systematic diagram showing evolution of the dolomitization front with time. (**a**) Initiation of the dolomitization front with RD1. (**b**) With the continued Mg influx and advancement of the front, this recrystallizes RD1 to euhedral RD2. (**c**) With further advancement, euhedral RD2 is recrystallized into subhedral RD2. Note how the fabric of the rock evolved along the length of the front with deleting and cooling fluid. Modified after [6].

5.4. Calcitisation

The ongoing India–Asia collision gradually uplifted the rock to shallower depths and produced fractures in lithified rock. Such shallow burial settings introduced low Mg undersaturated meteoric fluids which resulted in the transformation of dolomite to calcite through calcitisation. Depleted $\delta^{13}C$ values also suggest a likewise association (Figure 10). On the other hand, pronounced negative values of $\delta^{18}O$ negate the pure meteoric water influx and instead indicate the mixing of surface-derived waters with hot burial fluids. Frequent occurrence of FC with the dolomitisation fronts indicates that calcitising fluids flowed preferentially through the enhanced intercrystalline porosity of dolostones (Figure 8e,f). [6,7] reported similar calcitisation events where transformation from RD1 to calcite increased local porosity within dolostone. This suggests the process of calcitisation consists of two steps, an initial dissolution that enhances porosity around the periphery of dolostone bodies followed by a contemporaneous calcite cement (FC) precipitation. The cross-cutting relationship and replacement of dolomite crystals along calcite fractures clearly indicates that these fractures postdate dolomitization.

5.5. Diagenetic Model

The Kawagarh Formation underwent four phases of diagenesis (Figure 12). In the first phase, early calcite cementation occurred during shallow marine burial, resulting in the formation of microspar cement by replacing micritic limemud matrix and bounding the skeletal grain. This also involves the dissolution of aragonite from limemud of marls into calcites of limestone. The later stage of the first phase corresponds to the neomorphim where the carbonate grain, particularly skeletal grains, were transformed into coarse-grained microspar calcite. The second phase corresponds to intermediate burial diagenesis, where mechanical and chemical compaction dominated with stylolization in the later stages. Stylolites formed at burial depths ranging between 300–800 m. Deep burial diagenesis corresponds to the third phase, which was characterized by dolomitization fronts, and the formation of RD1 and RD2 dolomite. The fourth and final phase corresponds to uplift, resulting in large-scale fractures and veins. Low Mg fresh fluids became incorporated into the system causing calcitisation (dedolomizttion) and the precipitation of fracture calcite.

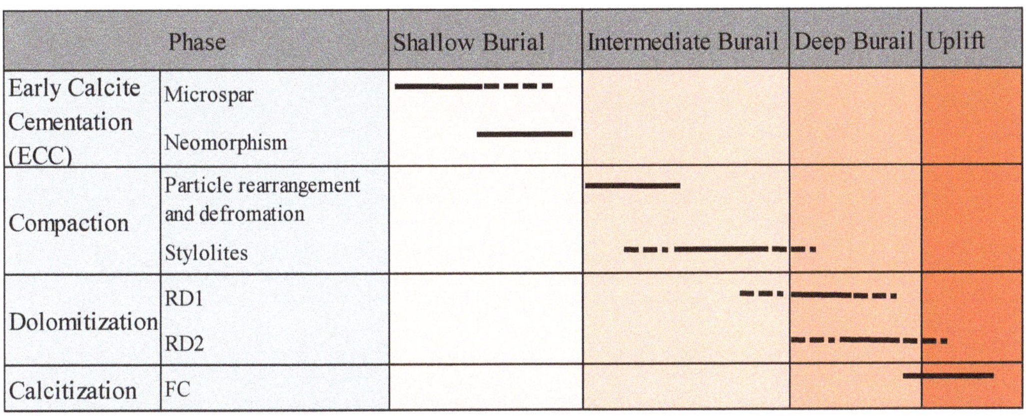

Figure 12. Paragenetic history of the Kawagarh Formation showing the sequence of key diagenetic phases along with the time of formation.

6. Conclusions

The Kawagarh Formation was deposited during High Stand Tract (HST) under open shallow marine ramp environments, bounded by Low Stand (LST) clastics. The regional tectonic settings suggest that the depositional trend of the Kawagarh Formation was terminated by the initial collision of India with the KLA at the end of Cretaceous. The following conclusions can be drawn in context with the complex diagenetic history of the Kawagarh Formation by interpreting a multiproxy dataset integrating outcrop, petrographic, stable isotope, and geochemical data.

1. Early calcite cementation happened well before mechanical compaction which also inhibited later compaction in the Kawagarh Formation.
2. Predominantly, the dissolution of aragonite from lime mud in marls of the Kalachitta and high Mg calcite/aragonite from mudstones of the Hazara basin provided low Mg calcite for the precipitation of microspar.
3. The fault-related hydrothermal dolomitization model claims to explain the source of Mg by the convection of seawater along rift-related open fault systems and basal clastic aquifers for the Kawagarh formation.

4. The uplift related to the India–Asia collision introduced a fracture system and fresh low Mg fluid percolating the fracture system mixed with subsurface fluid caused the calcitisation in the Kawagarh Formation.

The present study provides significant insights about burial depths, fluid influx, the geochemical gradients, and the evolution of reservoir properties (porosity, permeability) of the formation as well as the basin itself, such as the depth and pressure at which it was buried, the fluid flow, and the geochemical gradients. This appraisal can be used to better understand the geologic history of the region and to develop better exploration strategies for hydrocarbon reservoirs.

Author Contributions: Conceptualization, S.U.R., N.A. and M.J.M.; methodology, S.U.R. and M.J.M.; software, M.J.M.; validation, M.M.S., N.A. and M.K.; formal analysis, M.J.M.; investigation, V.L. and G.K.; resources, S.U.R. and N.A.; data curation, M.J.M. and H.T.J.; writing—original draft preparation, S.U.R. and M.J.M.; writing—review and editing, N.A., M.M.S. and H.T.J.; visualization, M.J.M. and G.K.; supervision, N.A., M.M.S., V.L. and G.K.; project administration, S.U.R. All authors have read and agreed to the published version of the manuscript.

Funding: This research received no external funding.

Data Availability Statement: The data used in this work are available on request to the corresponding author(s).

Acknowledgments: This work benefited from numerous discussions and collaboration with Mohammad Al Suwaidi and Aleem Zahid. The first author would thank Meng Fanwei for technical support and experiments. The authors also acknowledge contributions from the University of Sargodha for field and logistics support.

Conflicts of Interest: The authors declare no conflict of interest.

References

1. Tomašových, A. Microfacies and depositional environment of an Upper Triassic intra-platform carbonate basin: The Fatric Unit of the West Carpathians (Slovakia). *Facies* **2004**, *50*, 77–105. [CrossRef]
2. Gomez-Rivas, E.; Martín-Martín, J.D.; Bons, P.D.; Koehn, D.; Griera, A.; Travé, A.; Llorens, M.-G.; Humphrey, E.; Neilson, J. Stylolites and stylolite networks as primary controls on the geometry and distribution of carbonate diagenetic alterations. *Mar. Pet. Geol.* **2022**, *136*, 105444. [CrossRef]
3. Whitaker, F.F.; Smart, P.L.; Jones, G.D. Dolomitization: From conceptual to numerical models. *Geom. Petrog. Dolomite Hydrocarb. Reserv.* **2004**, *235*, 99. [CrossRef]
4. Moore, C.H.; Wade, W.J. Carbonate Porosity and Introduction to Diagenesis. In *Carbonate Reservoirs*; Elsevier: Amsterdam, The Netherlands, 2013; p. 49.
5. Peyrotty, G.; Brigaud, B.; Martini, R. δ18O, δ13C, trace elements and REE in situ measurements coupled with U–Pb ages to reconstruct the diagenesis of upper triassic atoll-type carbonates from the Panthalassa Ocean. *Mar. Pet. Geol.* **2020**, *120*, 104520. [CrossRef]
6. Koeshidayatullah, A.; Corlett, H.; Stacey, J.; Swart, P.K.; Boyce, A.; Hollis, C. Origin and evolution of fault-controlled hydrothermal dolomitization fronts: A new insight. *Earth Planet Sci. Lett.* **2020**, *541*, 116291. [CrossRef]
7. Koeshidayatullah, A.; Corlett, H.; Stacey, J.; Swart, P.K.; Boyce, A.; Robertson, H.; Whitaker, F.; Hollis, C. Evaluating new fault-controlled hydrothermal dolomitization models: Insights from the Cambrian Dolomite, Western Canadian Sedimentary Basin. *Sedimentology* **2020**, *67*, 2945–2973. [CrossRef]
8. Alsuwaidi, M.; Mohamed, A.A.I.; Mansurbeg, H.; Morad, S.; Alsuwaidi, A.; Al-Shalabi, E.W.; Gomes, J.; Al-Ramadan, K.; Mohammed, I.Q.; Farouk, S. Depositional and diagenetic controls on reservoir quality of microporous basinal lime mudstones (Aptian), United Arab Emirates. *Sediment. Geol.* **2021**, *420*, 105925. [CrossRef]
9. Ehrenberg, S.N.; Nadeau, P.H. Sandstone vs. carbonate petroleum reservoirs: A global perspective on porosity-depth and porosity-permeability relationships. *Am. Assoc. Pet. Geol. Bull.* **2005**, *89*, 435–445. [CrossRef]
10. Morse, J.W.; Wade, W.J. Early Marine Diagenesis of Shoal-Water Carbonate Sediments. In *Geochemistry of Sedimentary Carbonates*; Elsevier: Amsterdam, The Netherlands, 1990; pp. 241–276.
11. Tucker, M.E.; Bathurst, R.G. Carbonate diagenesis. In *International Association of Sedimentologists Series*, 1st ed.; Wiley–Blackwell: Hoboken, NJ, USA, 1990.
12. Moore, C.H.; Wade, W.J. Porosity Evolution in the Marine, Meteoric, and Burial Realms. In *Carbonate Reservoirs*; Elsevier: Amsterdam, The Netherlands, 2013; p. 91.
13. Regnet, J.B.; David, C.; Robion, P.; Menéndez, B. Microstructures and physical properties in carbonate rocks: A comprehensive review. *Mar. Pet. Geol.* **2019**, *103*, 366–376. [CrossRef]

14. Choquette, P.W.; Pray, L.C. Geologic Nomenclature and Classification of Porosity in Sedimentary Carbonates. *Am. Assoc. Pet. Geol. Bull.* **1970**, *54*, 207–250.
15. Munnecke, A.; Wright, V.P.; Nohl, T. The origins and transformation of carbonate mud during early marine burial diagenesis and the fate of aragonite: A stratigraphic sedimentological perspective. *Earth-Sci. Rev.* **2023**, *239*, 104366. [CrossRef]
16. Land, L.S. Limestone diagenesis-some geochemical considerations. *Geol. Surv. Bull.* **1986**, *5*, 129–137.
17. Choquette, P.W.; James, N.P. DIAGENESIS #12. Diagenesis in Limestones—3. The Deep Burial Environment. *Geosci. Can.* **1987**, *14*, 3–35.
18. Dutta, N.C. Deepwater geohazard prediction using prestack inversion of large offset P-wave data and rock model. *Lead. Edge* **2002**, *21*, 193–198. [CrossRef]
19. Ahsan, N. Facies Modeling, Depositional and Diagenetic Environments of Kawagarh Formation, Hazara Basin, Pakistan. Ph.D. Thesis, Punjab University, Lahore, Pakistan, 2008; Unpublished.
20. Rehman, S.; Ahsan, N.; Shah, M.; Munawar, M.; Miraj, M.; Rehman, F.; Mahmood, K. Depositional Environments and Microfacies of the Upper Turonian–Maastrichtian Kawagarh Formation, Kalachitta Range, Lesser Himalayas, Pakistan. *Aust. J. Earth Sci.* **2021**, *68*, 1017–1030. [CrossRef]
21. Latif, M. Explanatory Notes on the Geology of Southern Hazara to Accompany the Revised Geological Map. *Wein Jahrb Der Geol. Bundesanst.* **1970**, *15*, 5–20.
22. Calkins, J.A.; Offield, T.W.; Abdullah, S.K.M.; Ali, S.T. Geology of the southern Himalaya in Hazara, Pakistan and Adjoining Areas. Geol Surv Prof Pap 716-C, C1-C29 Geol. *Surv. Pakistan* **1975**, *19*, 93–132.
23. Aitchison, J.C.; Ali, J.R.; Davis, A.M. When and where did India and Asia collide? *J. Geophys. Res. Solid Earth* **2007**, *11*, 112–127. [CrossRef]
24. Ahsan, N.; Chaudhry, M.N. Geology of Hettangian to middle Eocene rocks of Hazara and Kashmir basins, Northwest lesser Himalayas, Pakistan. *Geol. Bull. Punjab Univ.* **2008**, *43*, 131–152.
25. Burg, J.-P.; Célérier, B.; Chaudhry, M.N.; Ghazanfar, M.; Gnehm, F.; Schnellmann, M. Fault analysis and paleostress evolution in large strain regions: Methodological and geological discussion of the southeastern Himalayan fold-and-thrust belt in Pakistan. *J. Asian Earth Sci.* **2005**, *24*, 445–467. [CrossRef]
26. Searle, M.P. Geological evidence against large-scale pre-Holocene offsets along the Karakoram Fault: Implications for the limited extrusion of the Tibetan Plateau. *Tectonics* **1996**, *15*, 171–186. [CrossRef]
27. Acton, G.D. Apparent polar wander of India since the Cretaceous with implications for regional tectonics and true polar wander, in the Indian Subcontinent and Gondwana: A Paleomagnetic and Rock Magnetic prospective. *Mem. Geol. Soc. India* **1999**, *44*, 129–175.
28. Ali, J.R.; Aitchison, J.C. Greater India. *Earth-Sci. Rev.* **2005**, *72*, 169–188. [CrossRef]
29. Najman, Y. The detrital record of orogenesis: A review of approaches and techniques used in the Himalayan sedimentary basins. *Earth-Sci. Rev.* **2006**, *74*, 1–72. [CrossRef]
30. Yin, A. Cenozoic tectonic evolution of the Himalayan orogen as constrained by along-strike variation of structural geometry, exhumation history, and foreland sedimentation. *Earth-Sci. Rev.* **2006**, *76*, 1–131. [CrossRef]
31. Gansser, A. *Geology of the Himalayas*; Interscience: New York, NY, USA, 1964; Volume 289.
32. Ghazanfar, M. Petrotectonic Elements and Tectonic Framework of Northwest Himalaya. Ph.D. Thesis, University of the Punjab, Punjab, Pakistan, 1993; Volume 1 & 2, pp. 1–380.
33. Powell, C.M. A speculative tectonic history of Pakistan and surroundings: Some constraints from the Indian Ocean. *Geodyn. Pak.* **1979**, *13*, 5–24.
34. Yeats, R.; Lawrence, R. Tectonics of the Himalayan thrust belt in northern Pakistan. In Proceedings of the US-Pakistan Workshop on Marine Sciences in Pakistan, Karachi, Pakistan, 11–16 November 1982; 39p.
35. Qureshi, K.A.; Ahmad, M. *Geological Map of Kalachitta Range, Northern Punjab, Pakistan*; Geological Survey of Pakistan: Quetta, Pakistan, 2001.
36. Latif, M.A.; Yasin, A.R.; Shafique, N.A.; Ashraf, M. Late mesozoic sedimentary megacycle in the rifted Haro Trough, Hazara, Pakistan and its hydrocarbon implications in the Northern Rim of the North West Himalayan Basin. *Pak. J. Hydrocarb. Res.* **1995**, *7*, 31–52.
37. Shah, S.M. *Stratigraphy of Pakistan*, 22nd ed.; Geological Survey of Pakistan: Quetta, Pakistan, 2009.
38. Rehman, S.U. Sedimentology of Turonian-Maastrichtian Kawagarh Formation, Attock Hazara Fold and Thrust Belt, Northwestern Lesser Himalayas, Pakistan. Doctoral Dissertation, University of Sargodha, Sargodha, Pakistan, 2017.
39. Garzanti, E.; Sciunnach, D.; Gaetani, M.; Corfield, R.; Watts, A.; Searle, M. Discussion on subsidence history of the north Indian continental margin, Zanskar–Ladakh Himalaya, NW India. *J. Geol. Soc. Lond.* **2005**, *162*, 889–892. [CrossRef]
40. Van Hinsbergen, D.J.; Lippert, P.C.; Dupont-Nivet, G.; McQuarrie, N.; Doubrovine, P.V.; Spakman, W.; Torsvik, T.H. Greater India Basin hypothesis and a two-stage Cenozoic collision between India and Asia. *Proc. Natl. Acad. Sci. USA* **2012**, *109*, 7659–7664. [CrossRef]
41. Rajkakati, M.; Bhowmik, S.K.; Ao, A.; Ireland, T.R.; Avila, J.; Clarke, G.L.; Aitchison, J.C. Thermal history of Early Jurassic eclogite facies metamorphism in the Nagaland Ophiolite Complex, NE India: New insights into pre-Cretaceous subduction channel tectonics within the Neo-Tethys. *Lithos* **2019**, *346*, 105166. [CrossRef]

42. Ahsan, N.; Rehman, S.U.; Shah, M.M. Kawagarh Formation (Turonian to Lower Maastrichtian)-A Homoclinal Ramp Deposit in Hazara Basin on Northern Margin of the Indian Plate. In Proceedings of the 15th Bathurst Meeting, Edinburgh, UK, 13–16 July 2015; University of Edenburgh: Edinburgh, UK, 2015.
43. Haq, B.U.; Hardenbol, J.; Vail, P.R. Mesozoic and Cenozoic chronostratigraphy and cycles of sea-level change. In *Society of Economic Paleontologists and Mineralogists*; Special Publication, 42; American Association of Petroleum Geoscience: Tulsa, OK, USA, 1988.
44. Rehman, S.U.; Mahmood, K.; Ahsan, N.; Shah, M. Microfacies and depositional environments of upper cretaceous Kawagarh Formation from Chinali and Thoba sections Northeastern Hazara Basin, lesser Himalayas, Pakistan. *J. Himal. Earth Sci.* **2016**, *49*, 1–16.
45. Chaudhry, M.N.; Mahmood, T.; Ahmed, R.; Ghazanfar, M. Microfacies, diagenesis and environment of deposition of Datta Formation from jaster Gali, Abbottabad, Hazara, Pakistan. *Geol. Bull. Punjab Univ.* **1992**, *27*, 47–62.
46. Ding, L.; Kapp, P.; Wan, X. Paleocene–Eocene record of ophiolite obduction and initial India-Asia collision, south central Tibet. *Tectonics* **2005**, *24*, 1–18. [CrossRef]
47. Klootwijk, C.T.; Gee, J.; Peirce, J.; Smith, G. Constraints on the India-Asia convergence: Paleomagnetic results from Ninetyeast Ridge. In *Proceedings of the Ocean Drilling Program*; Scientific Results; Ocean Drilling Program: College Station, TX, USA, 1991; Volume 121.
48. Rehman, H.U.; Seno, T.; Yamamoto, H.; Khan, T. Timing of collision of the Kohistan–Ladakh Arc with India and Asia: Debate. *Isl. Arc* **2011**, *20*, 308–328. [CrossRef]
49. Spencer, D.A. *Tectonics of the Higher-and Tethyan Himalaya, Upper Kaghan Valley, NW Himalaya, Pakistan: Implications of an Early Collisional, High Pressure (Eclogite Facies) Metamorphism to the Himalayan Belt*; Eidgenoessische Technische Hochschule Zuerich: Zürich, Switzerland, 1993.
50. Yin, A.; Harrison, T.M. Geologic evolution of the Himalayan-Tibetan orogen. *Annu. Rev. Earth Planet. Sci.* **2000**, *28*, 211–280. [CrossRef]
51. Dickson, J.A.D. A Modified Staining Technique for Carbonates in Thin Section. *Nature* **1965**, *205*, 587. [CrossRef]
52. Adams, A.; MacKenzie, I. *Carbonate Sediments and Rocks under the Microscope: A Colour Atlas*, 1st ed.; CRC Press: Boca Ratona, FL, USA, 1998.
53. Robinson, P. Determination of calcium, magnesium, manganese, strontium, sodium and iron in the carbonate fraction of limestones and dolomites. *Chem. Geol.* **1980**, *28*, 135–146. [CrossRef]
54. Folk, R.L. Some Aspects of Recrystallization in Ancient Limestones[1]. In *Dolomitization and Limestone Diagenesis*; SEPM Society for Sedimentary Geology: Tulsa, OK, USA, 1965; Volume 13.
55. Bathurst, R. Chapter 12 Neomorphic Processes in Diagenesis. In *Carbonate Sediments and Their Diagenesis; Developments in Sedimentology*; Elsevier: Amsterdam, The Netherlands, 1972; pp. 475–516.
56. James, N.P.; Choquette, P.W. Diagenesis 5. Limestones: Introduction. *Geosci. Can.* **1983**, *10*. Available online: https://journals.lib.unb.ca/index.php/GC/article/view/3352 (accessed on 1 November 2023).
57. Tucker, M.E.; Wright, V.P. Diagenetic Processes, Products and Environments. *Carbonate Sedimentol.* **1990**, *314*, 364.
58. Neumeier, U. Experimental modelling of beachrock cementation under microbial influence. *Sediment. Geol.* **1999**, *126*, 35–46. [CrossRef]
59. Hillgärtner, H.; Dupraz, C.; Hug, W. Microbially induced cementation of carbonate sands: Are micritic meniscus cements good indicators of vadose diagenesis? *Sedimentology* **2001**, *48*, 117–131. [CrossRef]
60. Flügel, E. (Ed.) *Diagenesis, Porosity, and Dolomitization BT—Microfacies of Carbonate Rocks: Analysis, Interpretation and Application*; Springer: Berlin/Heidelberg, Germany, 2004; pp. 267–338.
61. Boggs, S. *Principles of Sedimentology and Stratigraphy*; Pearson Prentice Hall: Hoboken, NJ, USA, 2006.
62. Reinhold, C.; Kaufmann, B. Sea-level changes as controlling factor of early diagenesis: The reefal limestones of Adnet (Late Triassic, Northern Calcareous Alps, Austria). *Facies* **2010**, *56*, 231–248. [CrossRef]
63. Flügel, E. (Ed.) *Karbonatdiagenese BT—Mikrofazielle Untersuchungsmethoden von Kalken*; Springer: Berlin/Heidelberg, Germany, 1978; pp. 51–78.
64. Smith, J.E. The dynamics of shale compaction and evolution of pore-fluid pressures. *J. Int. Assoc. Math. Geol.* **1971**, *3*, 239–263. [CrossRef]
65. Koehn, D.; Renard, F.; Toussaint, R.; Passchier, C.W. Growth of stylolite teeth patterns depending on normal stress and finite compaction. *Earth Planet Sci. Lett.* **2007**, *257*, 582–595. [CrossRef]
66. Ebner, M.; Piazolo, S.; Renard, F.; Koehn, D. Stylolite interfaces and surrounding matrix material: Nature and role of heterogeneities in roughness and microstructural development. *J. Struct. Geol.* **2010**, *32*, 1070–1084. [CrossRef]
67. Toussaint, R.; Aharonov, E.; Koehn, D.; Gratier, J.P.; Ebner, M.; Baud, P.; Rolland, A.; Renard, F. Stylolites: A review. *J. Struct. Geol.* **2018**, *114*, 163–195. [CrossRef]
68. Nelson, R.A. Significance of fracture sets associated with stylolite zones: Geologic notes. *Am. Assoc. Pet. Geol. Bull.* **1981**, *65*, 2417–2425.
69. Railsback, L.B. Lithologic controls on morphology of pressure-dissolution surfaces (stylolites and dissolution seams) in Paleozoic carbonate rocks from the mideastern United States. *J. Sediment. Res.* **1993**, *63*, 513–522. [CrossRef]

70. Ben-Itzhak, L.L.; Aharonov, E.; Karcz, Z.; Kaduri, M.; Toussaint, R. Sedimentary stylolite networks and connectivity in limestone: Large-scale field observations and implications for structure evolution. *J. Struct. Geol.* **2014**, *63*, 106–123. [CrossRef]
71. Folk, R.L. Practical Petrographic Classification of Limestones1. *Am. Assoc. Pet. Geol. Bull.* **1995**, *43*, 1–38. [CrossRef]
72. Lasemi, Z.; Sandberg, P.A. Transformation of aragonite-dominated lime muds to microcrystalline limestones. *Geology* **1984**, *12*, 420–423. [CrossRef]
73. Beales, F.W. Conditions of Deposition of Palliser (Devonian) Limestone of Southwestern Alberta1. *Am. Assoc. Pet. Geol. Bull.* **1956**, *40*, 848–870. [CrossRef]
74. James, N.P.; Choquette, P.W. Diagenesis 9. Limestones—The Meteoric Diagenetic Environment. *Geosci. Can.* **1984**, *11*, 161–194.
75. Bathurst, R.G.C. Microfabrics in Carbonate Diagenesis: A Critical Look at Forty Years in Research BT. In *Carbonate Microfabrics*; Rezak, R., Lavoie, D.L., Eds.; Springer: New York, NY, USA, 1993; pp. 3–14.
76. Munnecke, A.; Westphal, H.; Reijmer, J.; Samtleben, C. Microspar development during early marine burial diagenesis: A comparison of Pliocene carbonates from the Bahamas with Silurian limestones from Gotland (Sweden). *Sedimentology* **1997**, *44*, 977–990. [CrossRef]
77. Canfield, D.E.; Raiswell, R. Carbonate Precipitation and Dissolution: Its Relevance to Fossil Preservation. *Geography* **1991**, 411–453. Available online: https://www.semanticscholar.org/paper/Carbonate-Precipitation-and-Dissolution%3A-Its-to-Canfield-Raiswell/3bc334ef1bc6aa986a81739738ddcb5071324a9b (accessed on 1 November 2023).
78. Munnecke, A.; Westphal, H. Variations in primary aragonite, calcite, and clay in fine-grained calcareous rhythmites of Cambrian to Jurassic age—An environmental archive? *Facies* **2005**, *51*, 592–607. [CrossRef]
79. Munnecke, A.; Westphal, H.; Elrick, M.; Reijmer, J. The mineralogical composition of precursor sediments of calcareous rhythmites: A new approach. *Int. J. Earth Sci.* **2001**, *90*, 795–812. [CrossRef]
80. Westphal, H.; Böhm, F.; Bornholdt, S. Orbital frequencies in the carbonate sedimentary record: Distorted by diagenesis? *Facies* **2004**, *50*, 3–11. [CrossRef]
81. Westphal, H.; Munnecke, A.; Böhm, F.; Bornholdt, S. Limestone-marl alternations in epeiric sea settings—Witnesses of environmental changes, or of rhythmic diagenesis? Geol. *Assoc. Can.* **2008**, *48*, 389–406.
82. Shinn, E.A.; Halley, R.B.; Hudson, J.H.; Lidz, B.H. Limestone compaction: An enigma. *Geology* **1977**, *5*, 21–24. [CrossRef]
83. Shinn, E.A.; Robbin, D.M. Mechanical and chemical compaction in fine-grained shallow-water limestones. *J. Sediment. Res.* **1983**, *53*, 595–618. [CrossRef]
84. Gnoli, M. The amount of the taphonomic/tectonic compaction in the Fluminimaggiore Formation (SW Sardinia, Italy) with discussion of a new paleontological method for estimating compaction. *Palaeontol. Electron.* **2002**, *4*, 18–37.
85. Ricken, W. *Diagenetic Bedding: A Model for Marl-Limestone Alternations (Lecture Notes in Earth Sciences #6)*; Springer: New York, NY, USA, 1986.
86. Bathurst, R. The integration of pressure solution with mechanical compaction and cementation. In *Stylolites and Associated Phenomena. Relevance to Hydrocarbon Reservoirs*; Abu Dhabi Natl Reserv Found: Abu Dhabi, United Arab Emirates, 1984; pp. 41–55.
87. Lasemi, Z.; Sandberg, P.A.; Boardman, M.R. New microtextural criterion for differentiation of compaction and early cementation in fine-grained limestones. *Geology* **1990**, *18*, 370–373. [CrossRef]
88. Monaco, C.; Tortorici, L.; Nicolich, R.; Cernobori, L.; Costa, M. From collisional to rifted basins: An example from the southern Calabrian arc (Italy). *Tectonophysics* **1996**, *266*, 233–249. [CrossRef]
89. Westphal, H.; Munnecke, A. Mechanical compaction versus early cementation in fine-grained limestones: Differentiation by the preservation of organic microfossils. *Sediment. Geol.* **1997**, *112*, 33–42. [CrossRef]
90. Morrow, D.W. Diagenesis 2. Dolomite—Part 2 Dolomitization Models and Ancient Dolostones. *Geosci. Can.* **1982**, *9*, 95–107.
91. Land, L.S. The Origin of Massive Dolomite. *J. Geol. Educ.* **1985**, *33*, 112–125. [CrossRef]
92. Machel, H.G. Concepts and models of dolomitization: A critical reappraisal. *Geom. Petrog. Dolomite Hydrocarb. Reserv.* **2004**, *235*, 7–63. [CrossRef]
93. Martín-Martín, J.; Travé, A.; Gomez-Rivas, E.; Salas, R.; Sizun, J.-P.; Vergés, J.; Corbella, M.; Stafford, S.; Alfonso, P. Fault-controlled and stratabound dolostones in the Late Aptian–earliest Albian Benassal Formation (Maestrat Basin, E Spain): Petrology and geochemistry constrains. *Mar. Pet. Geol.* **2015**, *65*, 83–102. [CrossRef]
94. Hollis, C.; Bastesen, E.; Boyce, A.; Corlett, H.; Gawthorpe, R.; Hirani, J.; Rotevatn, A.; Whitaker, F. Fault-controlled dolomitization in a rift basin. *Geology* **2017**, *45*, 219–222. [CrossRef]
95. Hirani, J.; Bastesen, E.; Boyce, A.; Corlett, H.; Eker, A.; Gawthorpe, R.; Hollis, C.; Korneva, I.; Rotevatn, A. Structural controls on non-fabric-selective dolomitization within rift-related basin-bounding normal fault systems: Insights from the Hammam Faraun Fault, Gulf of Suez, Egypt. *Basin Res.* **2018**, *30*, 990–1014. [CrossRef]
96. Al-Ramadan, K.; Koeshidayatullah, A.; Cantrell, D.; Swart, P.K. Impact of basin architecture on diagenesis and dolomitization in a fault-bounded carbonate platform: Outcrop analogue of a pre-salt carbonate reservoir, Red Sea rift, NW Saudi Arabia. *Pet. Geosci.* **2019**, *26*, 448–461. [CrossRef]
97. Lavoie, D.; Jackson, S.; Girard, I. Magnesium isotopes in high-temperature saddle dolomite cements in the lower Paleozoic of Canada. *Sediment. Geol.* **2014**, *305*, 58–68. [CrossRef]
98. Falk, E.S.; Kelemen, P.B. Geochemistry and petrology of listvenite in the Samail ophiolite, Sultanate of Oman: Complete carbonation of peridotite during ophiolite emplacement. *Geochim. Cosmochim. Acta* **2015**, *160*, 70–90. [CrossRef]

99. Robertson, H.; Corlett, H.; Hollis, C.; Kibblewhite, T.; Whitaker, F. Listwanitization as a Source of Mg for Dolomitization: Field Evaluation in Atlin, British Columbia. In Proceedings of the Goldschmidt Conference, Barcelona, Spain, 18–23 August 2019.
100. Garzanti, E.; Hu, X. Latest Cretaceous Himalayan tectonics: Obduction, collision or Deccan-related uplift? *Gondwana Res.* **2015**, *28*, 165–178. [CrossRef]
101. Humphrey, E.; Gomez-Rivas, E.; Neilson, J.; Martín-Martín, J.D.; Healy, D.; Yao, S.; Bons, P.D. Quantitative analysis of stylolite networks in different platform carbonate facies. *Mar. Pet. Geol.* **2020**, *114*, 104203. [CrossRef]
102. Yao, S.; Gomez-Rivas, E.; Martín-Martín, J.D.; Gómez-Gras, D.; Travé, A.; Griera, A.; Howell, J. A Fault-controlled dolostone geometries in a transgressive–regressive sequence stratigraphic framework. *Sedimentology* **2020**, *67*, 3290–3316. [CrossRef]
103. Davies, G.R.; Smith, L.B., Jr. Structurally controlled hydrothermal dolomite reservoir facies: An overview. *Am. Assoc. Pet. Geol. Bull.* **2006**, *90*, 1641–1690. [CrossRef]
104. Sharp, I.; Gillespie, P.; Morsalnezhad, D.; Taberner, C.; Karpuz, R.; Vergés, J.; Horbury, A.; Pickard, N.; Garland, J.; Hunt, D. Stratigraphic architecture and fracture-controlled dolomitization of the Cretaceous Khami and Bangestan groups: An outcrop case study, Zagros Mountains, Iran. In *Mesozoic and Cenozoic Carbonate Systems of the Mediterranean and the Middle East: Stratigraphic and Diagenetic Reference Models*; Geological Society: London, UK, 2010; Volume 329, pp. 343–396.

minerals

MDPI

Article

Petrological, Geochemical and Chronological Characteristics of Dolomites in the Permian Maokou Formation and Constraints to the Reservoir Genesis, Central Sichuan Basin, China

Xuejing Bai [1,2], Jianfeng Zheng [2,3,*], Kun Dai [4,*], Shuxin Hong [1,2], Junmao Duan [2,3] and Yunmiao Liu [1,2]

1 Exploration and Development Research Institute, PetroChina Daqing Oilfield Co., Ltd., Daqing 163712, China; bxuejing@petrochina.com.cn (X.B.); hshuxin@petrochina.com.cn (S.H.); liuyunmiao@petrochina.com.cn (Y.L.)
2 Key Laboratory of Carbonate Reservoirs CNPC, Hangzhou 310023, China; duanjm_hz@petrochina.com.cn
3 PetroChina Hangzhou Research Institute of Geology, Hangzhou 310023, China
4 School of Earth Sciences, China University of Petroleum (Beijing), Beijing 102249, China
* Correspondence: zhengjf_hz@petrochina.com.cn (J.Z.); thous_nite@foxmail.com (K.D.)

Abstract: The Middle Permian Maokou Formation in the Sichuan Basin has huge resources and is an important target for natural gas exploration. In recent years, significant exploration breakthroughs have been made in the dolomite field of member Mao-2 in central Sichuan, and the gas production of several wells has exceeded 1×10^6 m^3/d, indicating promising prospects for exploration. However, the origin of the dolomite reservoir in member Mao-2 remains ambiguous, which restricts the accurate prediction of favorable reservoirs. This study focuses on drilling in the Hechuan area as its research object, by using a detailed description of the cores from member Mao-2 of seven wells; samples were selected for tests of the degree of dolomite cation ordering, stable carbon and oxygen isotopic compositions, strontium isotopic composition, rare earth elements, LA-ICP-MS element mapping and U-Pb dating. It is clarified that: (1) The crystalline dolomite of member Mao-2 in the Hechuan area is the main reservoir rock, and the heterogeneous vugs and fractures are the main reservoir space. The dolomite in member Mao-2 has been characterized by a low degree of cation ordering value (avg. 0.59), with values of δ^{13}C (avg. 3.87‰), δ^{18}O (avg. −7.15‰) and ^{87}Sr/^{86}Sr (avg. 0.707474) having similar geochemical characteristics to Middle Permian seawater; the REEs normalized distribution patterns have similar characteristics to limestone; and the U-Pb age (261.0~262.0 Ma) corresponds to the age in the Capitanian stage of the Permian Guadalupian Series. (2) Petrological studies show that member Mao-2 has vertical karstification zonation characteristics; syngenetic karstification controls the formation of a large-scale fracture-cave system in the phreatic zone; the dolomitization of sediment in the fracture-cave system occurred during the penecontemporaneous period with locally restricted seawater. (3) The main controlling factors of the reservoir were syngenetic karstification, early dolomitization and hydrothermal dissolution related to Emei taphrogenesis. The research results are of great significance for dolomite reservoir prediction; the highlands of paleogeomorphology with syndepositional faults are favorable areas for dolomite reservoirs.

Keywords: dolomite; geochemistry; chronology; reservoir genesis; Permian Maokou Formation; central Sichuan Basin

Citation: Bai, X.; Zheng, J.; Dai, K.; Hong, S.; Duan, J.; Liu, Y. Petrological, Geochemical and Chronological Characteristics of Dolomites in the Permian Maokou Formation and Constraints to the Reservoir Genesis, Central Sichuan Basin, China. *Minerals* **2023**, *13*, 1336. https://doi.org/10.3390/min13101336

Academic Editors: Hamzeh Mehrabi, Vahid Tavakoli and János Haas

Received: 15 August 2023
Revised: 2 October 2023
Accepted: 11 October 2023
Published: 17 October 2023

1. Introduction

Sichuan Basin is a large superposed petroliferous basin in southwest China [1]; its resources consist primarily of natural gas, with vast reserves, of which the total Permian resources are approximately 1.51×10^{12} m^3, with proven reserves of 881.3×10^8 m^3, and an exploration rate of only 5.84%, showing a great exploration potential. In the past two decades, significant progress has been made in the exploration of the Middle Permian Qixia Formation and Maokou Formation in northern and central Sichuan, with the discovery of the Yuanba, Shuangyushi and Moxi gas fields [2,3]; among these, the exploration of the

Maokou Formation has been targeting the top unconformable limestone karst reservoirs. In recent decades, great breakthroughs in the Maokou Formation of central Sichuan have been successful in obtaining economically viable gas flow from wells NC1, MX39 and JT1, sequentially [4], among which the gas production in well JT1 is 112.8 × 10^4 m^3/d [5], making this one of the important areas of risk exploration in Sichuan Basin. The previous three years, with the continuous high production of economically viable gas flow obtained from well HS4 (test gas production 113 × 10^4 m^3/d), TS4 (205 × 10^4 m^3/d) and TS11 (233 × 10^4 m^3/d) in the Hechuan area of Maokou Formation in central Sichuan, have further revealed that this target is a good exploration prospect.

The dolomite reservoir of the Middle Permian Maokou Formation has gradually become a research hotspot, with the continuous discovery of topics for exploration. Previous studies on dolomite reservoirs have mainly focused on dolomitization models, including seepage reflux, sabkha, burial dolomitization and hydrothermal dolomitization. For the origin of the dolomite, through the study of drilling and outcrops in the central Sichuan area, Liu's [6] research indicates that the insufficient mantle-derived hydrothermal fluids mixed with the fluids in the early fracture-cave system, resulting in dolomitization of the surrounding rocks. According to Jiang's [7] research on drilling cores in central Sichuan, the dolomite is originated by structural hydrothermal. Liu [8], based on the research of the drilling core from central Sichuan, thinks that the dolomite originated from the mixing of residual seawater in the strata and hydrothermal fluids associated with magma activity. As for the main controlling factors of dolomite reservoir genesis, Hu's [9] studies show that the dolomite reservoir in southwest Sichuan and central Sichuan is dominated by vuggy porosity; the main controlling factors of the reservoir are granular beach facies, epidiagenetic dissolution and buried hydrothermal dissolution. Hu [10] thought that, in the east Sichuan region, the dolomite reservoir is porous and fractured; the bioclastic beach, formation water subjected to magmatism and basement faults commonly control the genesis of reservoirs. The study of the Huaying Mountain outcrop area in eastern Sichuan by Li [11] shows that the dolomite reservoir is dominated by cracks and vuggy porosity, and the bioclastic beach and syndepositional faults are identified as the main factors controlling the genesis of reservoirs. However, it still fails to reach a consensus on the genesis of the dolomite reservoir of the Maokou Formation in the central Sichuan area. In addition, the lack of drilling data leads to difficulties in the prediction of favorable dolomite reservoirs. Therefore, the study of the diagenesis and genetics of dolomites is essential, to better deal with effective reservoir prediction and further exploration deployment in this area.

This paper discusses the genesis of the dolomite and the reservoir control factors in the Maokou Formation in the Hechuan area of the central Sichuan Basin, by using the cores of seven wells, through description and thin-section observations, tests of degree of dolomite cation ordering, stable carbon and oxygen isotopic compositions, strontium isotopic composition, rare earth elements, laser-ablation inductively-coupled-plasma mass spectrometry (LA-ICP-MS) element mapping, and U-Pb dating, so as to explore the genesis of the dolomite and the main factors controlling the reservoirs in the Maokou Formation, and thus identify the genesis and distribution of reservoirs in the Maokou Formation. The research findings provided effective guidance for the prediction of dolomite reservoir distribution in the Maokou Formation Hechuan Block, and are of great significance for dolomite reservoir evaluation for exploration in the central Sichuan Basin and even the entire basin.

2. Geological Setting

The Sichuan Basin is located in the eastern part of Sichuan Province, China (Figure 1a); it is a large superposed oil and gas basin developed on the basis of craton [12], with an area of about 19 × 10^4 km^2, the basin can be divided into five structural units: the north Sichuan low-steep structural belt, west Sichuan low-steep structural belt, south Sichuan low-steep structural belt, central Sichuan gentle structural belt and east Sichuan high-steep structural belt (Figure 1b). The Neoproterozoic to Middle Triassic basin evolution was influenced by the extension-convergence cycles of the Rodinia and Pangaea continents, and it experienced four tectonic cycles [13]: the Yangtze cycle, Caledonian cycle, Hercynian cycle and Indochina cycle. Under the influence of the "Guangxi Movement" in the late Caledonian period, the Leshan-Longnusi paleo-uplift was formed, which placed the tectonic slope break belt on the periphery of the uplift depression of central and northern Sichuan in the highland of paleogeomorphology for a long time, and controlled the sedimentary pattern of the Early Permian [14,15]; and the "Dongwu Movement" in the late Middle Permian (sedimentary period of Maokou Formation) caused the tectonic differentiation of the basin [16]. It strongly promoted the maturity of the Lower Paleozoic to Lower Permian source rocks and the generation of oil and gas [17]. Emei taphrogenesis entered a high-incidence period, under a background of extensional and tensile gravity; the basement fault in the basin was revived, more tensile fractures were formed, and a series of NW-SE trending platform internal taphrogeneses were developed. At this time, the mantle plume uplifted in the Emeishan large igneous province (ELIP), resulting in the overall uplift of the Upper Yangtze region and large-scale uplift and denudation of the Maokou Formation in the basin. With the increase of mantle plume uplift in the later period, rapid subsidence occurred in the central and northern parts of the basin, while the southwest inherited higher paleogeomorphological characteristics, providing conditions for the karst at the top of the Maokou Formation [18,19].

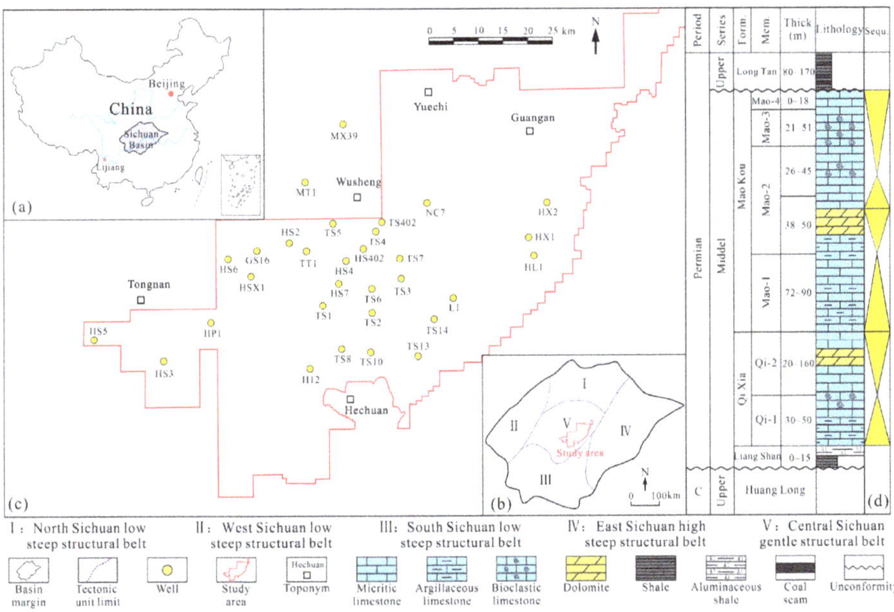

Figure 1. Geological setting map of the study area: (**a**) Location of the Sichuan Basin in eastern Sichuan Province, China is shown in the purple zone; (**b**) Tectonic location of the study area indicated with a red line polygon; (**c**) Well location in the study area; (**d**) Stratigraphic sequence of Middle Permian Maokou Formation.

The Hechuan area belongs to the gentle structural belt in central Sichuan (Figure 1b,c), which experienced a gradual transition from clastic rock deposition to carbonate rock deposition in the Middle Permian, and developed the Liangshan Formation, Qixia Formation and Maokou Formation (Figure 1d) from bottom to top. The Liangshan Formation was in parallel unconformable contact with the Carboniferous strata. In the early Middle Permian, there was extensive transgression in the Sichuan Basin, and the Liangshan Formation mainly developed shore swamp facies of sandy mudstone [20]; with a step of transgression, the study area gradually evolved into a carbonate gentle slope sedimentary system. The Qixia Formation is composed of a third-order sequence; during the transgression period of the Qixia Formation, it mainly developed medium gentle slope argillaceous limestone and wackestone; in the highstand, inner gentle slope facies of packstone, grainstone and crystalline dolomite are mainly developed. The Maokou Formation is composed of two third-order sequences, which can be divided into four members from bottom to top: Mao-1, Mao-2, Mao-3, and Mao-4. The member Mao-1 has the structural characteristics of "eyelid and eyeball" (Figure 2a,b), with a thickness of about 72 m to 90 m. The "eyelid limestone" is mainly dark gray-black argillaceous limestone with high organic matter content, and the "eyeball limestone" is mainly wackestone/packstone [21]. The member Mao-2 mainly develops dark gray layered argillaceous limestone, gray wackestone/packstone (Figure 2c) and dark gray dolomite/dolomite with limestone (Figure 2d); it contains a lot of foraminifers, green algae, coral, brachiopods and other biological debris, with an overall thickness of about 60~90 m. Among these, dolomite is distributed in the lower middle part of member Mao-2, which is relatively continuous horizontally, but the thickness varies greatly (2~22 m). The dolomite crystals are mainly fine-grained to medium-finely crystalline dolomite (Figure 2e). The restoration of the dolomite protolith structure shows that the protolith of crystalline dolomite is bioclastic limestone (Figure 2f,g); member Mao-3 is 26–45 m thick, mainly developed with gray bioclastic packstone; member Mao-4 is generally absent, and mainly develops dark gray bioclastic packstone, which is in parallel unconformity with the overlying Longtan Formation (Figure 1d).

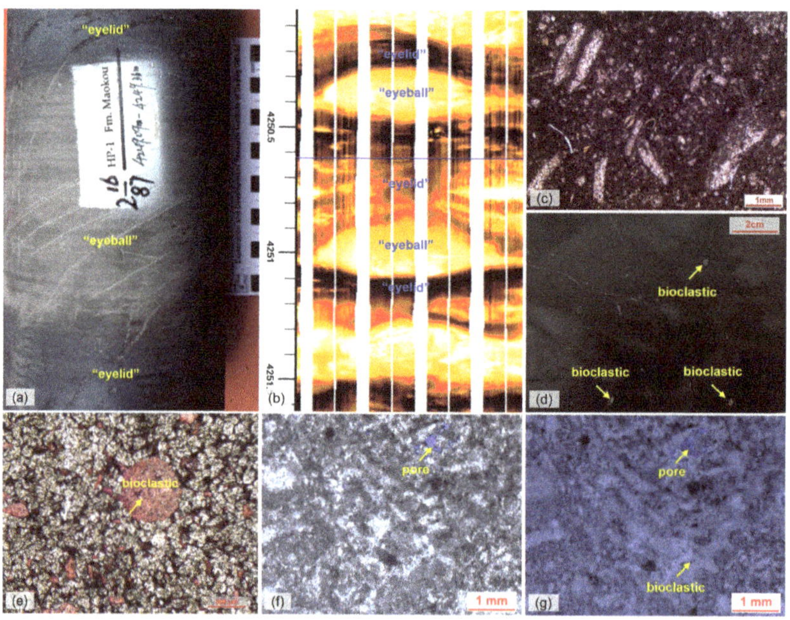

Figure 2. Photographs of carbonate rocks in the Maokou Formation: (**a**) "Eyeball-eyelid" structure, well HP1, member Maokou-1, core; (**b**) "Eyeball-eyelid" structure, well HP1, member Maokou-1, core; (**c**) Bioclastic limestone, well HS4, member Mao-2, thin section PPL image; (**d**) Dolomite with undolomitized bioclastics, well TS4, member Mao-2, slabbed coe; (**e**) Crystalline dolomite with alizarin red stain, the component stained red is calcite bioclastic, well HS4, member Mao-2, thin section PPL image; (**f**) Crystalline dolomite, well TS4, member Mao-2, thin section PPL image; (**g**) Same view as (**f**), the bioclastic profile is clear by protolith reconstruction, thin section PPL image.

3. Petrology

The member Mao-2 dolomite development in the Hechuan area has zonation characteristics (Figure 3a). The dolomite is darker in color and develops fractures filled with saddle-shaped dolomite (Figure 3b), gray irregularly rounded limestone "breccia" commonly in the upper and lower of the dark gray dolomite zone (Figure 3c); near the top and bottom of the dolomite zone is a dolomitic limestone zone. Dark gray dolomite has the characteristics of "groove" distribution, which "cuts" the gray limestone into a "breccia" (Figure 2d). A dark gray "network" calcareous dolomite can be found in the gray limestone zone (Figure 2e). Overall, the dolomite and limestone boundaries of member Mao-2 are clear and can be recognized with the naked eye.

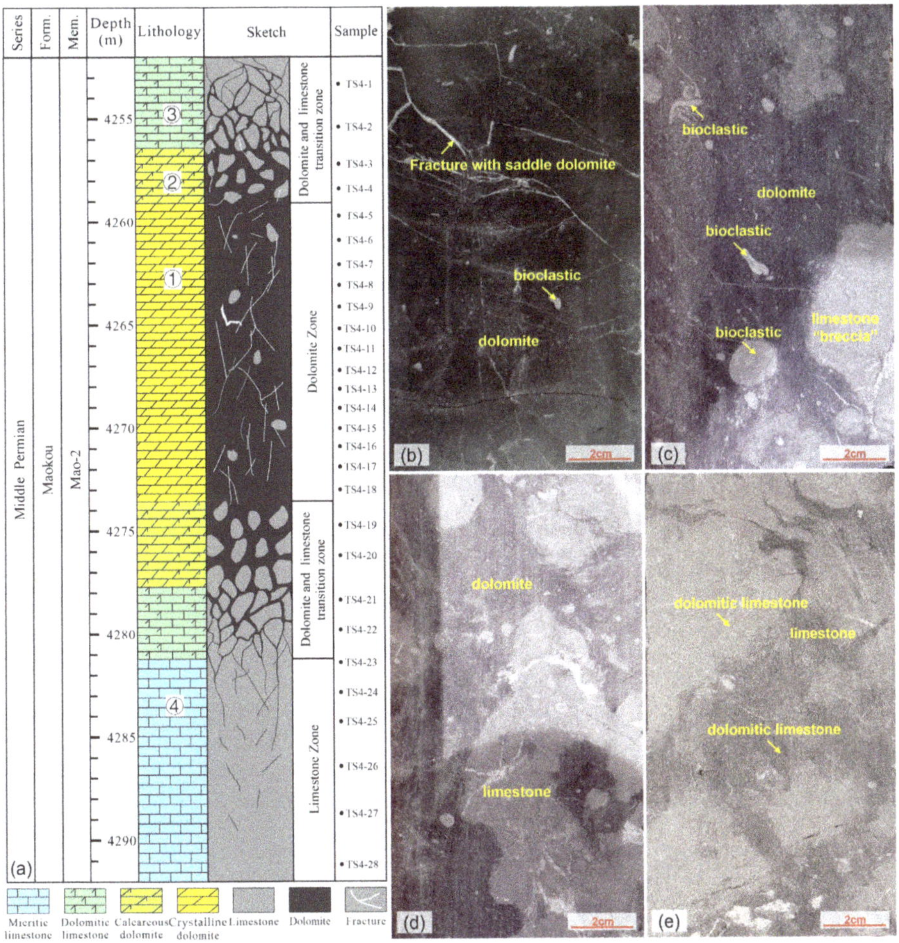

Figure 3. Core characteristics of member Mao-2 in well TS4: (**a**) Core column; (**b**) Photograph of position ① in dolomite zone; (**c**) Photograph of position ② in dolomite and limestone transition zone; (**d**) Photograph of position ③ in dolomite and limestone transition zone; (**e**) Photograph of position ④ in limestone.

Pores are found almost exclusively in the dolomite of member Mao-2; the types of pores include vugs (Figure 4a,b), intercrystalline (dissolved) pores (Figure 4c,d), and fractures (Figure 4e,f), among which vugs are the most common and are mainly distributed unevenly along the fractures (Figure 4b,h). White saddle dolomite is the most common in the vugs and fractures (Figure 4b,f–k), which has the characteristics of undulatory extinction under the microscope (Figure 4i), and asphalt (Figure 4j,k) is common above the saddle dolomite. In addition, the vugs were filled with quartz, fluorite (Figure 4k) and sparry calcite (Figure 4j,k). In conclusion, the vugs and fractures are usually filled with one or more minerals; the whole is characterized by sequentially cemented saddle dolomite-pitch-quartz/fluorite-calcite from the edge of the vug to the center.

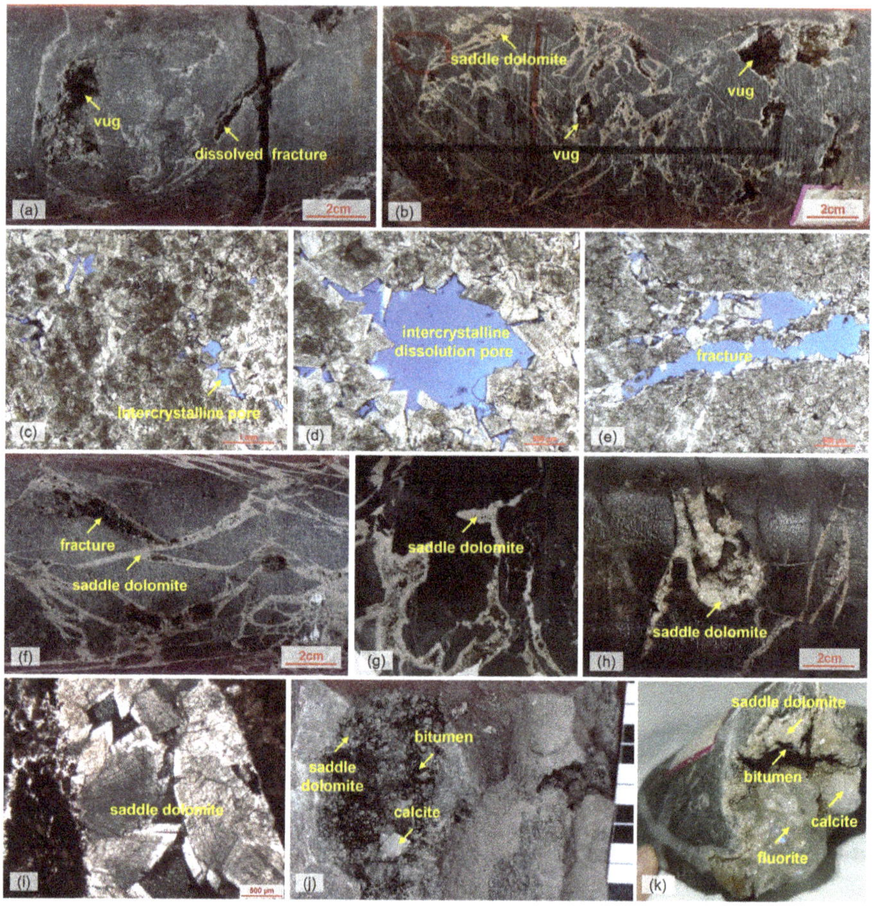

Figure 4. Photographs of dolomite reservoir in the Maokou Formation: (**a**) Dolomite with vugs and dissolved fractures, well TS402, member Mao-2, core; (**b**) Dolomite with vugs and fractures, the rock broke up and became brecciated, well TS4, member Mao-2, core; (**c**) Mesocrystalline dolomite with pores, well TS3, member Mao-2, blue casting thin section PPL image; (**d**) Mesocrystalline dolomite with pores, well TS3, member Mao-2, blue casting thin section PPL image; (**e**) Mesocrystalline dolomite with pores, well HS3, member Mao-2, blue casting thin section PPL image; (**f**) Dolomite with fractures, well HS402, member Mao-2, core; (**g**) Dolomite with fractures, well TS4, member Mao-2, core; (**h**) Dolomite with vugs and fractures, well TS4, member Mao-2, core; (**i**) Saddle dolomite in the fracture, well TS4, member Mao-2, thin section XPL image; (**j**) Dolomite with vugs, the vugs are filled with saddle dolomite, bitumen and sparry calcite, well TS402, member Mao-2, core; (**k**) Dolomite with vugs, the vugs are filled with saddle dolomite, bitumen, fluorite and sparry calcite, well TS13, member Mao-2, core.

4. Samples and Methods

A total of 38 carbonate rock samples were collected through research, all of which came from the Maokou Formation of TS4 Well in the Hechuan area, Sichuan Basin. Member Mao-2 of the Maokou Formation of this well encountered dolomite of about 22 m, which is the well with the largest thickness of dolomite in the study area. All samples were made into cast thin sections. Based on core characteristics and thin-section microscopy analysis, 28 dolomite samples were selected for whole-rock carbon and oxygen isotopes ($\delta^{13}C$ and $\delta^{18}O$), 19 samples for strontium isotopes, 14 samples for dolomite ordering, and 16 samples for rare earth element detection and analysis. In order to minimize the interaction between carbonate cements and host rocks, single structural components were drilled from all samples using a small micro-sampling drill, ground to a powder of 200 mesh with an agate marl. Then, the sample powder was packaged with transparent drawing paper. In addition, 1 sample was selected for elemental mapping (Figure 5a; Table 1), and 3 samples for laser U-Pb dating detection (Figure 5b–d). Before the LA-ICP-MS trace elemental mapping and laser U-Pb dating detection, the samples were made into a 100 μm thick sections, and then polished on a single side, and purified in a super-clean laboratory.

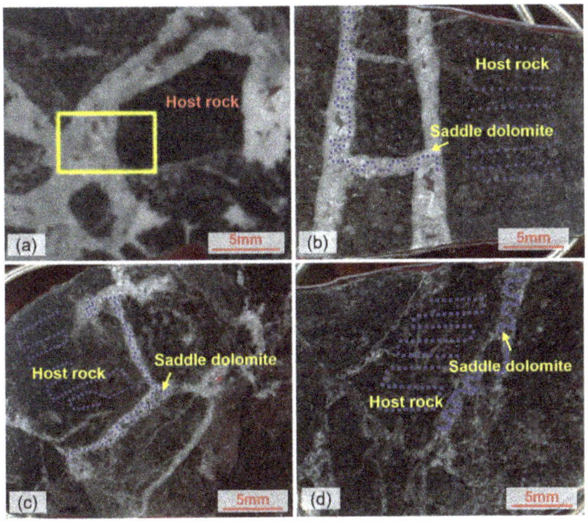

Figure 5. Photographs of the Maokou Formation test sample ((**a**) is an elemental mapping sample, where the yellow rectangle is element mapping area; (**b**–**d**) are U-Pb dating samples): (**a**) sample TS4-8; (**b**) Sample TS4-6; (**c**) Sample TS4-10; (**d**) Sample TS4-17.

Table 1. Message of the sample.

Lithology	Member	Sample	Depth/m	Lithology	Member	Sample	Depth/m
Limestone	Mao-2	TS4-1	4253.3	Dolomite	Mao-2	TS4-15	4270
Limestone	Mao-2	TS4-2	4255.4	Dolomite	Mao-2	TS4-16	4270.9
Dolomite	Mao-2	TS4-3	4257.2	Dolomite	Mao-2	TS4-17	4271.9
Dolomite	Mao-2	TS4-4	4258.4	Dolomite	Mao-2	TS4-18	4273
Dolomite	Mao-2	TS4-5	4259.7	Dolomite	Mao-2	TS4-19	4274.7
Dolomite	Mao-2	TS4-6	4260.9	Limestone	Mao-2	TS4-20	4276.2
Dolomite	Mao-2	TS4-7	4262.1	Limestone	Mao-2	TS4-21	4278.4
Dolomite	Mao-2	TS4-8	4263.1	Limestone	Mao-2	TS4-22	4279.8
Dolomite	Mao-2	TS4-9	4264.2	Limestone	Mao-2	TS4-23	4281.4

Table 1. *Cont.*

Lithology	Member	Sample	Depth/m	Lithology	Member	Sample	Depth/m
Dolomite	Mao-2	TS4-10	4265.2	Limestone	Mao-2	TS4-24	4282.8
Dolomite	Mao-2	TS4-11	4266.2	Limestone	Mao-2	TS4-25	4284.3
Dolomite	Mao-2	TS4-12	4267.2	Limestone	Mao-2	TS4-26	4286.4
Dolomite	Mao-2	TS4-13	4268.1	Limestone	Mao-2	TS4-27	4288.7
Dolomite	Mao-2	TS4-14	4269.1	Limestone	Mao-2	TS4-28	4291.2

Determination of the order degree of dolomite, and whole-rock C, O and Sr isotope measurement, REEs and LA-ICP-MS trace elemental mapping and U-Pb dating, were all completed in the Key Laboratory of Carbonate rock Reservoir of CNPC (Hangzhou). Order degree of dolomite was measured using an X′pert Pro X-ray diffractometer, with a relative error <10%; the X-ray diffraction peak ratio (015)/(110) was used to calculate the order degree [22]. The whole-rock C and O isotope detection instrument was a Delta V Advantage isotope ratio mass spectrometer. GBW4405 and GBW4406 standard samples were used in the test process. The test results were standardized using the Cretaceous Vienna PeeDee Belemnite (VPDB) [23]. The test accuracy of $\delta^{13}C$ is ±0.06‰ and the test accuracy of $\delta^{18}O$ is ±0.08‰. Whole-rock Sr isotope measurement was conducted with a Triton Plus IRMS (Isotope Ratio Mass Spectrometry), when the carbonate standard reference material of GBW04411 was used, and a precision better than 0.01% was obtained. The REEs detection instrument model was a Thermo iCAP TQ ICP-MS. W-2a and BHVO-2 international standard samples were used in the test process, and the analysis accuracy and accuracy are better than 5%. LA-ICP-MS trace elemental mapping used an ASI RESOlution LR laser ablation system; erosion using square beam spot is 40 microns in length, beam spot movement rate 0.04 mm/s. The test instrument model was a Thermo iCAP TQ ICP-MS; an NIST612 standard sample was used in the test process, original data were processed by Iolite 3.6 (University of Melbourne, Parkville, VIC, Australia) to generate element distribution images. Laser U-Pb dating detection used a COHERENT GeoLasHD laser ablation system, denudation is 160 microns in diameter circular beam spot. The test instrument model was an Element XR ICP-MS. Two international standard samples, NIST614 and WC-1, and DaMY-1 laboratory standard samples were used in the test process [24]. After the data were processed by Iolite3.6, Isoplot3.0 (University of California Berkeley, Berkeley, CA, USA) was used to calculate the age and draw the Tera-Wasserburg concordia diagram.

5. Results

5.1. Degree of Dolomite Cation Ordering

Dolomite is a tripartite crystal system mineral, and its lattice parameters are affected by composition, temperature and pressure [25]; the ideal crystal structure of dolomite is where Ca^{2+} and Mg^{2+} are arranged alternately along the c axis, and the molar percentage of Ca^{2+} and Mg^{2+} is the same, but in disordered dolomite, Ca^{2+} and Mg^{2+} are randomly distributed, similar to the structure of calcite. The degree of dolomite cation ordering is an important index for measuring the crystallization speed, crystallization temperature and evolution degree of dolomite; the slower the crystallization speed and the higher the crystallization temperature, the higher the degree of dolomite cation ordering, and vice versa [26]. In general, the degree of dolomite cation ordering greater than 0.8 is defined as high order, 0.6~0.8 is defined as secondary order, 0.4~0.6 is defined as low order, and less than 0.4 is defined as disorder [27]. According to the test results of 14 dolomite samples (Table 2), the dolomite degree of dolomite cation ordering of member Mao-2 of the TS4 well in the Hexhuan area ranges from 0.51 to 0.71, with an average value of 0.59, which belongs to the low order degree.

Table 2. Order degree, $\delta^{13}C$, $\delta^{18}O$ and $^{87}Sr/^{86}Sr$ values of dolomite (whole rock) in the Maokou Formation.

Sample	Lithology	Order Degree	$\delta^{13}C$ ‰ (PDB)	$\delta^{18}O$ ‰ (PDB)	$^{87}Sr/^{86}Sr$
TS4-1	Limestone	/	4.54	−7.03	0.707266
TS4-2	Limestone	/	4.30	−7.31	/
TS4-21	Limestone	/	4.28	−7.45	0.707317
TS4-22	Limestone	/	4.14	−6.82	/
TS4-23	Limestone	/	4.61	−7.27	0.707223
TS4-24	Limestone	/	4.01	−6.94	/
TS4-25	Limestone	/	4.46	−7.59	0.707215
TS4-26	Limestone	/	4.65	−6.94	0.707330
TS4-27	Limestone	/	4.17	−7.64	/
TS4-28	Limestone	/	4.25	−6.95	0.707245
TS4-3	Dolomite	0.58	3.93	−7.62	0.707278
TS4-4	Dolomite	0.55	3.98	−6.74	0.707341
TS4-5	Dolomite	0.54	3.77	−7.19	0.707452
TS4-6	Dolomite	0.69	4.09	−7.33	/
TS4-7	Dolomite	0.71	3.90	−7.57	0.707380
TS4-8	Dolomite	0.59	3.93	−6.94	0.707415
TS4-9	Dolomite	/	3.82	−7.48	/
TS4-10	Dolomite	0.60	3.80	−7.25	0.707324
TS4-11	Dolomite	/	4.03	−6.79	/
TS4-12	Dolomite	0.59	3.63	−6.71	0.707458
TS4-13	Dolomite	/	4.07	−6.84	/
TS4-14	Dolomite	0.55	3.93	−6.73	0.707497
TS4-15	Dolomite	0.63	3.78	−7.00	0.707593
TS4-16	Dolomite	0.51	3.95	−7.46	0.707676
TS4-17	Dolomite	/	3.71	−7.03	0.707639
TS4-18	Dolomite	0.56	3.76	−7.41	/
TS4-19	Dolomite	0.60	3.52	−7.55	0.707587
TS4-20	Dolomite	0.60	4.03	−7.07	0.707533

5.2. Stable Carbon and Oxygen Isotopic Compositions

Stable carbon and oxygen isotopic compositions of dolomitization are related to the stable carbon and oxygen isotopic compositions of dolomitization objects and the fluids that cause dolomitization, and are mainly affected by the salinity and temperature of the fluids [28]. Generally, the carbon and oxygen isotopes of sea water migrate in a positive direction due to evaporation, while the oxygen isotopes of underground brine will migrate in a negative direction due to high temperatures under buried conditions [29]. According to the whole-rock carbon and oxygen isotope test results of 28 carbonate samples (10 limestone samples, 18 dolomite samples) from member Mao-2 of well TS4 in the Hechuan area (Table 2) and the binary scatter diagram (Figure 6), there is no obvious correlation between $\delta^{13}C$ and $\delta^{18}O$ values, indicating that the samples are weakly affected by late diagenetic transformation. The basic information on diagenetic fluid is basically preserved. The $\delta^{13}C$ and $\delta^{18}O$ values of the dolomite range from 3.52‰ to 4.09‰ and −7.64‰ to −6.71‰, respectively, with average values of 3.87‰ and −7.15‰, the $\delta^{13}C$

and $\delta^{18}O$ values of the limestone range from 4.01‰ to 4.65‰ and −7.64‰ to −6.82‰, respectively, with average values of 4.34‰ and −7.19‰.

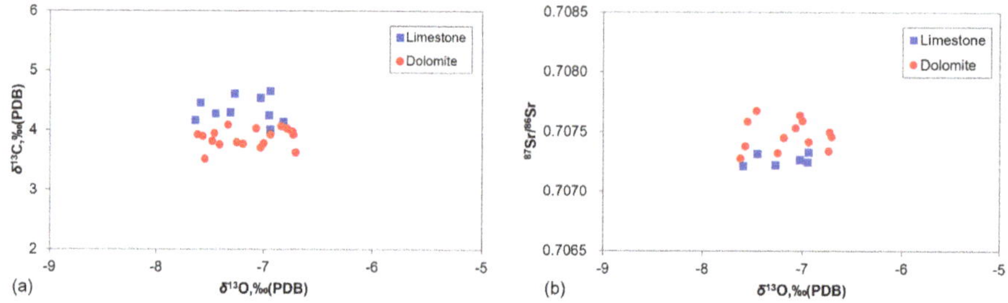

Figure 6. Binary plots of geochemistry of carbonate rocks of member Mao-2; (**a**) $\delta^{13}C$ versus $\delta^{18}O$; (**b**) $^{87}Sr/^{86}Sr$ versus $\delta^{18}O$.

5.3. Stable Strontium Isotopic Composition

Stable strontium isotopes are important parameters indicating paleoclimate and diagenetic fluid properties [30]. According to the whole-rock strontium isotope test results (Table 2) and the $^{87}Sr/^{86}Sr$ and $\delta^{18}O$ binary scatter diagram (Figure 6) of 19 carbonate samples from the TS4 well of member Mao-2, Hexhuan area, it can be seen that the $^{87}Sr/^{86}Sr$ value of dolomite ranges from 0.707278 to 0.707676; the average value is 0.707474. The $^{87}Sr/^{86}Sr$ values of limestone range from 0.707215 to 0.707330, and the average value is 0.707266. Overall, the 87Sr/86Sr value of dolomite is slightly higher than that of limestone, which conforms to the general law that the 87Sr/86Sr value of dolomite samples is usually higher than that of limestone samples [31].

5.4. Rare Earth Elements

The relative abundance of rare earth elements in carbonate minerals mainly depends on the content of rare earth elements in the fluid [32]. Rare earth elements enter the carbonate framework mainly through Ca^{2+} of metasomatic carbonate minerals, and their content is very weakly affected during the diagenetic process. Therefore, it can better indicate the sedimentary environment and the source of dolomitization fluid [33]. According to the rare earth element test results of 13 carbonate samples from the TS4 well of member Mao-2 in the Hechuan area (Table 3), ΣREEs of dolomite ranges from 1.284 ppm to 4.168 ppm, with an average value of 2.9 ppm. The ΣREEs of limestone ranges from 0.917 ppm to 1.822 ppm, with an average value of 1.37 ppm. In general, dolomite ΣREEs are higher than limestone. The measured results were standardized for the Australian Post-Archean mean shale (PAAS) [34], and the standardized (SN) element anomalies were calculated using the following methods: $\delta Ce = 2 \times Ce_{SN}/(La_{SN} + Pr_{SN})$, $\delta Eu = 2 \times Eu_{SN}/(Sm_{SN}+)$ [25]; if δCe and δEu are greater than 1.2, they are judged as positive anomalies; if δEu and δCe are less than 0.8, they are judged as negative anomalies. As can be seen from the standardized rare earth element partitioning pattern (Figure 7), both dolomites and limestone have the characteristics that the content of light rare earth elements (La~Eu) is smaller than that of heavy rare earth elements (Gd~Lu), showing no "tilt upward" pattern. There are no δCe and δEu anomalies in dolomite and limestone, but there is a positive δEu anomaly in one dolomite sample.

Table 3. Rare earth elements content values of dolomite in member Mao-2.

Sample	Lithology	La (ppm)	Ce (ppm)	Pr (ppm)	Nd (ppm)	Sm (ppm)	Eu (ppm)	Gd (ppm)	Tb (ppm)	Dy (ppm)	Ho (ppm)	Er (ppm)	Tm (ppm)	Yb (ppm)	Lu (ppm)	ΣREEs (ppm)
TS4-1	Limestone	0.277	0.494	0.059	0.227	0.042	0.009	0.045	0.007	0.045	0.011	0.034	0.005	0.034	0.005	1.294
TS4-22	Limestone	0.195	0.358	0.041	0.158	0.031	0.006	0.031	0.005	0.032	0.007	0.024	0.004	0.022	0.003	0.917
TS4-25	Limestone	0.375	0.692	0.083	0.32	0.066	0.013	0.068	0.011	0.068	0.016	0.049	0.008	0.046	0.007	1.822

Table 3. *Cont.*

Sample	Lithology	La (ppm)	Ce (ppm)	Pr (ppm)	Nd (ppm)	Sm (ppm)	Eu (ppm)	Gd (ppm)	Tb (ppm)	Dy (ppm)	Ho (ppm)	Er (ppm)	Tm (ppm)	Yb (ppm)	Lu (ppm)	ΣREEs (ppm)
TS4-27	Limestone	0.324	0.526	0.066	0.257	0.052	0.01	0.056	0.009	0.058	0.014	0.043	0.006	0.035	0.006	1.462
TS4-4	Dolomite	0.489	0.903	0.107	0.395	0.076	0.014	0.074	0.012	0.08	0.018	0.055	0.008	0.053	0.009	2.293
TS4-6	Dolomite	0.346	0.633	0.077	0.285	0.06	0.011	0.065	0.01	0.066	0.015	0.048	0.007	0.044	0.007	1.674
TS4-7	Dolomite	0.554	1.030	0.123	0.461	0.089	0.016	0.089	0.014	0.086	0.021	0.065	0.011	0.067	0.01	2.636
TS4-8	Dolomite	0.828	1.570	0.177	0.651	0.128	0.025	0.122	0.02	0.128	0.030	0.084	0.013	0.079	0.013	3.868
TS4-9	Dolomite	0.705	1.390	0.166	0.638	0.138	0.026	0.135	0.022	0.138	0.032	0.097	0.016	0.095	0.015	3.613
TS4-10	Dolomite	0.751	1.413	0.166	0.617	0.12	0.021	0.115	0.018	0.117	0.027	0.083	0.013	0.082	0.013	3.556
TS4-11	Dolomite	0.871	1.640	0.198	0.739	0.142	0.024	0.134	0.021	0.138	0.031	0.1	0.016	0.098	0.016	4.168
TS4-12	Dolomite	0.651	1.178	0.140	0.521	0.101	0.019	0.098	0.016	0.100	0.023	0.071	0.011	0.068	0.011	3.008
TS4-14	Dolomite	0.270	0.471	0.060	0.234	0.046	0.012	0.048	0.008	0.049	0.012	0.034	0.005	0.030	0.005	1.284

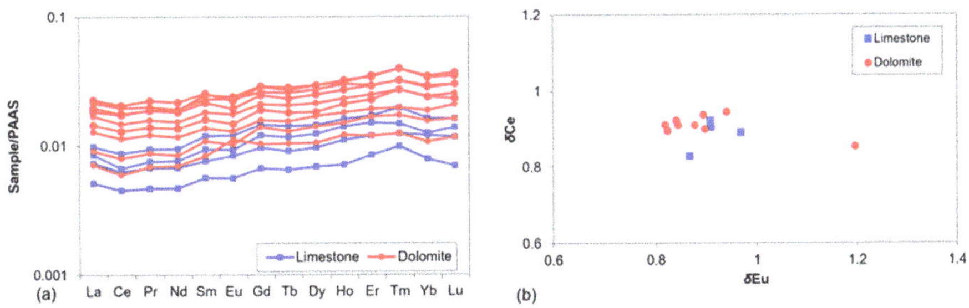

Figure 7. Rare earth element map of carbonate rocks of member Mao-2 in Hechuan area: (**a**) REEs normalized distribution patterns; (**b**) REEs binary plots of geochemistry: δCe versus δEu.

5.5. Element Mapping

The element composition of multi-phase cements in carbonate fractures and vugs is non-uniform. Element mapping technology based on LA-ICP-MS is used to directly display the characteristics of the element plane changes in the millimeter–centimeter region of the sample, which is helpful for analyzing the origin of the sub-cements of different periods [35]. The plane distribution characteristics of Mg and Ca can judge the type of mineral. Trace elements Mn, Fe, Sr, Ba, Th, U and rare earth elements Y, La, Ce, Eu are relatively sensitive to the characteristics of diagenetic fluid, and the source and environment of diagenetic fluid can be analyzed according to their plane distribution characteristics [27].

According to the element mapping of sample TS4-8 (Figure 8), the area (red) with high and uniform distribution of major elements Mg and Ca is dolomite, while the area (black) without Mg and Ca is noncarbonate rock mineral. Combined with the rock characteristics in Figure 5a, it can be ascertained that the host rock is dolomite, and two phases of cements can be identified in the fracture: phase I is dolomite, which is symmetrically distributed along the edge of both sides of the fracture; phase II is quartz (judging by the absence of calcium in the mineral and the quartz seen in the core vugs and fractures). Based on the distribution characteristics of trace and rare earth elements, it can be seen that the host rock has high Sr, Ce, Y, Th, Fe, La, U and low Eu, Mn, Ba values. The host rock edge has higher Mn and lower Ce, Y, Th, Fe, La, U values, and almost no Ba element; there is a difference between them. The distribution of trace and rare earth elements in the host rock edge is similar to that of dolomite in phase I.

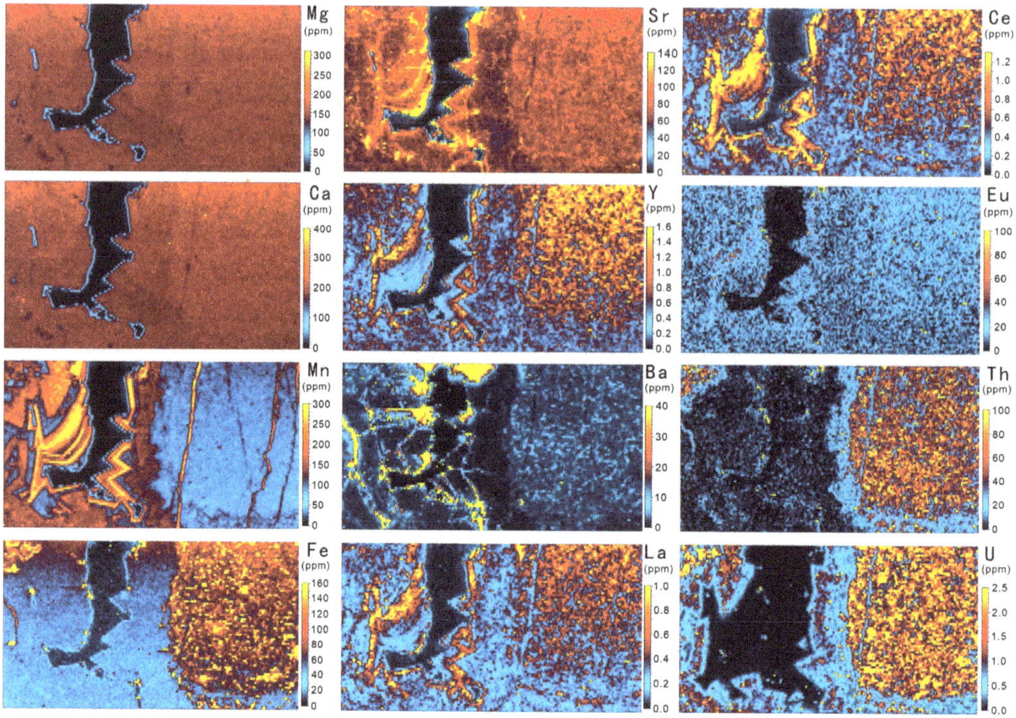

Figure 8. Element mapping based on LA-ICP-MS, sample TS4-8.

5.6. U-Pb Dating

An U-Pb dating technique of carbonate rocks based on LA-ICP-MS can be used to determine the relative diagenetic ages of carbonate minerals [24,36]. According to the test results (Table 4, Figure 9), the age of the host rock of sample TS4-6 is 261 ± 6.1 Ma with the MSWD (mean square of weighted deviates) value of 2.2, and the age of the saddle dolomite in the fracture is 256 ± 10 Ma with the MSWD value of13. The age of host rock of sample TS4-10 is 261.0 ± 4 Ma with the MSWD value of 2, and that of saddle dolomite in the fracture is 258.9 ± 5.1 Ma with the MSWD value of 7.1. The age of the host rock of sample TS4-15 is 262 ± 16 Ma with the MSWD value of 9.9, and the age of saddle dolomite in the fracture is 260 ± 2.4 Ma with the MSWD value of 3. It can be seen that the host rock of the three dolomite samples in member Mao-2 are similar in age, and the corresponding stratigraphic age is the Capitanian stage of the Permian Guadalupian Series (259.51~264.28 Ma). The age of saddle dolomite of the three samples is also close, and the corresponding stratigraphic age is from the Wuchiapingian stage to Changhsingian Stage (251.9~259.51 Ma) of the Permian Lopingian Series. In addition, the age of the host rock of the three samples is greater than that of the saddle dolomite in the fracture, which reveals that it is consistent with the rock sequence, and therefore reflects that the measurement results are relatively reliable.

Table 4. The result of U-Pb dating in member Mao-2 samples.

Sample	Components	U-Pb Age (Ma)	Components	U-Pb Age (Ma)
TS4-6	Host rock (dolomite)	261 ± 6.1	Saddle dolomite	256.0 ± 10

Table 4. *Cont.*

Sample	Components	U-Pb Age (Ma)	Components	U-Pb Age (Ma)
TS4-10	Host rock (dolomite)	261.0 ± 4.0	Saddle dolomite	258.9 ± 5.1
TS4-15	Host rock (dolomite)	262.0 ± 16	Saddle dolomite	260.0 ± 2.4

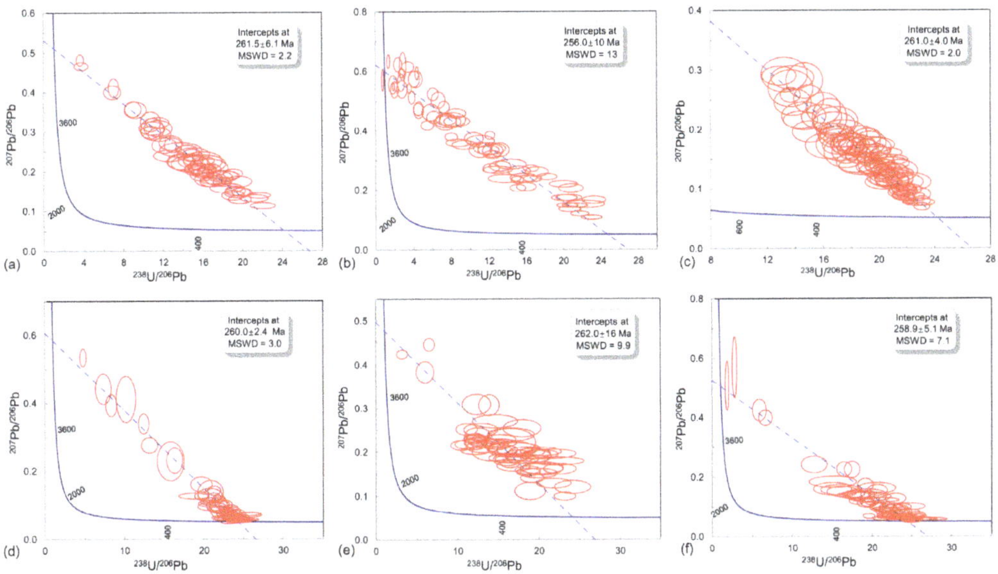

Figure 9. LA-ICP-MS U-Pb concordia diagram of dolomite in the Maokou Formation (The blue line represents the theoretical age curve when the sample U-Pb system is always closed. The blue dashed line represents the theoretical age curve with certain initial Pb in the sample U-Pb system, but it has been closed since its formation. The lower intersection point between the blue line and the sample U-Pb system is the rock formation age. The red circle represents the margin of error): (**a**) Host rock of sample TS4-6; (**b**) Saddle dolomite of sample TS4-6; (**c**) Host rock of sample TS4-10; (**d**) Saddle dolomite of sample TS4-10; (**e**) Host rock of sample TS4-15; (**f**) Saddle dolomite of sample TS4-15.

6. Discussion

6.1. Genesis of Dolomite

Dolomitization can improve the compressive properties of rocks; the dolomitization of the early diagenetic stage plays an important constructive role in the preservation of primary pores and early dissolved pores; however, dolomitization in the late burial stage had little effect on early pore preservation. In addition, large-scale pressure dissolution not only causes compaction of pore space, but also the calcium fluid which forms the pressure dissolution can fill primary pores. It has a destructive effect on the reservoir [37,38], so it is significant to judge the time of dolomitization.

According to the geochemical data obtained by analysis and testing, the $\delta^{13}C$ (avg. 3.87‰) and $\delta^{18}O$ (avg. 7.15‰) values of dolomite in member Mao-2 are within the variation range of seawater in the Middle Permian ($\delta^{18}O$ ranges from 7‰ to 4‰ and $\delta^{13}C$ ranges from 3‰ to 5‰) [30,39]. The range of $\delta^{18}O$ values of dolomite and limestone is similar, but the $\delta^{13}C$ of limestone is higher than that of dolomite, which is related to the rise and fall of sea levels. During the rise of sea levels, the rate of organic carbon burial increases, resulting in a corresponding increase in the $\delta^{13}C$ value of limestone, whereas a large amount of ^{12}C enters the seawater during the fall of sea levels. The $\delta^{13}C$ value of the dolomite decreases during the same period [30]. The $^{87}Sr/^{86}Sr$ values (average 0.707474) are also close to the coeval seawater [30,39], indicating that the dolomitization fluid in member Mao-2 was seawater [40]. Dolomite and limestone in member Mao-2 have similar REEs normalized distribution patterns, which reveals that dolomite and limestone have similar diagenetic fluids (seawater). In addition, the low ΣREEs content of dolomite (avg. 2.9 ppm) and the unobvious overall anomalies of δCe and δEu also indicate that dolomite did not undergo transformation by meteoric water and hydrothermal fluid during diagenesis. Member Mao-2 has a low degree of dolomite cation ordering (avg. 0.59), indicating that the dolomitization process was characterized by rapid replacement and crystal growth. The U-Pb age (259.51~264.28 Ma) is within the age range of the Capitanian stage of the Permian Guadalupian Series, indicating that dolomite was formed in a penecontemporaneous period. In a word, member Mao-2 dolomites were formed in a penecontemporaneous period with seawater as the dolomitization fluid.

As mentioned above, the vertical karst zonation characteristics of member Mao-2 in the study area are displayed. The dark gray dolomite at the transition zone of dolomite and limestone is a "network of cut" limestone, which is "breccia" (Figure 4), the "brecciform" limestone without edges and corners, and the sedimentary particles in the "network" have the characteristics of a vadose zone, which is similar to the vertical vadose zone in a kast fracture-cave system. Below the transition zone of dolomite and limestone is the dolomite zone; it has irregular limestone "breccia" at the top and bottom, not touching, and a few isolated "breccia" in the middle; this zone has the same characteristics as a horizontal phreatic zone in kast. Below the dolomite zone is limestone with dolomitic, and the "network fracture" with dark grey gradually changes downward into a high angle "fracture" until it disappears. These characteristics indicating the fracture-cave system may develop in member Mao-2. In addition, from the geological background, there is a sequence boundary in member Mao-2 at the study area, which indicates that it had a long exposure and karstification during the sedimentary period [20]. According to the drilling results, there are leakage and venting zones in member Mao-2 near the study area, indicating there is a fracture-cave system in the phreatic zone. From the outcrop, a large cave with a height of 6~10 m has developed about 20~30 m from the top of member Mao-2; From the drilling in the study area, the dolomite of member Mao-2 in the Hechuan area is stably distributed in the middle and lower part of member Mao-2, but the thickness changes greatly (in the range of 2~22 m), which is similar to the cave distribution characteristics in horizontal phreatic zone in a karst system. To sum up, the dolomite in member Mao-2 is controlled by syngenetic karstification which is known for karst features, including caves, that form within a soft, porous, soluble sediment at the same time as it is being cemented into a rock. Speleogenesis and lithogenesis are concurrent [41], and there is a fracture-cave system in the phreatic zone.

Early deposition of the Maokou Formation in the Hechuan area was in the inner gentle ramp of the carbonate ramp, controlled by a NW trending strike-slip fault within the carbonate platform; a localized formation of alternating uplifted and depression paleogeomorphological patterns was established and maintained until member Mao-2 [14,16]. In this geological setting, the bioclastic banks developed in the highlands of paleogeomorphology, and in the highstand stage, the sea level dropped; the beach would emerge from the water surface for a long time, and be leached by meteoric water, forming a syngenetic karstification fracture-cave system. At the same time, the local topography of uplift and depression also makes the local formation of a relatively limited water environment. In this environment, the evaporation of seawater is relatively strong, the salinity of seawater is increased, and the content of Mg^{2+} is also relatively increased. During the transgression period, due to wave action, seabed bioclastics and soft sediment will fill the fractures and caves; with the gradual rise of sea level, it also leads to the continuous replenishment of normal sea water, so that Mg^{2+} is supplemented, and when the sea level is relatively reduced, it will be subject to evaporation, the content of Mg^{2+} in sea water will be increased

(Figure 10). In the shallow burial stage, the particles and marl filled in the fractures and vugs have rapid dolomitization with Mg^{2+} rich seawater. This explains the clear boundary between dolomites and limestone, the absence of dolomitization in limestone "breccia" in dolomites, and the occurrence of dolomitization in limestone only in the fracture (Figure 3).

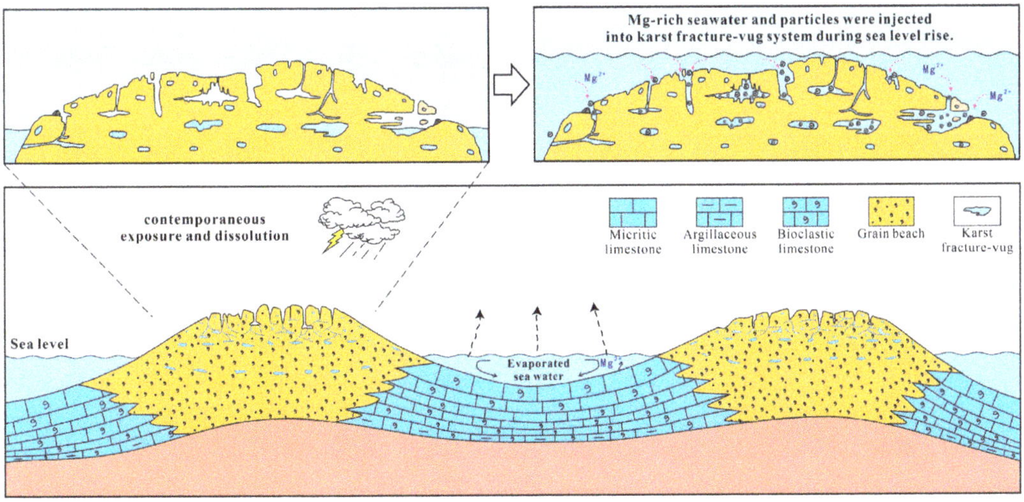

Figure 10. Dolomitization model of the Maokou Formation in Hechuan area.

6.2. Reservoir Main Control Factors

The pore types of the dolomite reservoir in the study area are dominated by heterogeneous vugs and fractures, and locally homogeneous intercrystalline (dissolution) pores are developed. Therefore, it is of great significance to identify the genesis of fractures and vugs for the understanding of reservoir-controlling factors.

According to the tectonic setting of the Sichuan Basin, in the late Middle Permian, Emei taphrogenesis led to the revival of the basement fault in the basin [14], and was subjected to NW stretching, resulting in a large number of syndepositional faults [42]. At this point, tectonic faulting was active in the Hechuan area, and a number of NW trending syndepositional faults were formed; because of the brittleness of dolomite, it is easier to form fractures in dolomite than limestone, so the fractures in the Hechuan area are mainly in the dolomite. As mentioned above, the age of the dolomite in member Mao-2 is 261.0~262.0 Ma, the age of the saddle dolomite in the fractures is 256.0~260.0 Ma; according to the characteristic that the fractures are mainly developed in the dolomite in member Mao-2, it is clear that these fractures were formed after the dolomitization and before the saddle dolomite was filled, which further reveals that the fractures in the dolomite in member Mao-2 are mainly controlled by syndepositional faults. Fractures are not only an important reservoir space, but also a channel for fluid migration in subsequent diagenetic processes.

According to the core characteristics (Figure 4), the vugs in the dolomite of member Mao-2 are mostly developed along the fractures, with the characteristics of non-fabric selectivity and heterogeneous distribution, which shows that these vugs are not controlled by sedimentary facies, but closely related to fractures; after the formation of the dolomite, they underwent modifications due to fluid dissolution along fractures. The Emeishan large igneous province (ELIP) began to erupt around 260 Ma, and the main eruption period was about 259 to 257 Ma [43]. The eruption center is located in the southwest outside the Sichuan Basin, which gradually weakens from west to east, and forms an unusually high heat flow field with a large regional scale [44,45]. The central Sichuan region is in a weakened zone of plume activity in the ELIP, unaffected directly by magma activity, with minimal melt from mantle plumes; However, it is affected by high heat flow by extension of deep mantle plume materials. In the outer regions of the eruption area, the Mg^{2+} rich strata on the basin floor are driven by an ancient thermal flow, migrating along fractures and developing vugs [8,43,46,47]. In addition, the results of rare earth elements in the dolomite of member Mao-2 at the study area show that there are local Eu anomalies in the dolomite, and element mapping results show significant differences between the host rock and its elements; the host rock edge shows similarity to the elements of the phase I saddle dolomite cement. Therefore, it can be inferred that the host rock has been eroded by hydrothermal fluids, forming an erosion, which subsequently led to the annular-shaped saddle dolomite cement through continuous precipitation. To sum up, it shows that partial dolomite in the study area has been transformed by hydrothermal solution, and the formation of vugs and fractures is related to hydrothermal solution.

Vugs and fractures in the dolomite of member Mao-2 are filled with multi-phase minerals. According to the core observation, saddle dolomite, asphalt, quartz/fluorite and calcite can be identified in turn from the edge of fractures to the center, but not all fractures have a complete filling sequence; some are only filled with one or more of them (Figure 4j,k). According to the results of elemental mapping, it can also be seen that saddle dolomite and quartz fill the vugs in turn, indicating that the pores have been transformed by fluids in different periods. According to the results of previous studies, the Middle-Late Triassic in the study area was the early stage of crude oil cracking, forming an ancient oil reservoir, and the middle stage of the late Jurassic to Middle Cretaceous was the stage of crude oil cracking [48], so it can be inferred that the formation time of asphalt was the early-middle Cretaceous. According to the burial history of central Sichuan Basin [8], in the Late Cretaceous, due to the uplift of the Qinghai-Tibet Plateau and the southeast compression, some early faults in central Sichuan were activated, which led to the similar migration of deep silicon-rich and fluorine-rich hydrothermal solutions along the faults, and to their precipitation into hydrothermal minerals such as quartz and fluorite in the vugs [41], so it can be inferred that quartz/fluorite in the vugs of dolomite in member Mao-2 was formed in the deep burial stage of the late Yanshanian. The Hercynian fault activity once again entered the high-incidence period, and the multi-stage faults activity caused the early strike-slip fault activation. Based on this, it is inferred that the calcite that finally filled in the vugs of member Mao-2 was formed in this period.

Based on this analysis of the origin, pore formation and evolution of dolomite in member Mao-2 of the Maokou Formation the in Hechuan area, and combined with the geological background, the diagenetic sequence and pore evolution curve (Figure 11) of each important stage of the Maokou Formation in the study area are established. Referring to Clyde H. Moore [49], the initial porosity value and the distribution area of residual pores and cement under the microscope were accumulated, and the initial porosity was selected as 30%. After a different diagenesis, the appropriate increase or decrease value was selected to indicate the evolution of porosity. The early diagenetic stage mainly experienced the micritization of sediment mud in the mixed environment of seawater, seawater cementation, the dissolution of meteoric water (forming fracture-cave systems), the collapse of fracture-caves, karst sedimentary filling and dolomitization. The early burial stage mainly experienced mechanical compaction, tectonic fracturing I, hydrothermal dissolution and saddle dolomite cementation. The middle diagenetic stage mainly experienced burial recrystallization and oil-gas charging. In the late diagenetic stage, it mainly experienced tectonic fracturing II, quartz/fluorite filling, tectonic fracturing III and calcite cementation. Among all the main diageneses, syngenetic karstification, dolomitization, tectonic fracture I and hydrothermal dissolution are the most important constructive diageneses, in which syngenetic karstification controls the development of a large fracture-cave system in the subsurface zone, which lays the foundation for subsequent dolomitization. The pre-existing pores are inherited and preserved by penecontemporaneous dolomitization, and the brittle characteristics of dolomite create conditions for a large number of fractures to develop in the later period. The tectonic fracturing I has control of the reservoir, which is not only an important reservoir space, but also a channel for fluid migration, providing conditions for the formation of vugs. The reservoir has undergone multi-phase hydrothermal transformation, and the early hydrothermal dissolution related to Emei taphrogenesis is the key to the development of dolomite reservoir in member Mao-2, and the porous reservoir is heterogeneous.

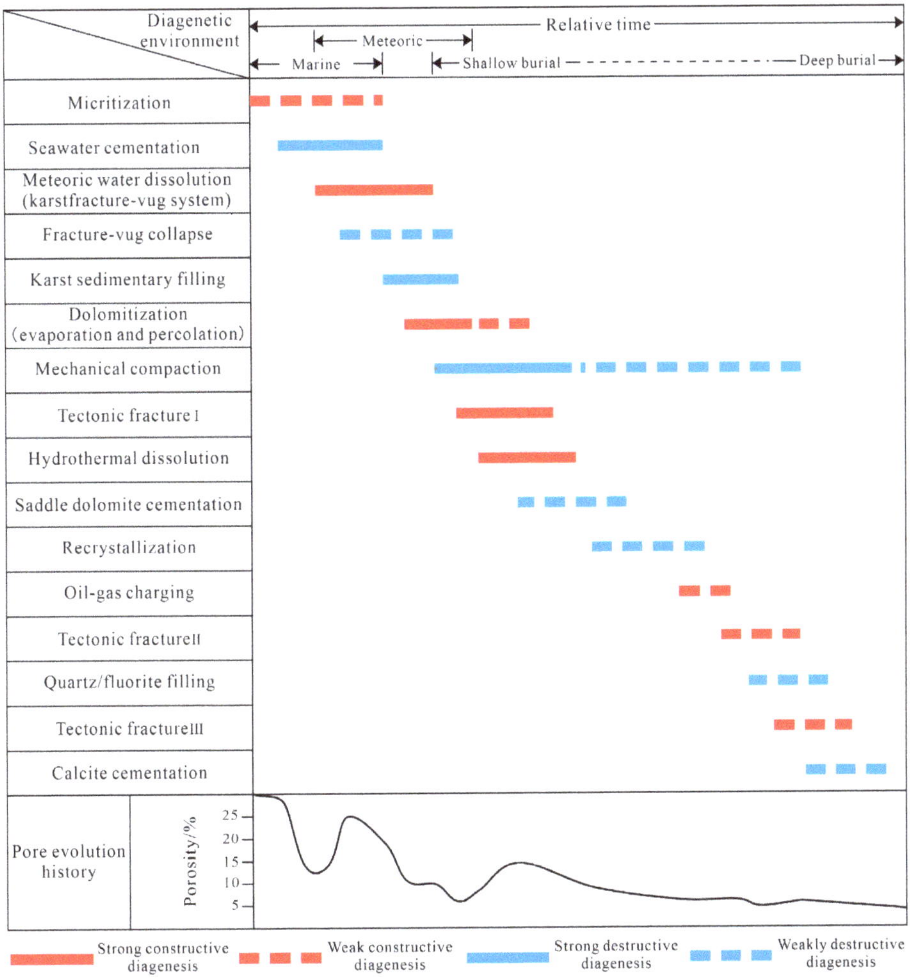

Figure 11. Diagenetic sequence of Maokou Formation.

To sum up, syngenetic karstification, early dolomitization and hydrothermal dissolution related to faults are the main controlling factors for the development of dolomite reservoirs in member Mao-2 of the Hechuan area. Syndepositional faults play a crucial role in the development of dolomite reservoirs. The development of faults not only provides migration channels for hydrothermal fluids, but also serves as an important reservoir space. Syndepositional faults control the distribution law of high-quality reservoirs to a certain extent, playing an important role in the generation, migration, accumulation and accumulation of oil and gas. Accordingly, the prediction of dolomite reservoir development in this study can be achieved by incorporating highlands of paleogeomorphology with syndepositiona faults, which provides a reliable basis for exploration strategies of the Maokou Formation in the Hechuan region.

7. Conclusions

Crystalline dolomite is the main carrier of the reservoir in member Mao-2 in the Hechuan area of Sichuan Basin, and the heterogeneous vugs and fractures are the main reservoir space. The dolomite in member Mao-2 has been characterized by a low degree of cation ordering value (avg. 0.59), with values of $\delta^{13}C$ (avg. 3.87‰), $\delta^{18}O$ (avg. −7.15‰) and $^{87}Sr/^{86}Sr$ (avg. 0.707474); the rare earth elements (REEs) normalized distribution patterns have no obvious outliers, the age determined by laser ablation U-Pb dating range from 261.0 to 262.0 Ma.

Member Mao-2 is controlled by syngenetic karstification to form a large fracture-cave system. The sediment within the fracture-cave system underwent dolomitization during the penecontemporaneous period with local seawater as the dolomitization fluid.

The fracture-cave system formed under control of syngenetic karstification laid the foundations for subsequent dolomitization. Penecontemporaneous dolomitization inherits and preserves the pre-existing pores, and the brittle characteristics of dolomite played a pivotal role in the extensive development of later-stage fractures. The NW trending faults associated with Emei taphrogenesis have a significant control over the reservoir, serving as important reservoir spaces and fluid migration pathways. Reservoirs have undergone multi-phase hydrothermal transformation, most of the vugs exit along the fractures in a heterogeneous way, so the highlands of paleogeomorphology with syndepositional faults are favorable areas for dolomite reservoirs.

Author Contributions: Conceptualization, J.Z. and X.B.; methodology, J.Z. and K.D.; software, K.D. and J.D.; validation, J.Z. and X.B.; formal analysis, S.H. and Y.L.; investigation, J.Z. and K.D.; resources, X.B. and S.H.; data curation, J.D. and Y.L.; writing—original draft preparation, J.Z. and K.D.; writing—review and editing, J.Z.; visualization, K.D. and J.D.; supervision, J.Z.; project administration, J.Z. and X.B.; funding acquisition, J.Z. and X.B. All authors have read and agreed to the published version of the manuscript.

Funding: This research was funded by the Scientific Research and Technology Development Project of PetroChina Company Limited (Grant No. 2021DJ0501).

Data Availability Statement: Data are available upon reasonable request. The data that support the findings of this study are available on request from the corresponding author. The data are not publicly available due to privacy or ethical restrictions.

Acknowledgments: We would thank Anjiang Shen and Jingao Zhou for their valuable suggestions, Xianying Luo and Feng Liang for their guidance in the experiments. We would also like to thank Mao Zhu, Yi Hao, Qianying Yao and Kedan Zhu for the help in data collection.

Conflicts of Interest: The authors declare no conflict of interest.

References

1. Ma, Y.S.; Guo, X.S.; Guo, T.L.; Huang, R.; Cai, X.; Li, G.X. The Puguang gas field: New giant discovery in the mature Sichuan Basin, southwest China. *AAPG Bull.* **2007**, *91*, 627–643. [CrossRef]
2. Guo, X.S.; HU, D.F.; Li, Y.P.; Duan, J.B.; Ji, C.H.; Duan, H. Discovery and theoretical and technical innovations of Yuanba gas field in Sichuan Basin, SW China. *Pet. Explor. Dev.* **2018**, *45*, 14–26. [CrossRef]
3. Du, J.H.; Zhou, C.N.; Xu, C.C.; He, H.Q.; Shen, P.; Yang, Y.M.; Li, Y.L.; Wei, G.Q.; Wang, Z.C.; Yang, Y. Theoretical and technical innovations in strategic discovery of a giant gas field in Cambrian Longwangmiao Formation of central Sichuan paleo-uplift, Sichuan Basin. *Pet. Explor. Dev.* **2014**, *41*, 268–277. [CrossRef]
4. Yang, Y.M.; Yang, Y.; Wen, L.; Zhang, X.H.; Chen, C.; Chen, K.; Zhang, Y.; Di, G.L.; Wang, H.; Xie, C. New progress and prospect of Permian natural gas exploration in Sichuan Basin. *Nat. Gas Ind.* **2020**, *40*, 10–22. [CrossRef]
5. Yang, Y.; Xie, J.R.; Zhao, L.Z.; Huang, P.H.; Zhang, X.H.; Chen, C.; Zhang, B.J.; Wen, L.; Wang, H.; Gao, Z.L.; et al. Breakthrough of natural gas exploration in the beach facies porous dolomite reservoir of Middle Permian Maokou Formation in the Sichuan Basin and its implications: A case study of the tridimensional exploration of Well JT1 in the central-northern Sichuan Basin. *Nat. Gas Ind.* **2021**, *41*, 1–9. [CrossRef]
6. Liu, H.; Ma, T.; Tan, X.C.; Zeng, W.; Hu, G.; Xiao, D.; Luo, B.; Shan, S.J.; Su, C.P. Origin of structurally controlled hydrothermal dolomite in epigenetic karst system during shallow burial: An example from Middle Permian Maokou Formation, central Sichuan Basin, SW China. *Pet. Explor. Dev.* **2016**, *43*, 916–927. [CrossRef]

7. Jiang, Y.Q.; Gu, Y.F.; Li, K.H.; Li, S.; Luo, M.S.; He, B. Spatial types and genesis of Middle Permian hydrothermal dolomite reservoirs in central Sichuan Basin. *Nat. Gas Ind.* **2018**, *38*, 16–24. [CrossRef]
8. Liu, J.Q.; Zheng, H.F.; Liu, B.; Liu, H.G.; Shi, K.B.; Guo, R.T.; Zhang, X.F. Characteristics and genetic mechanism of the dolomite in the Middle Permian Maokou Formation, central Sichuan area. *Acta Pet. Sin.* **2017**, *38*, 386–398. [CrossRef]
9. Hu, A.P.; Pan, L.Y.; Hao, Y.; Shen, A.J.; Gu, M.F. Origin, Characteristics and Distribution of Dolostone Reservoir in Qixia Formation and Maokou Formation, Sichuan Basin, China. *Marin. Orig. Petrl. Geol.* **2018**, *23*, 39–52. [CrossRef]
10. Hu, D.F.; Wang, L.J.; Huang, R.C.; Duan, J.B.; Xu, Z.X.; Pan, L. Characteristics of dolomite reservoirs in the middle Permian Maokou Formation in the eastern Sichuan Basin and their main controlling factors. *Nat. Gas Ind.* **2019**, *39*, 13–21. [CrossRef]
11. Li, R.B.; Duan, J.B.; Pan, L.; Li, H. Genetic mechanism and main controlling factors of the Middle Permian Maokou Formation dolomite reservoirs in the eastern Sichuan Basin. *Nat. Gas Geosci.* **2021**, *32*, 1347–1357. [CrossRef]
12. Zhao, W.Z.; Shen, A.J.; Hu, S.Y.; Zhang, B.M.; Pan, W.Q.; Zhou, J.G.; Wang, Z.C. Geological conditions and distributional features of large-scale carbonate reservoirs onshore China. *Pet. Explor. Dev.* **2012**, *39*, 1–12. [CrossRef]
13. Liu, S.G.; Yang, Y.; Deng, B.; Zhong, Y.; Wen, L.; Sun, W.; Li, Z.W.; Jansa, L.; Li, J.X.; Song, J.M.; et al. Tectonic evolution of the Sichuan Basin, Southwest China. *Earth-Sci. Rev.* **2021**, *213*, 103470. [CrossRef]
14. Ma, B.S.; Liang, H.; Wu, G.H.; Tang, Q.S.; Tian, W.Z.; Zhang, C.; Yang, S.; Zhong, Y.; Zhang, X.; Zhang, Z.L. Formation and evolution of the strike-slip faults in the central Sichuan Basin, SW China. *Pet. Explor. Dev.* **2023**, *50*, 333–345. [CrossRef]
15. Gu, M.F.; Liu, R.; Zhang, H.; He, Y.; Li, K.Y.; Hao, Y.; Jiang, H.; Li, W.Z. Characteristics of Caledonian tectonic movement and its geological response in Sichuan Basin. *Nat. Gas Ind.* **2023**, *43*, 32–43.
16. Zhong, Y.; Yang, Y.M.; Wen, L.; Luo, B.; Xiao, D.; Tan, X.C.; Zhao, L.K.; Li, M.L. Sedimentary environments controlled by tectonics and induced differential subsidence: A perspective in the Permian Liangshan and Qixia Formations, northwestern Sichuan Basin, China. *J. Cent. South Univ.* **2020**, *27*, 3398–3416. [CrossRef]
17. Huang, S.P.; Jiang, Q.C.; Jiang, H.; Tang, Q.S.; Zeng, F.Y.; Lu, W.H.; Hao, C.G.; Yuan, M.; Wu, Y. Genetic and source differences of gases in the Middle Permian Qixia and Maokou formations in the Sichuan Basin, SW China. *Org. Geochem.* **2023**, *178*, 104574. [CrossRef]
18. Zhao, Z.J.; Zhou, H.; Chen, X.; Liu, Y.H.; Zhang, Y.B.; Liu, Y.E.; Yang, Y. Sequence lithofacies paleogeography and favorable exploration zones of the Permian in Sichuan Basin and adjacent areas, China. *Acta Pet. Sin.* **2012**, *33*, 35–51. [CrossRef]
19. Li, D.J.; Chen, H.; Chen, H.D.; Liang, H.; Peng, C.; Xia, M.; Duan, H.Z. Relationship between reservoir development in the Middle Permian Maokou Formation and paleostructure evolution in the Sichuan Basin. *Oil Gas Geol.* **2016**, *37*, 756–763. [CrossRef]
20. Hu, M.Y.; Hu, Z.G.; Wei, G.Q.; Yang, W.; Liu, M.C. Sequence lithofacies paleogeography and reservoir potential of the Maokou Formation in Sichuan Basin. *Pet. Explor. Dev.* **2012**, *39*, 45–55. [CrossRef]
21. Tian, X.B.; Shi, J.B.; Dong, J.H.; Tan, W.C.; Yin, C.H.; Li, Q.; Song, Z.Y.; Zhang, K.; Xiao, C.T. New understanding of sedimentary facies of the first member of Maokou Formation based on the latest drilling in central Sichuan Basin. *Nat. Gas Geosci.* **2021**, *32*, 1646–1655.
22. Goldsmith, J.R.; Graf, D.L. Structural and compositional variations in some natural dolomites. *J. Geol.* **1958**, *66*, 678–693. [CrossRef]
23. Helie, J.F.; Adamowicz-Walczak, A.; Middlestead, P.; Chartrand, M.M.G.; Mester, Z.; Meija, J. Discontinuity in the Realization of the Vienna Peedee Belemnite Carbon Isotope Ratio Scale. *Anal. Chem.* **2021**, *93*, 10740–10743. [CrossRef]
24. Shen, A.J.; Zhao, W.Z.; Hu, A.P.; Wang, H.; Liang, F.; Wang, Y.S. The datingand temperature measurement technologies for carbonate minerals and their application in hydrocarbon accumulation research in the paleo-uplift in central Sichuan Basin, SW China. *Pet. Explor. Dev.* **2021**, *48*, 476–487. [CrossRef]
25. Merlini, M.; Sapelli, F. High-temperature and high-pressure behavior of carbonates in the ternary diagram $CaCO_3$-$MgCO_3$-$FeCO_3$. *Am. Mineral.* **2006**, *101*, 1423–1430. [CrossRef]
26. Deng, M.; Qian, G.R.; Tang, M.S. Ordered index and dedolomitization of dolomite crystals. *J. Nanjing Univ. Chem. Technol.* **2001**, *23*, 1–5. [CrossRef]
27. Zheng, J.F.; Wang, H.; Shen, A.J.; Luo, X.Y.; Zhao, C.; Dai, K. Genesis of Dolomite Reservoir in Ediacaran Chigbrak Formation of Tarim Basin, NW China: Evidence from U–Pb Dating, Isotope and Element Geochemistry. *Am. Mineral.* **2023**, *13*, 725. [CrossRef]
28. Northrop, D.A.; Clayton, R.N. Oxygen-isotope fractionations in systems containing dolomite. *J. Geol.* **1966**, *74*, 174–196. [CrossRef]
29. Saller, A.H.; Dickson, J.A.D. Partial dolomitization of a Pennsylvanian limestone buildup by hydrothermal fluids and its effect on reservoir quality and performance. *AAPG Bull.* **2011**, *95*, 1745–1762. [CrossRef]
30. Veizer, J.; Ala, D.; Azmy, K.; Bruckschen, P.; Buhl, D.; Bruhn, F.; Carden, G.A.; Diener, A.; Ebneth, S.; Godderis, Y. 87Sr/86Sr, δ13C and δ18O evolution of Phanerozoic seawater. *Chem. Geol.* **1999**, *161*, 59–88. [CrossRef]
31. Baud, A.; Magaritz, M.; Holser, W.T. Permian-Triassic of the Tethys: Carbon isotope studies. *Geol. Rundsch.* **1989**, *78*, 649–677. [CrossRef]
32. Burns, S.J.; Haudenschild, U.; Matter, A. The strontium isotopic composition of carbonates from the late Precambrian (∼560–540 Ma) Huqf Group of Oman. *Chem. Geol.* **1994**, *111*, 269–282. [CrossRef]
33. Lottermoser, B.G. Rare earth elements and hydrothermal ore formation processes. *Ore Geol. Rev.* **1992**, *7*, 25–41. [CrossRef]
34. Zheng, J.F.; Shen, A.J.; Qiao, Z.F.; Wu, X.N.; Zhang, T.F. Characteristics and pore genesis of dolomite in the Penglaiba Formation in Keping-Bachu outcrop area. *Acta Pet. Sin.* **2014**, *35*, 664–672. [CrossRef]
35. McLennan, S.M.; Taylor, S.; McGregor, V. Geochemistry of Archean metasedimentary rocks from West Greenland. *Geochim. Cosmochim. Acta* **1984**, *48*, 1–13. [CrossRef]

36. Drost, K.; Chew, D.; Petrus, J.A.; Scholze, F.; Woodhead, J.D.; Schneider, J.W.; Harper, D.A.T. An Image Mapping Approach to U-Pb LA-ICP-MS Carbonate Dating and Applications to Direct Dating of Carbonate Sedimentation. *Geochem. Geophys. Geosyst.* **2018**, *19*, 4631–4648. [CrossRef]
37. Zheng, J.; Zhu, Y.J.; Huang, L.L.; Guo, Y.; Hu, F.J. Geochemical Characteristics and Their Geological Significance of Lower Cambrian Xiaoerblak Formation in Northwestern Tarim Basin, China. *Am. Minerals.* **2022**, *12*, 781. [CrossRef]
38. Shen, A.J.; Luo, X.Y.; Hu, A.P.; Qiao, Z.F.; Zhang, J. Dolomitization evolution and its effects on hydrocarbon reservoir formation from penecontemporaneous to deep burial environment. *Pet. Explor. Dev.* **2022**, *49*, 637–647. [CrossRef]
39. Korte, C.; Jasper, T.; Kozur, H.W.; Veizer, J. δ18O and δ13C of Permian brachiopods: A record of seawater evolution and continental glaciation. *Palaeogeogr. Palaeoclimatol. Palaeoecol.* **2005**, *224*, 333–351. [CrossRef]
40. Zheng, J.F.; Huang, L.L.; Yuan, W.F.; Zhu, Y.J.; Qiao, Z.F. Geochemical features and its significance of sedimentary and diagenetic environment in the Lower Cambrian Xiaoerblak Formation of Keping area, Tarim Basin. *Nat. Gas Geosci.* **2020**, *31*, 698–709. [CrossRef]
41. Zheng, J.F.; Shen, A.J.; Liu, Y.F.; Chen, Y.Q. Multi-parameter comprehensive identification of the genesis of Lower Paleozoic dolomite in Tarim Basin, China. *Acta Pet. Sin.* **2012**, *33*, 145–153. [CrossRef]
42. Grimes, K. Syngenetic Karst in Australia: A Review. *Helictite* **2003**, *39*, 27–38.
43. Fu, X.D.; Zhang, B.J.; Wang, Z.C.; Lei, M.; Zhang, J.Y.; Guang, S.W.; Li, W.Z.; Zhong, Y.; Gu, M.F.; Chen, T. Strike-Slip Faults in Central and Western Sichuan Basin and Their Control Functions on Hydrocarbon Accumulation. *Earth Sci.* **2023**, *48*, 2221–2237.
44. Sun, Z.M.; Bian, C.R.; Liu, G.X. Advances on Emeishan Mantle Plume and Discussions on Dynamic Mechanism of Formation of Sichuan Basin in Permian. *Geosciences* **2023**, 1–15. [CrossRef]
45. Liang, X.Y.; Li, H.; Zhang, D.D.; Yang, K.; Zhou, D.W.; Zheng, T.Y.; Dong, Y.K.; Zhai, L.G. Geochemical characteristics and genetic analysis of Huayingshan Emeishan basalt in Sichuan Basin. *Sci. Geol. Sin.* **2022**, *56*, 288–302. [CrossRef]
46. Zhu, C.Q.; Xu, M.; Yuan, Y.S.; Zhao, Y.Q.; Shan, J.N.; He, Z.G.; Tian, Y.T.; Hu, S.B. Palaeo-geothermal response and record of the effusing of Emeishan basalts in Sichuan basin. *Chin. Sci. Bull.* **2010**, *55*, 474–482. [CrossRef]
47. Li, H.B.; Zhang, Z.C.; Lv, L.S.; Wang, Y.F.; Kou, C.H.; Li, Y.S.; Liao, B.L. Isopach maps of the Qixia and Maokou formations: Implication for mantle plume model of the Emeishan large igneous province. *Acta Pet. Sin.* **2011**, *27*, 2963–2974.
48. He, W.Y.; Meng, Q.A.; Yin, C.H.; Wang, X.D.; Zhang, H.J.; Shi, J.B. Geological characteristics and favorable exploration plays of gas in Qixia Formation dolomite in Hechuan-Tongnan area of Sichuan Basin. *Petroleum Pet. Geol. Oilfield Dev. Daqing.* **2022**, *41*, 1–11. [CrossRef]
49. Clyde, H.M. Carbonate reservoirs-porosity evolution and diagenesis in a sequence stratigraphic framework. *Mar. Pet. Geol.* **2001**, *19*, 293–298. [CrossRef]

minerals

MDPI

Review

Deposition, Diagenesis, and Geochemistry of Upper Cretaceous Carbonates (Sarvak Formation) in the Zagros Basin and the Persian Gulf, Iran

Hamzeh Mehrabi

School of Geology, College of Science, University of Tehran, Tehran 1417935840, Iran; mehrabi.hamze@ut.ac.ir or hmehrabi.ut@gmail.com

Abstract: Upper Cretaceous carbonate rocks of the Sarvak Formation are among the most important hydrocarbon reservoirs of the Middle East. This paper summarizes the depositional history, diagenetic alteration, and geochemistry of this formation in south and southwest Iran. The combined effects of tectonic activities and the paleoclimate had remarkable control over the depositional–diagenetic evolution of this formation. Large changes in facies types, their frequency, and thickness were recorded across the study area and attributed to the tectonic activities. The complex diagenetic history of this formation includes minor marine, intensive meteoric, and burial diagenesis. Two pronounced paleoexposure events were recorded at the Cenomanian–Turonian boundary (CT-ES) and the middle Turonian (mT-ES), which are associated with remarkable changes in facies and diagenetic characteristics and special geochemical trends. Elemental (Mg, Fe, Rb, Mn, Sr), isotopic (carbon, oxygen), and $^{87}Sr/^{86}Sr$ ratios provide useful tools for delineating the paleoexposure surfaces, their associated diagenetic alterations, and chemo-sequence stratigraphic interpretation. Variable exposure durations resulted in immature to over-matured karstified profiles below the CT-ES and mT-ES, respectively. Meteorically dissolved rudist-dominated facies provided the best reservoir zone of the Sarvak Fm. in the regressive systems tract of the Cenomanian sequence, just below the CT-ES. Microporous and dolomitized mud-dominated facies formed the second important reservoir intervals, especially within the transgressive systems tracts. The results of this study revealed that the reservoir characteristics of the Sarvak Fm. were considerably controlled by paleoexposure events and, therefore, it can be considered as a typical paleokarst hydrocarbon reservoir in Iran.

Keywords: stable isotopes; paleokarst; trace elements; strontium isotope; cretaceous

Citation: Mehrabi, H. Deposition, Diagenesis, and Geochemistry of Upper Cretaceous Carbonates (Sarvak Formation) in the Zagros Basin and the Persian Gulf, Iran. *Minerals* **2023**, *13*, 1078. https://doi.org/10.3390/min13081078

Academic Editor: Stephen E Laubach

Received: 16 July 2023
Revised: 9 August 2023
Accepted: 12 August 2023
Published: 13 August 2023

1. Introduction

Upper Cretaceous carbonate successions (Sarvak and Ilam formations) are important petroliferous rock units in the Middle East, including the Zagros Basin [1–3]. They contain large volumes of hydrocarbons in several giant to supergiant oil and gas fields in SW Iran, including the Dezful Embayment (e.g., Ahwaz, Marun, Gachsaran, Rag-e-Sefid, Abteymour, Mansuri, etc.), the Abadan Plain (e.g., Azadegan, Jufair, Yadavaran, Darquain, Kushk, etc.), the Interior Fars (e.g., Sarvestan and Saadat Abad), and the Persian Gulf (e.g., Sirri, Hendijan, Bahregansar, Binak) (Figure 1). The Sarvak Formation is the second most important oil reservoir in Iran, after the Oligocene–Miocene Asmari Formation.

Since 2012, these reservoir rocks have been the subject of several studies focusing on their sedimentological [1,4–10], palaeontological [11–13], geochemical [2,14–19], and petrophysical [20–22] characteristics. Most of these studies were conducted by the author of this paper and his colleagues at the University of Tehran.

This study aims to bring the main results of all the previous studies together and present an overview of the depositional settings, diagenetic alterations, and geochemical compositions of Upper Cretaceous shallow-marine carbonates in the Zagros Basin. This large-scale synopsis will help to better understand the controlling factors of the depositional–diagenetic evolution of the Zagros Basin and, on a wider scale, the Middle East during the Late Cretaceous. It will also be useful for the exploration and development planning of hydrocarbon fields supplied by Upper Cretaceous reservoirs in various parts of the Zagros Basin.

2. Geological Setting and Stratigraphy

The Late Cretaceous (Cenomanian–Turonian: C–T) was a time of remarkable tectonic activity on the NE margin of the Arabian Plate (AP,) including the Zagros Basin [23–25]. At that time, a previous passive tectonic margin of the AP evolved into an active margin, following the initiation of the Neo-Tethys subduction beneath the Central Iranian Plate (CIP) at the end of the Cenomanian [16,24,26]. This huge change in the tectonic regime resulted in highly variable depositional settings for the C–T carbonates throughout the Zagros Basin [7,12].

The Cretaceous was the time of the highest eustatic sea level and, consequently, the maximum development of shallow-marine carbonate platforms over the continental plate margins, such as the AP [27]. During the C–T, the Zagros Basin was part of the NE margin of the AP, located at paleolatitudes of 0–5° north of the equator [28,29] (Figure 1). As a result, C–T carbonates were deposited under a warm and humid (tropical) climate [13,18,19]. The combined effects of tectonic activities, eustatic sea level changes, and paleoclimatic condition resulted in an intense and highly variable diagenetic evolution of the sediments deposited in various parts of the Zagros Basin [5,9]. Such remarkable changes in depositional–diagenetic characteristics are reflected by conspicuous variations in the geochemical compositions of C–T carbonates [2,18,19,30].

C–T carbonate rocks from the Zagros Basin have been included into the Sarvak Formation [31]. Similar rocks from the adjacent areas have been included into the Mishrif Formation in Iraq, UAE, and Qatar [3], the Derdere Formation (Mardin Group) in Turkey [32,33], and the Natih Formation in Oman [29] (Figure 2). The Sarvak Fm. is dominantly composed of rudist-dominated and foraminiferal limestone and dolomitic to argillaceous limestone. The Sarvak Formation is underlain by the Kazhdumi Fm. (shale and marl) along a sharp contact and overlain by the Ilam, Surgah, Laffan, or Gurpi formations with a disconformable boundary (Figure 2). In this study, the lower part of the Ilam Fm. is also included for delineating the Sarvak–Ilam boundary that corresponds to the mid-Turonian disconformity (abbreviated as mT-ES).

3. Materials and Methods

The dataset in this study includes results of thin-section, scanning electron microscope (SEM), X-ray diffraction (XRD), carbon and oxygen isotope, trace element, and strontium isotope analyses of the Sarvak Fm. and the lower parts of the Ilam Fm. from 23 wells studied in 14 oilfields located in various parts of the Zagros Basin, including the Dezful Embayment, Abadan Plain, Fars, and the Persian Gulf (Table 1). Macroscopic and microscopic petrographic studies of 3186 m of cores and 7452 thin sections are integrated with the results of SEM (147 samples) and XRD (116 samples) analyses and geochemical analysis of 701 samples for stable isotopes (C&O), 490 samples for elemental analysis, and 274 samples for strontium isotope measurements. Geochemical data from the Dezful Embayment and Abadan Plain were previously published [2,16,18,19], and the rest of the data in this study are new.

The textural classification schemes of Dunham [34] and Embry and Klovan [35] were used along with the standard microfacies model of Flügel [36] for the identification of sedimentary facies and the interpretation of the depositional environment. SEM analysis was implemented using the MIRA3 TESCAN apparatus at the Razi Metallurgical Research Center, Tehran, Iran. The geochemical samples were drilled from the micrite, cements, unaltered shell fragments, and whole rocks using a 0.5 mm tungsten carbide bit under a binocular microscope. The carbon and oxygen isotopic measurements were conducted at the stable isotope laboratory of the Institute of Geology, Leibniz University Hannover, Germany, using a Thermo Fisher Scientific Gas bench II mass spectrometer. The external reproducibility values were $\leq$0.06‰ for δ^{13}C and $\pm$ 0.08‰ for δ^{18}O (see [17] for the isotopic values of standards).

Strontium isotope composition was analyzed using a Thermo Scientific TRITON thermal ionization mass spectrometer (TIMS) at the Boston TIMS Facility, USA. The ^{87}Sr/^{86}Sr values of all samples were corrected for in situ Rb decay incorporated by clay minerals [37] using the equations by Faure and Mensing [38]. Elemental concentrations were measured using the Analytik Jena ICP-OES device in the Central Lab of Arak University and using atomic absorption (AA) and inductively coupled plasma (ICP) spectrometers at the University of Tehran.

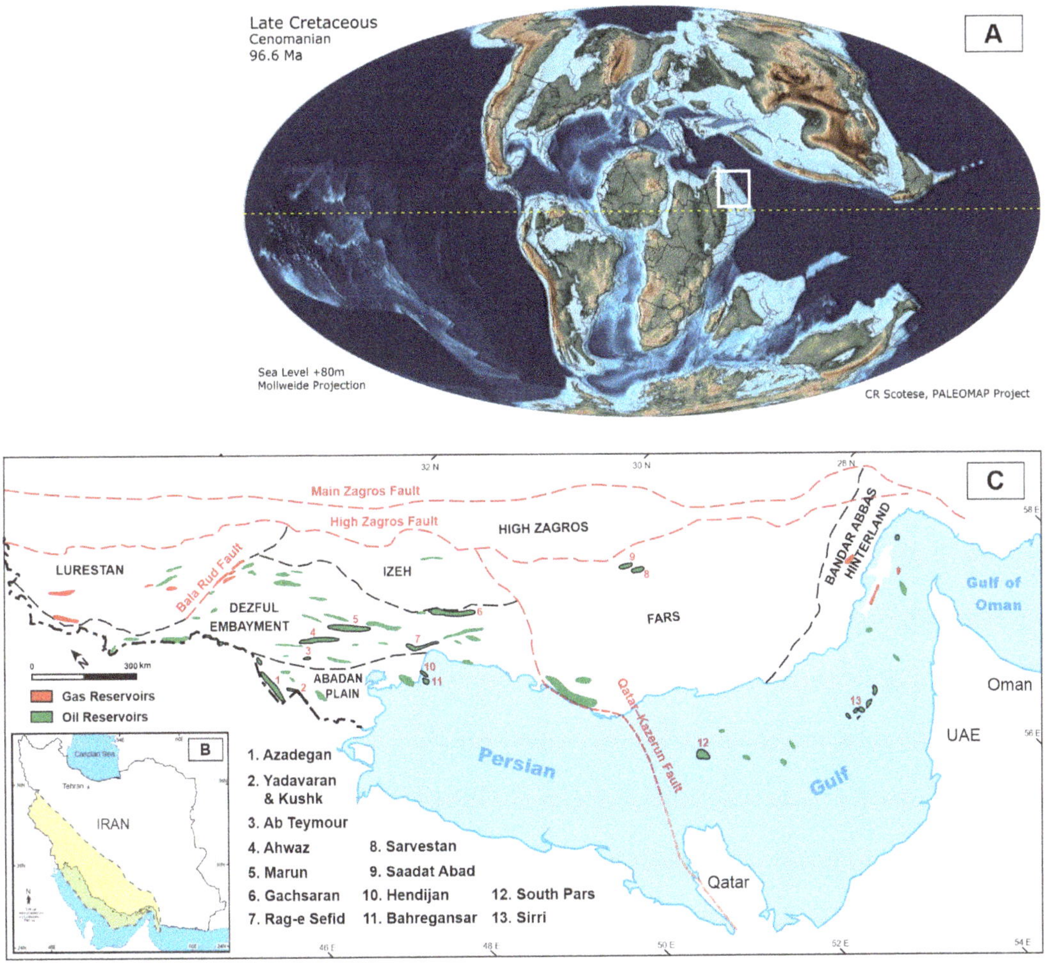

Figure 1. (**A**) Paleogeographic map of the world during the Late Cretaceous (modified after [39]). Study area on the northeastern margin of the Arabian Plate is marked by the white rectangle on this map. (**B**) General map of Iran (inset) and (**C**) location map of the studied fields in the Zagros and Persian Gulf basins (modified after [31]).

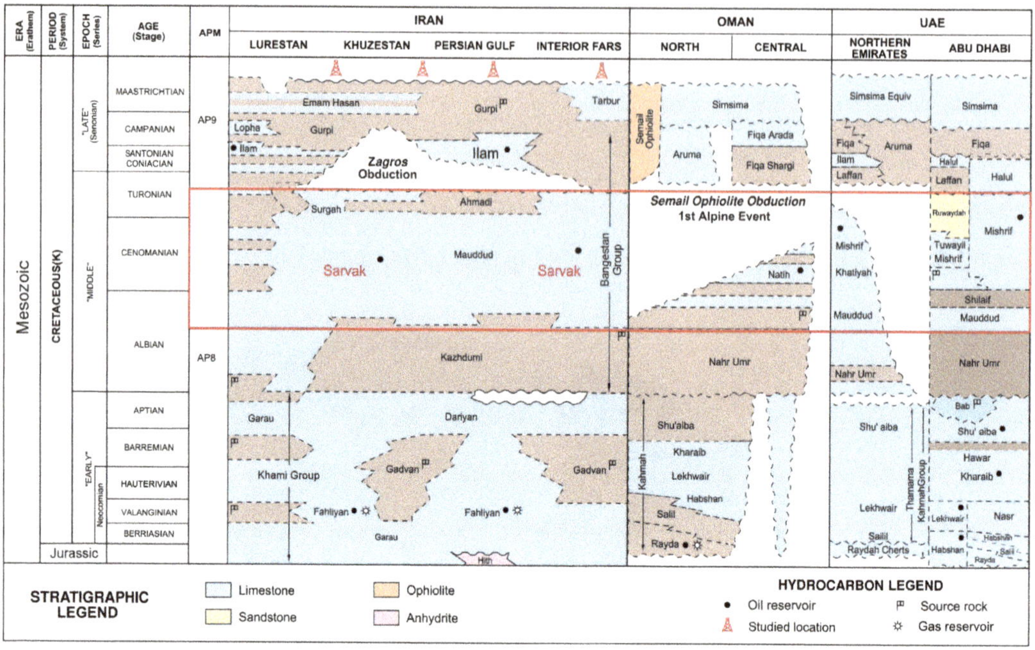

Figure 2. Stratigraphic chart of Cretaceous in Iran and neighboring countries (modified after petroleum geo-service; http://www.pgs.com; accessed on 13 August 2023). The Sarvak Formation and its stratigraphic equivalents are marked by the red box on this column.

Table 1. Summary of datasets used in this study.

Zone Name	Field Name	Well No.	Core (m)	Thin Section (No.)	XRD (No.)	SEM (No.)	C and O Isotopes (No.)	Trace Elements (No.)	Sr Isotopes (No.)
Dezful Embayment	Ahwaz	B	207	345	8	-	111	110	56
	Marun	C	120	170	5	-	69	66	31
	Abteymour	A	125	240	17	-	108	109	70
	Gachsaran	E2	62	84	5	-	57	55	22
	Gachsaran	E1	99	185	11	-	64	64	28
	Rag-e-Sefid	D	165	330	14	-	70	69	52
Abadan Plain	Azadegan	X03	269	807	9	12	25	-	12
	Azadegan	X13	105	315	6	10	14	-	-
	Azadegan	X23	56	168	4	5	6	-	-
	Azadegan	X21	186	558	5	10	20	-	3
	Kushk	X1	256	768	-	16	111	-	-
	Kushk	X2	69	207	-	-	46	-	-
	Yadavaran	X9	158	474		54	-	-	-
	Yadavaran	X20	146	438		10	-	-	-
Fars	Sarvestan	X5	63	147	5	3	-	-	-
	Sarvestan	X6	55	165	7	3	-	7	-
	Saadat Abad	X4	73	220	7	4	-	10	-

Table 1. *Cont.*

Zone Name	Field Name	Well No.	Core (m)	Thin Section (No.)	XRD (No.)	SEM (No.)	C and O Isotopes (No.)	Trace Elements (No.)	Sr Isotopes (No.)
Persian Gulf	Sirri	X06	60	195	8	-	-	-	-
	Hendijan	X7	133	160	-	-	-	-	-
	Bahregansar	X1	252	356	-	-	-	-	-
	Bahregansar	X5	129	258	-	-	-	-	-
	Bahregansar	X6	296	592	-	-	-	-	-
	South Pars	X	102	270	5	20	-	-	-
Total			**3186**	**7452**	**116**	**147**	**701**	**490**	**274**

4. Results

4.1. Microfacies

Microscopic petrographic analyses indicate that the Sarvak Fm. is composed of eight microfacies types in various parts of the Zagros Basin. The general characteristics of these facies are presented in Table 2. Representative photomicrographs of the microfacies are summarized in Figure 3. Short descriptions of the facies are presented below.

Table 2. Summary of facies characteristics of the Sarvak Formation in the Zagros Basin.

Facies Code	Facies Name	Allochems	Sedimentary Structures	Depositional Setting	Encountered Locations
F1	Benthic foraminifera mudstone, wackestone	Miliolids, Alveolinids, *Nezzazata* sp., *Dicyclina* sp., peloid, green algae, calcareous sponges, cortoids, oncoids, bivalve, gastropod	Bioturbation	Lagoon	Dezful, Abadan Plain, Persian Gulf
F2	Rudist-algal-coral bafflestone, rudstone, floatstone	Rudists, Dasycladacean algae, coral	-	Patch reefs and taluses	Dezful, Abadan Plain, Fars, Persian Gulf
F3	Bioclast, peloid, intraclast packstone, grainstone	Rudist, bivalve, algae, coral, echinoderm, benthic foraminifera	Cross-bedding, lamination	Shoal	Dezful, Abadan Plain, Fars, Persian Gulf
F4	Foraminifera (benthic and planktic), bioclast (echinoid) wackestone, packstone	Echinoderm, bivalve, benthic and planktic foraminifera, intraclast, rudist	-	Middle ramp	Dezful, Abadan Plain, Fars
F5	Microbioclast packstone, grainstone	Fine to very fine bioclasts such as echinoderms, rudist, bivalves, fine peloids, glauconite, small benthic and planktic foraminifera	Hummocky cross stratification	Storm wave base (Tempestites)	Fars, Persian Gulf
F6	Oligosteginids wackestone, packstone	Oligostegina, planktic foraminifera	Bioturbation	Outer ramp	Dezful, Fars, Persian Gulf
F7	Planktic foraminifera mudstone, wackestone	Planktonic foraminifera (*Rotalipora* sp., *Hedbergella* sp., *Globigerinelloides* sp., *Whiteinella* sp.), oligostegina	Lamination	Outer ramp	Dezful, Fars, Persian Gulf

Table 2. *Cont.*

Facies Code	Facies Name	Allochems	Sedimentary Structures	Depositional Setting	Encountered Locations
F8	Calcisiltite with sponge spicules	Siliceous sponge spicules, planktonic foraminifera and oligosteginids, fine peloids	Lamination	Basin	Dezful, Fars, Persian Gulf

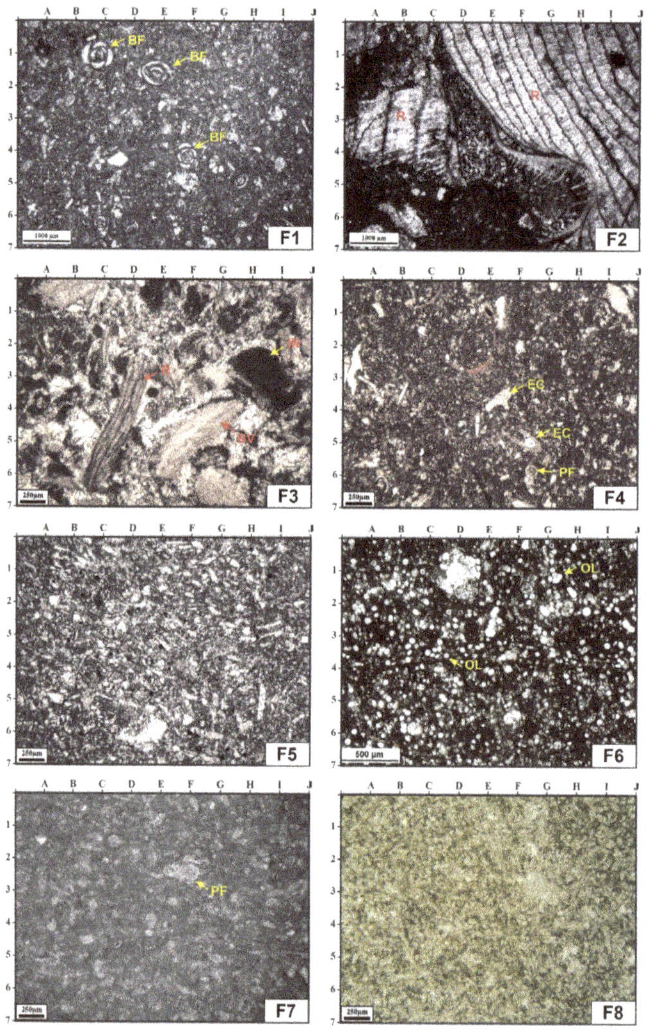

Figure 3. Photomicrographs of depositional facies of the Sarvak Fm. in the studied fields from the Zagros Basin and the Persian Gulf. (**F1**): benthic foraminifera (BF) mudstone, wackestone; (**F2**): rudist (R)-algal-coral bafflestone, rudstone, floatstone; (**F3**): bioclast (bivalve: BV), peloid, intraclast (IN) packstone, grainstone; (**F4**): foraminifera (benthic and planktic: PF), bioclast (echinoid: EC) wackestone, packstone; (**F5**): microbioclast packstone, grainstone; (**F6**): oligosteginids (OL) wackestone, packstone; (**F7**): planktic foraminifera (PF) mudstone, wackestone; (**F8**): calcisiltite with sponge spicules. All photos in plane-polarized light.

F1: Benthic Foraminifera Mudstone and Wackestone

This microfacies is marked by mud-supported textures in which benthic foraminifera such as miliolids, alveolinids, *Nezzazata* sp., and *Dicyclina* sp. are notable allochems (Figure 3). Peloids, oncoids, bivalves, and gastropods are subordinate constituents. It shows bioturbation as the only sedimentary feature. In the Abadan Plain and Dezful Embayment, green algae and calcareous sponge spicules are recorded within the facies.

F2: Rudist-Algal-Coral Bafflestone, Rudstone, and Floatstone

This facies is dominantly composed of reef-builder biota including rudists, green algae, and corals (Figure 3). In almost all cases, in situ forms of these fossils are not recognized in subsurface areas of the Zagros Basin. Instead, they commonly occur as allochthonous reefal fragments with floatstone (mud-supported) to rudstone (grain-dominated) textures. Boring and burrowing are frequently recorded in this facies.

F3: Bioclast, Peloid, Intraclast Packstone, and Grainstone

This microfacies encompasses all grain-dominated facies of the Sarvak Fm., with variable constituents including bioclasts (mostly rudists and echinoderms), peloids, and intraclasts (Figure 3). Cross-bedding is recognizable in the core samples of this facies.

F4: Foraminifera (Benthic and Planktic), Bioclast (Echinoid) Wackestone, and Packstone

This facies shows a mixture of small benthic and planktic foraminifera along with echinoderms, bryozoan, red algae, bivalves, and intraclasts as the main allochems (Figure 3). A range of mud-supported (wackestone) to grain-dominated (packstone) textures is distinguished within this facies. Bioturbation is the only notable sedimentary feature of F4.

F5: Microbioclast Packstone and Grainstone

This facies is marked by the dominance of fine to very fine sand-sized bioclasts indicating high levels of abrading. The bioclasts mostly include rudists and echinoderms (Figure 3). Pyrite and glauconite along with fine peloids are other notable constituents. This facies is dominantly recorded in the Fars area (Sarvestan and Saadat Abad fields) and the Persian Gulf (i.e., Sirri Field). Cross-bedding is a major sedimentary structure of F5.

F6: Oligosteginids Wackestone and Packstone

The microfacies is mainly composed of oligosteginids that provide a range of grain-dominated to mud-supported textures (Figure 3). Planktic foraminifers, fine peloids, and rare echinoderm fragments are other allochems of this facies. Pyrite and glauconite are also present.

F7: Planktic Foraminifera Mudstone and Wackestone

This facies contains various types of planktic foraminifera (*Globigerinoides* sp., *Hedbergella* sp., and *Rotalipora* sp.) in mud-dominated textures of mudstone to wackestone (Figure 3). Oligosteginids, fine peloids, and rare skeletal fragments (mostly echinoderms) are subordinate grains. Lamination and bioturbation are its important sedimentary features.

F8: Calcisiltite with Sponge Spicules

This facies is characterized by the presence of silt-sized (highly abraded) skeletal debris (echinoderms, thin-shelled bivalves) and siliceous sponge spicules (Figure 3). Fine peloids, clacisphere (oligosteginids), and planktic foraminifera are subordinate constituents of this microfacies. Pyritization and glauconitic grains are also recorded. It shows lamination in some cases. This microfacies is mainly recorded from the Dezful Embayment.

4.2. Diagenesis

Various types of diagenetic alteration are recorded from the Sarvak carbonates throughout the Zagros Basin. They include isopachous (aragonite) cementation, micritization, bioturbation (boring and burrowing), dissolution (karstification), collapse brecciation, development of paleosol, silicification, recrystallization, equant and drusy calcite cementation, mechanical to chemical compaction (stylolitization), dolomitization, blocky calcite cementation, pyritization, and fracturing. Core photos and photomicrographs of the major diagenetic processes of the Sarvak Fm. are presented in Figures 4 and 5. Brief descriptions of some major diagenetic alterations of this formation are presented below.

Collapse Brecciation

Dissolution–collapse-brecciated units are common features in the topmost part of the Sarvak Fm., especially in the Dezful Embayment (Figure 4A–C) and Abadan Plain (Figure 4G,H). In these units, carbonate breccia clasts are consolidated together with paleosol infiltrated sediments (Figure 4A–C). The most typical brecciated intervals of the Sarvak Fm. are recorded from the Rag-e-Sefid, Gachsaran, and Azadegan fields (see Table 1).

Figure 4. Core photos and photomicrographs of major diagenetic features of the Sarvak Fm. in the Zagros Basin and the Persian Gulf. **(A–C)** Solution collapse breccia with paleosol (ps) filling spaces

between breccia clasts (br); (**D**) vug/vuggy pores (vg); (**E,F**) pisolitic (pis) horizons; (**G**) solution collapse breccia with paleosol (ps) filling spaces between breccia clasts (br); (**H**) Fe concretions (Fe); (**I**) vug/vuggy pores (vg) in rudist-dominated facies.

Dissolution (Karstification)

Dissolution molds, vugs, channels, and caves are frequently recorded from the Sarvak Fm. in the Zagros Basin (Figure 4D,I). On a microscopic scale, dissolution forms considerable amounts (up to 35%) of pores in the form of separate to touching vugs, especially within the grain-supported facies of F2 and F3 (see Table 2 and Figure 4). Parts of these dissolution vugs are subsequently filled by calcite cements; however, a major portion of them remains unfilled.

Development of Paleosol

Iron- and aluminum-rich paleosols (bauxite and laterite) are common diagenetic features at the uppermost parts of the Sarvak Fm. in various parts of the Zagros Basin. They provide some economically important mineral deposits in the west and southwest of Iran (Hamedan, Kermanshah, Khuzestan, and Fars). In the subsurface, paleosols are recognized from the Gachsaran and Rag-e-Sefid oilfields in the Dezful Embayment (Figure 4A–C). In some localities, such as the Fars area and the Persian Gulf, pisolitic units are recorded at the top of the Sarvak Fm., beneath the disconformable surfaces (Figure 4E,F). They can be considered as stratigraphic equivalents of top-Sarvak paleosols in the neighboring areas. Paleosols show a close association with extensively dissolved and brecciated units of the Sarvak Fm. and, consequently, are known as the karst bauxite deposits in the Zagros Basin [40].

Cementation

Various types of cements are recorded in the Sarvak Fm., including isopachous, equant, drusy, syntaxial, blocky, pendant, and bladed circumgranular calcite (Figure 5A–F). Isopachous, pendant, and syntaxial cements are frequently recorded within the grain-dominated facies (F3). Equant, drusy, and blocky calcites filled dissolution vugs and molds as well as fractures, especially within the karstified rudist-dominated facies (F2) of the Sarvak Formation (Figure 5A,B).

Compaction

Effects of mechanical and chemical compaction are frequently recorded in the Sarvak Fm. in the form of deformed and broken grains and the development of solution seams and stylolites (Figure 5G–I). The latter features are among the most frequent and important diagenetic features of the Sarvak Fm. in the Zagros Basin. Solution seams are commonly recorded within the mud-dominated facies of F1 and F7 (Table 2). In contrast, stylolites are predominantly observed in the grain-supported facies, including F2 to F5 (see Figure 5I and Table 2).

Dolomitization

Dolomites of the Sarvak Fm. are mainly concentrated along the solution seams and stylolites (Figure 5J–L). They are made up of euhedral to subhedral crystals with a size ranging from silt to very fine sand (up to 8 μm). In some cases, they show a cloudy center surrounded by a clear margin. In stained thin sections, dolomites seem to be ferroan (Figure 5L). Dolomitization predominantly occurred within the mud-dominated facies (mud- to wackestone) of open marine and lagoonal settings. These replacement dolomites are commonly concentrated along the chemical compaction features (i.e., solution seams and stylolites) and their frequency ranges from >5% up to 65%. Dolomitization was mostly fabric-selective and fabric-retentive (Figure 5J–L). It is frequently recorded from the Sarvak Fm. in the Dezful Embayment and Abadan Plain. It is also recorded as a minor diagenetic feature in the Persian Gulf oilfields. Saddle/baroque dolomites are also present, with a very low frequency, in the mentioned areas [1]. It is commonly recorded as fracture-filling cement.

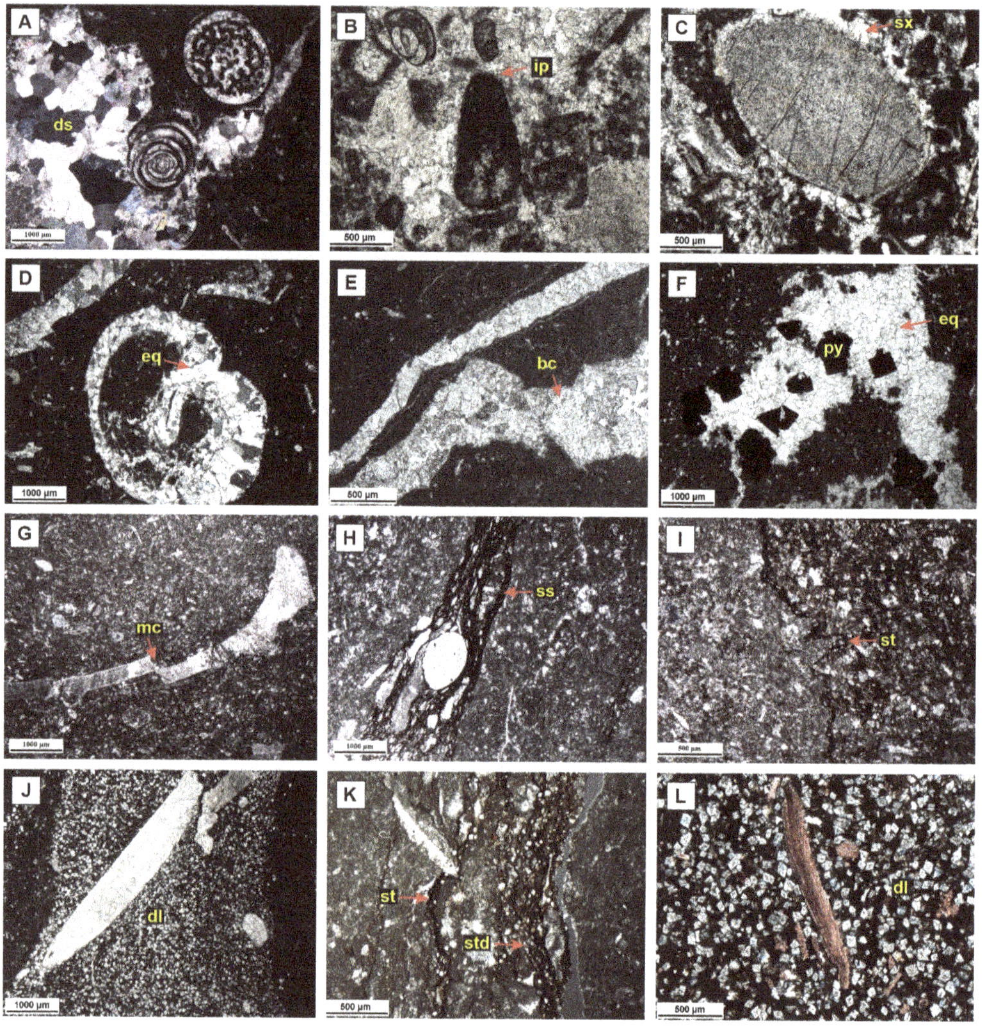

Figure 5. Photomicrographs of diagenetic features of the Sarvak Fm. in the Zagros Basin and the Persian Gulf. Various types of cements are recorded, including drusy (**A**), isopachous (**B**), syntaxial (**C**), equant (**D**,**F**), and blocky (**E**) calcite cements. (**G**) Mechanical compaction; (**H**) solution seams; (**I**) stylolitization; (**J**–**L**) burial (stylolite-related) dolomitization. (**A**,**D**,**H**,**I**) are shown in cross-polarized light, others are shown in plane-polarized light. Abbreviations—ds: drusy, ip: isopachous, sx: syntaxial, eq: equant, bc: blocky, py: pyrite, mc: mechanical compaction, ss: solution seam, st: stylolite, dl: dolomitization, std: stylolite-related dolomitization. (**L**) is a stained thin section in which calcitic components are stained and dolomites remain unstained. Dolomites seem to be ferroan.

4.3. Bio- and Sequence Stratigraphy

Cenomanian–Santonian sedimentary sequences in the Zagros Basin are marked by three well-known biozones defined based on the benthic foraminiferal assemblages within the shallow-marine carbonates of the Sarvak and Ilam formations (see [11] for more details). They include the following:

Biozone 1: *Nezzazata–Alveolinids* assemblage zone; biozone #25 of [41], (Cenomanian);
Biozone 2: *Nezzazatinella–Dicyclina* interval zone; biozone #29 of [41], (Turonian);
Biozone 3: *Rotalia sp.22*–algae assemblage zone; biozone #30 of [41], (Santonian).

Each of these biozones correspond to a third-order depositional sequence of Cenomanian (Sv_Cen), Turonian (Sv_Tur), and Santonian (IL_San) ages (Figure 6). Depositional sequences of the Sarvak Fm. are easily differentiable based on the presence of two regional disconformable surfaces at the Cenomanian–Turonian boundary (known as CT-ES) and middle Turonian (mT-ES). These paleoexposure surfaces are considered as type-I (erosional) sequence boundaries at the top of the Sv_Cen and Sv_Tur sequences (Figure 6).

Within these third-order sequences, the vertical accumulation of outer ramp, basin, and middle ramp facies formed the transgressive systems tracts (TSTs), and inner ramp facies (lagoon, shoal, and reef talus) formed the regressive systems tracts (RSTs). The deepest marine facies (F7 and F8) are considered as the maximum flooding surfaces (MFSs) (Figure 6).

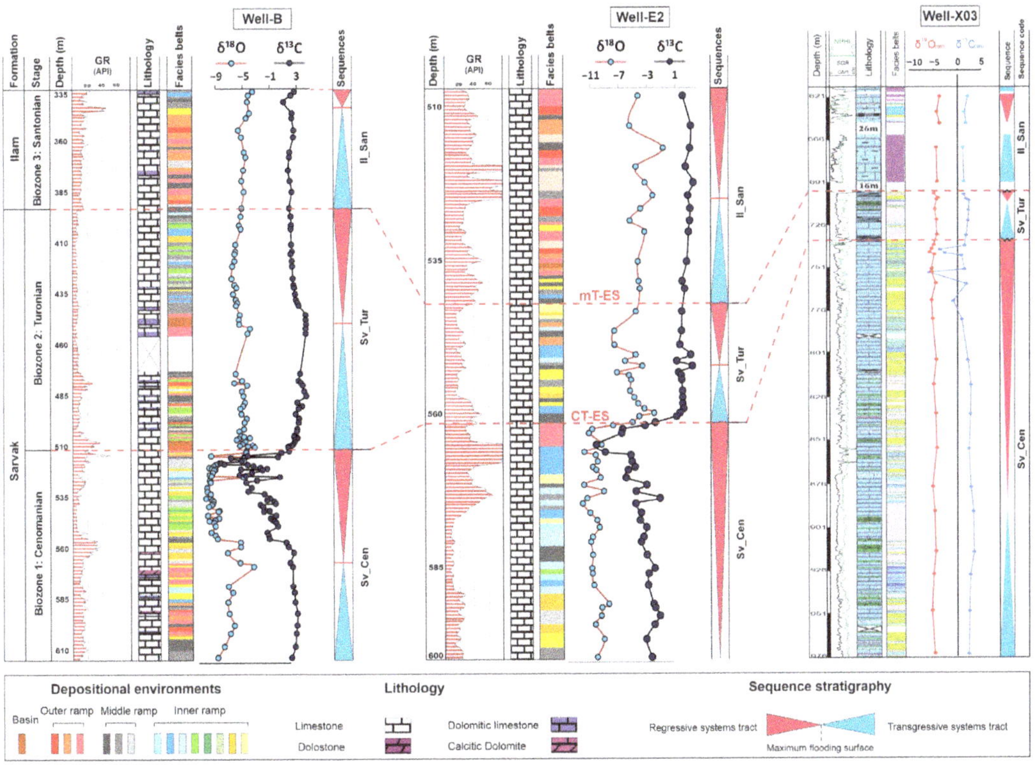

Figure 6. Chemo-sequence stratigraphic correlation of the Sarvak and Ilam formations across the studied wells. Three third-order sequences are shown that are dated according to the results of biostratigraphic studies. They include Cenomanian (Sv_Cen), Turonian (Sv_Tur), and Santonian (IL_San) sequences.

4.4. Geochemistry

4.4.1. Trace Elements

The elemental (Ca, Mg, Sr, Fe, Al, Mn, and Rb) analysis indicates some remarkable and meaningful variations in the Sarvak Fm. (Figure 7). Here, the results of the elemental analysis of Abteymour Well-A are presented as the key well, but evidence from other studied sections will be provided in the following sections.

Among the tested elements, Mg and Fe show a strong covariance across the studied intervals. The Mg content ranges from 1700 to 37,410 ppm (average: 5173 ppm) with maximum concentrations below the mT-ES (Figure 7) and within the Ilam Fm. (37,410 ppm). The Fe values are measured as 500 to 42,000 ppm (average: 2040 ppm) with similar trends of variations to Mg. The maximum Fe concentration is measured from the altered carbonates beneath the mT-ES. In contrast, no major changes are distinguished below the CT-ES (Figure 7).

The manganese (Mn) content of the analyzed samples shows remarkable fluctuations from 25 to 605 ppm (average: 188 ppm). The perturbations in Mn concentration are mainly recorded at two stratigraphic intervals around the CT-ES and mT-ES (Figure 7). However, all of the peaks of Mn are not related to the paleoexposures. Some Mn variations are also recorded within the Ilam Fm. (i.e., IL_San sequence).

The strontium (Sr) concentrations range from 150 ppm to 1250 ppm (average: 649 ppm) in the studied sequences (Figure 7). The lowest Sr content is measured from the altered carbonates below the CT-ES (i.e., Sv_Cen sequence). In the Sv_Tur sequence, the Sr content shows an upward increasing trend that reaches the maximum value of 1250 ppm beneath the mT-ES (Figure 7). Two different trends are visible in the Sr content of the Ilam Fm. (IL_San sequence), which include an upward decreasing trend within the TST that is followed by an increase in Sr content within the RST.

The rubidium (Rb) concentration of the studied samples varies between 0 and 42 ppm, with an average of 8 ppm. A sharp increase in Rb content is visible just below the CT-ES, while there is no detectable change around the mT-ES. The maximum variations in Rb are measured from the Ilam carbonates with a general increasing trend (range: 2–24 ppm, average: 9.5 ppm) (Figure 7).

Cross plots of Mn vs. $\delta^{18}O_{carb}$ and Mn vs. Sr/Ca are presented in Figure 8 for the Sarvak and Ilam formations in the Ahwaz Well-B and Rag-e-Sefid Well-D. The expected ranges for the Cretaceous marine carbonates are shown on the Mn vs. $\delta^{18}O_{carb}$ plot (Figure 8B,D). Deviations from the expected ranges are clearly visible for the Cenomanian and Turonian sequences of the Sarvak Fm. (Figure 8). These samples are marked by their decreased $\delta^{18}O_{carb}$ values (−4‰ to −7‰) and increased Mn contents (100 to 550 ppm).

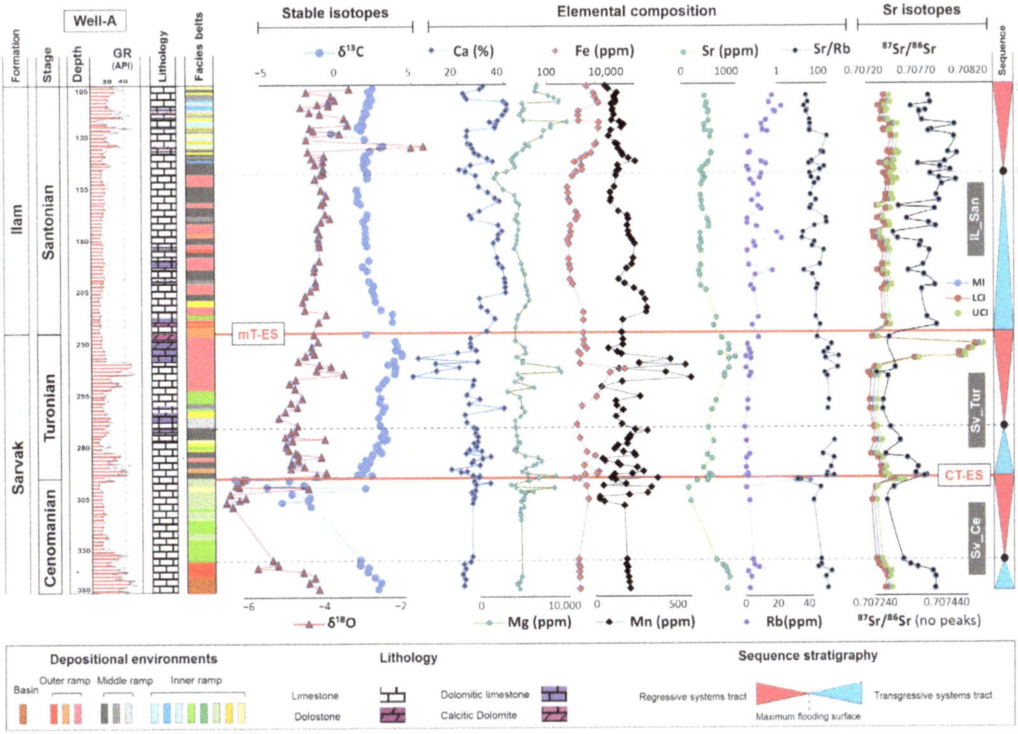

Figure 7. Integrated geochemical log of the Sarvak and Ilam formations in Abteymour Oilfield (Well-A). Elemental, carbon and oxygen, and $^{87}Sr/^{86}Sr$ ratios are shown along with the sedimentological characteristics (facies and sequences) of these formations. Paleoexposure surfaces are marked as CT-ES (Cenomanian–Turonian boundary exposure surface) and mT-ES (middle Turonian exposure surface).

Plots of Mn vs. Sr/Ca indicate remarkable deviations of analyzed samples compared to the expected ranges of the aragonite (indicated by A in Figure 8A,C), high-magnesium calcite (HMC), and low-magnesium calcite (LMC) (Figure 8A,C). They are marked by increased Mn contents. The Sr/Ca values of analyzed samples are all in the ranges of expected values for the HMC and LMC (Figure 8A,C).

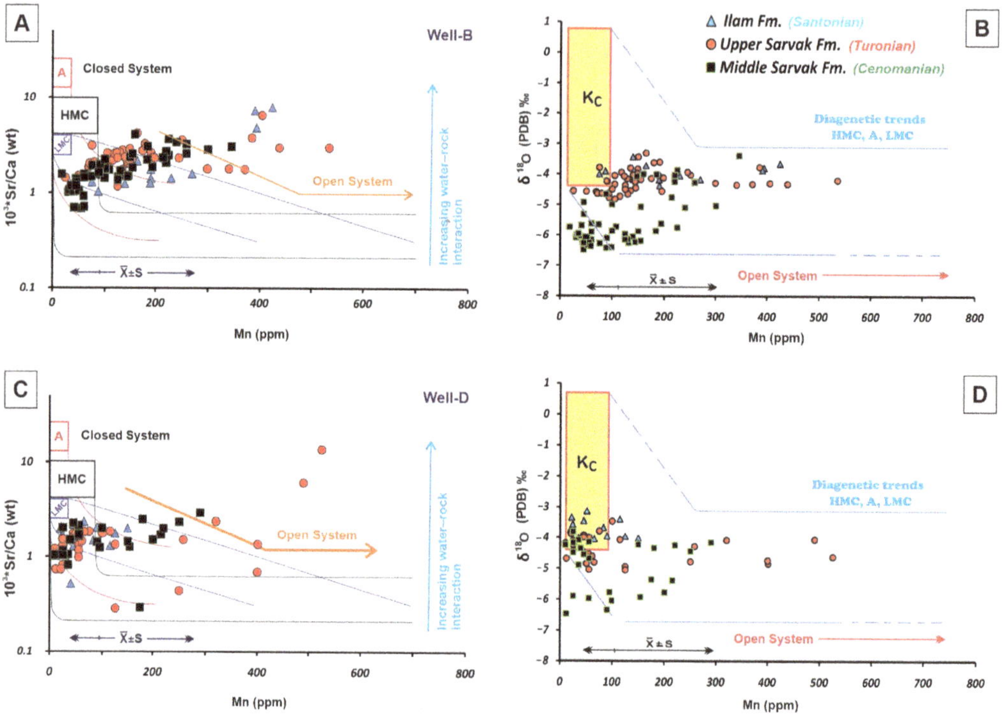

Figure 8. Cross plots of Mn vs. Sr/Ca (**A**,**C**) and Mn vs. $\delta^{18}O_{carb}$ (**B**,**D**) for analyzed samples from the Sarvak and Ilam formations in the Ahwaz (Well-B: plots (**A**,**B**)) and Rag-e-Sefid (Well-D: plots (**C**,**D**)) fields in the Dezful Embayment. A: aragonite; HMC: high-magnesium calcite; LMC: low-magnesium calcite; K_C: expected range for Cretaceous marine calcite [42]. X-bar: mean, S: standard deviation.

4.4.2. Carbon and Oxygen Isotopes

The $\delta^{13}C_{carb}$ and $\delta^{18}O_{carb}$ values are measured from Cenomanian–Santonian sequences in various parts of the Zagros Basin (Figures 7 and 9). The $\delta^{13}C_{carb}$ values vary between 4‰ and −8‰ in the studied wells. The lowest values of $\delta^{13}C_{carb}$ are measured from the upper parts of the Cenomanian (Sv_Cen) and Turonian (Sv_Tur) sequences in the Sarvak Fm. (Figure 9).

In the Cenomanian sequence, the positive $\delta^{13}C_{carb}$ values (2‰ to 3‰) change upwards to the negative values of −6‰ to −8‰. A similar upward decreasing trend is also visible in the Turonian sequence, where $\delta^{13}C_{carb}$ values of 3‰ to 4‰ decrease upward to 0‰ to 1‰ (Figure 9). In the Ilam Fm. (IL_San), a narrow variation in $\delta^{13}C_{carb}$ is measured between 2‰ and 3‰ (Figure 7).

The $\delta^{18}O_{carb}$ values of the analyzed samples range from −1‰ to −7‰ in the studied sections (Figure 9). Similar to $\delta^{13}C_{carb}$, the minimum values of $\delta^{18}O_{carb}$ are recorded from the samples taken from the uppermost parts of Sv_Cen sequence of the Sarvak Fm. (Figures 7 and 9). In the RST of this sequence, the $\delta^{18}O_{carb}$ values vary between −5‰ and −7‰ (Figure 7). A sharp decrease in $\delta^{18}O_{carb}$ values is not measured from the Sv_Tur and IL_San sequences. However, a small decrease in $\delta^{18}O_{carb}$ is noted just below the mT-ES in the Ahwaz (Well-B) and Rag-e-Sefid (Well-D) sections (Figure 9). The $\delta^{18}O_{carb}$ values of the Ilam Fm. show similar values around −4‰.

Cross plots of $\delta^{13}C_{carb}$ vs. $\delta^{18}O_{carb}$ are shown for the studied sequences in wells B and D in Figure 9. In these plots, analyzed samples are categorized into four classes marked as A to D.

Class A includes samples of the Ilam Fm. (IL_San sequence) that fall within the expected ranges of the Cretaceous marine carbonates. In this class, $\delta^{18}O_{carb}$ values range from −3‰ to −4‰, and $\delta^{13}C_{carb}$ varies between 0‰ and 4‰ (Figure 9).

Class B encompasses the samples of the Turonian sequence (Sv_Tur) with $\delta^{13}C_{carb}$ values similar to the expected ranges of Cretaceous marine carbonates (0‰ to 4‰) and decreased values of $\delta^{18}O_{carb}$ (−4‰ to −5.5‰) (Figure 9).

Class C incorporates the samples that belong to the RST of the Cenomanian sequence (Sv_Cen) with very low values of $\delta^{13}C_{carb}$ (0‰ to −9‰) and $\delta^{18}O_{carb}$ (−5‰ to −7‰). Samples of this class plotted far from the expected isotopic ranges for the Cretaceous marine carbonates (Figure 9).

Class D includes those samples taken from the TST of the Cenomanian sequence (Sv_Cen). The $\delta^{13}C_{carb}$ values of these samples are in the expected range of Cretaceous marine carbonates (0‰ to 4‰). A 1‰ decrease is recorded in the $\delta^{18}O_{carb}$ values of this class (Figure 9).

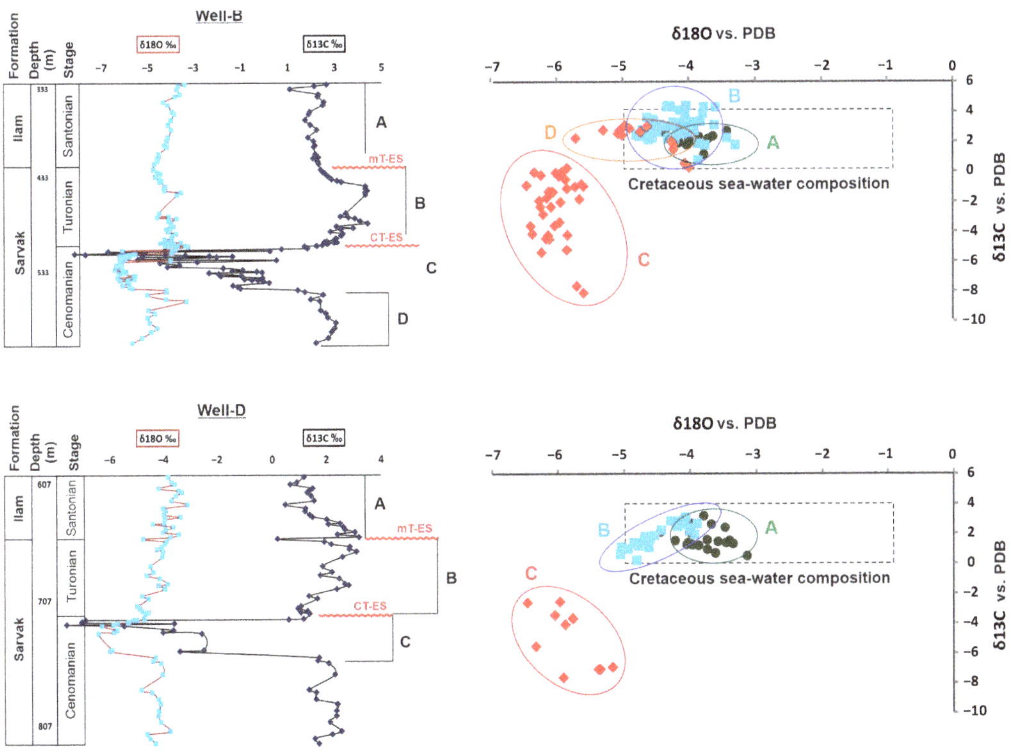

Figure 9. Plots of carbon and oxygen isotopic rations versus depth and their cross plots in the Ahwaz (Well-B) and Rag-e-Sefid (Well-D) fields. According to their stratigraphic positions, the samples were categorized into four classes of A to D. Paleoexposure surfaces are marked as CT-ES (Cenomanian–Turonian boundary exposure surface) and mT-ES (middle Turonian exposure surface).

4.4.3. Strontium Isotopes

The $^{87}Sr/^{86}Sr$ ratios of carbonate samples from the Sarvak and Ilam formations are plotted in stratigraphic order in Figure 10, along with the defined biozones for relative dating of the studied intervals. As shown, there are two remarkable increases in the $^{87}Sr/^{86}Sr$ ratios of the analyzed samples at the two paleoexposure surfaces (i.e., CT-ES and mT-ES) at the top of the Sv_Cen and Sv_Tur sequences (Figure 10). In the Abteymour Field (Well-A), the highest values of $^{87}Sr/^{86}Sr$ are recorded just below the CT-ES (0.707534 ± 24 ×

10^{-6}) and mT-ES (0.708259 $\pm$ 38 $\times$ 10^{-6}). Accordingly, clear increases in $^{87}Sr/^{86}Sr$ ratios are recorded in the topmost parts of the Cenomanian and Turonian sequences (Figure 10). The $^{87}Sr/^{86}Sr$ ratios of the Ilam Fm. show a narrow range with a relatively uniform distribution across the studied wells (Figure 10). In Well-A, the $^{87}Sr/^{86}Sr$ ratios of this formation range from 0.707282 $\pm$ 30 $\times$ 10^{-6} to 0.707393 $\pm$ 28 $\times$ 10^{-6} (Figure 10).

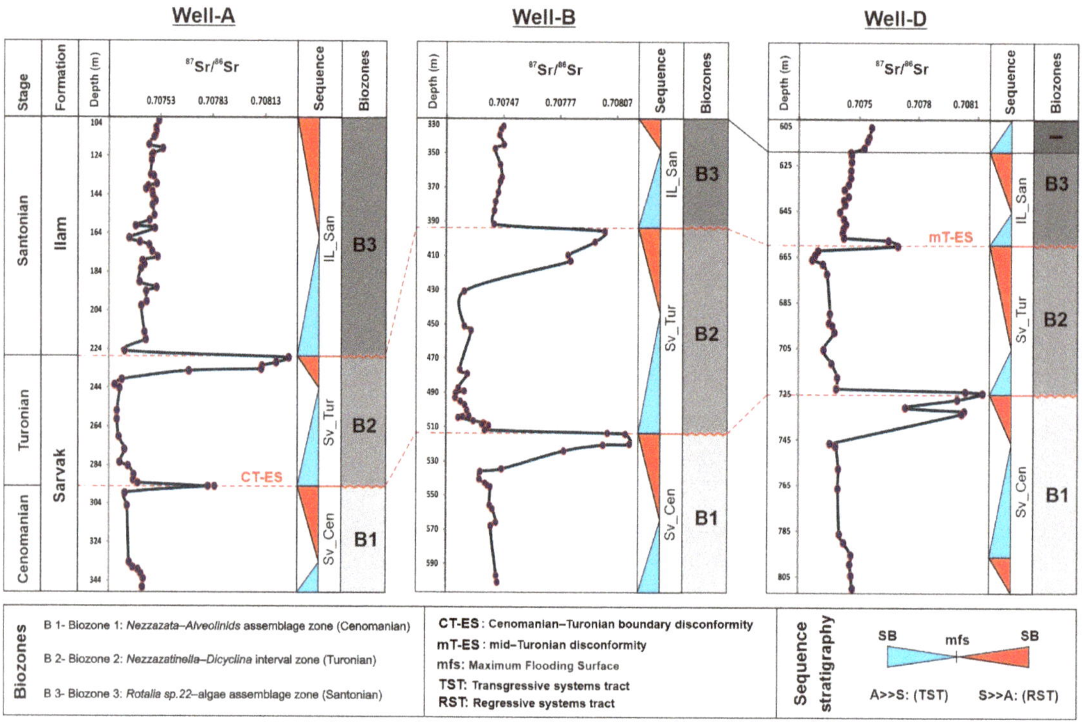

Figure 10. Strontium isotope stratigraphic (SIS) correlation of the Sarvak and Ilam formations across the Abteymour (Well-A), Ahwaz (Well-B), and Rag-e-Sefid (Well-D) fields in the Dezful Embayment. Sharp increases in $^{87}Sr/^{86}Sr$ ratios are measured around the paleoexposure surfaces. Third-order sequences and biozones are also shown.

For the absolute dating of the studied sequences, the $^{87}Sr/^{86}Sr$ ratios of the analyzed samples were plotted on the standard $^{87}Sr/^{86}Sr$ curves of Mc Arthur [37] for the Albian–Campanian time span (Figure 11). A rubidium correction was applied to the analyzed samples, and diagenetically altered samples were removed from the database. Then, the $^{87}Sr/^{86}Sr$ ratios of the remaining (unaltered) samples were plotted on the standard diagram. Lower and upper age limits along with mean age values were considered for each sample. Stratigraphic hiatuses were determined, and their associated time gaps were calculated (see [16] for full methodology). Accordingly, the duration of the paleoexposure event at the CT-ES was calculated as 0.53 Myr for Well-A and 0.39 Myr for Well-D. Comparatively longer time gaps were estimated for the mT-ES as 2.7 Myr in Well-A and 4.5 Myr in Well-D (Figure 11).

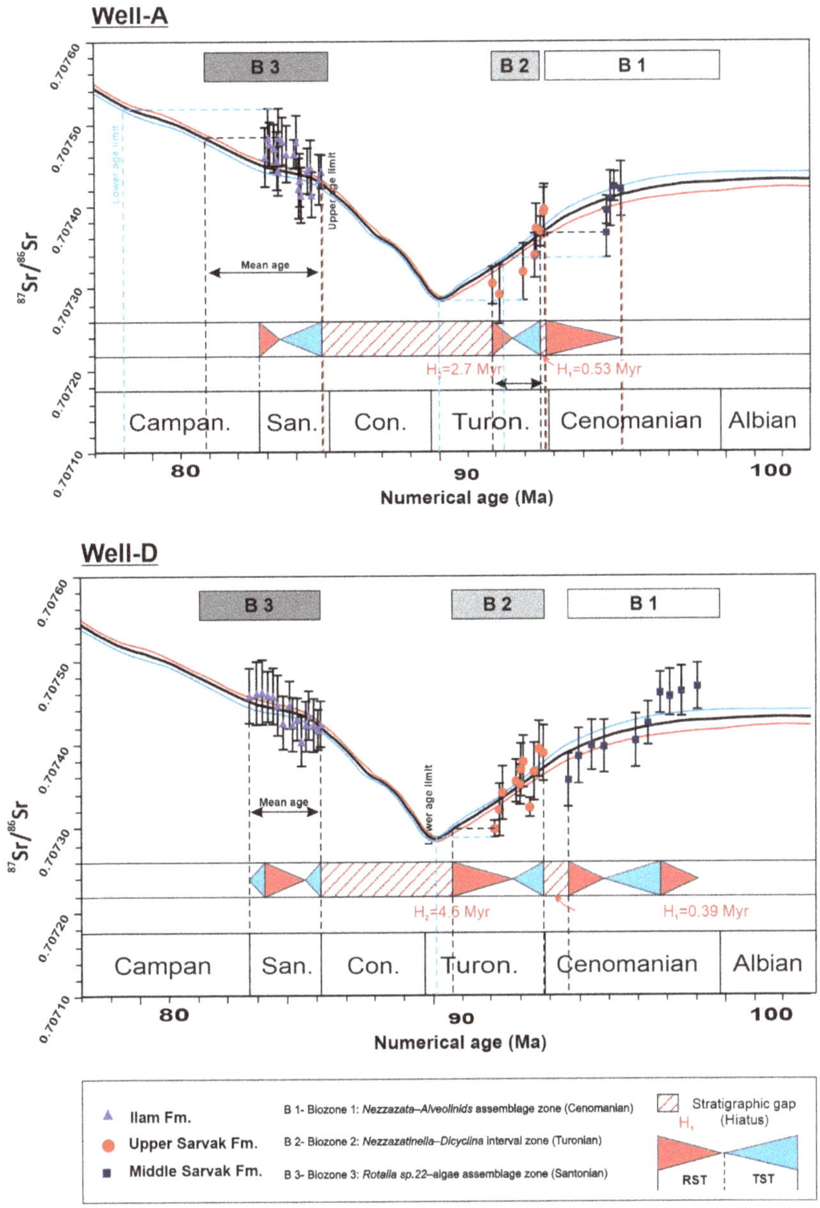

Figure 11. Plots of $^{87}Sr/^{86}Sr$ ratios of selected samples for absolute dating of the Sarvak and Ilam formations on the standard curve of Albian–Campanian [37]. Diagrams are plotted for the Abteymour (Well-A) and Rag-e-Sefid (Well-D) fields. Third-order sequences and biozones are also shown along with the calculated hiatus durations at the Cenomanian–Turonian (CT-ES) and middle Turonian (mT-ES) paleoexposure surfaces.

5. Discussion

5.1. Depositional Model

The results of the microfacies analysis of the Sarvak Fm. in 13 hydrocarbon fields show that it is mainly composed of eight facies types (see Table 2 and Section 4.1). Facies characteristics and their correlation with standard models (e.g., [36]) and previous studies indicated that they can be classified into four main facies associations, including inner ramp, middle ramp, outer ramp, and basin. Furthermore, inner ramp facies can be subdivided into three categories as lagoon (F1: benthic foraminifera mudstone, wackestone), patch reef taluses (F2: rudist-algal-coral bafflestone, rudstone, floatstone), and shoal (F3: bioclast, peloid, intraclast packstone, grainstone). Middle ramp facies include mud- to grain-supported facies with dual textural characteristics and fossil content.

In these facies, a mixture of neritic and pelagic biota was recorded (F4: foraminifera, bioclast wackestone, packstone). Extremely abraded and cross-bedded grain-dominated facies that are mainly composed of echinoderm debris and planktic foraminifers (F5: micro-bioclast packstone, grainstone) are attributed to the distal mid-ramp setting, around the storm wave base (SWB), as tempestites. Mud-dominated facies containing oligosteginids (F6) and planktic foraminifers (F7) are interpreted as outer ramp facies. Basinal facies include calcisiltite with frequent siliceous sponge spicules (F8).

Accordingly, the absence of continuous reefs and evidence of remarkable differences in depositional depth (e.g., reworked sediments, debris flows, or turbidites) all indicate that the Sarvak Fm. was deposited in a ramp-like carbonate platform in most parts of the Zagros Basin and the Persian Gulf. A similar model was also proposed previously for this formation in the other parts of the Zagros Basin, such as the Izeh Zone [43,44] and the Lorestan Zone [45]. In some localities (e.g., Lorestan and southern parts of the Dezful Embayment), the development of intra-platform basins and seaways provided a general open (non-rimmed) shelf configuration. Recently, Moeini et al. [46] proposed an isolated platform model for the Cenomanian sequences in the Lorestan Zone.

The conceptual depositional model, frequencies of facies associations, and paleogeographic setting of the Sarvak Fm. are presented in Figure 12. As shown, during the Cenomanian–Turonian, the shallow-marine carbonate platforms were developed on the northeastern margin of the Arabian Plate [3,28,29]. These platforms were dominantly ramp-like and surrounded some intra-platform basins in various parts of the Middle East [5,10,22,27]. Rudist bioherms (patch reefs) developed in the shallow (inner) parts of these ramps and provide important reservoir units in the Sarvak Fm. in Iran [21,47] and its stratigraphically equivalent strata in Iraq [48,49], UAE [50], Qatar [51], and Oman [52,53].

The results of the frequency analysis show that there are remarkable differences in the frequency of facies associations among the studied fields from different parts of the Zagros Basin (Figure 12B). Accordingly, nearly 50% of the Sarvak Fm. is composed of inner ramp (lagoon, shoal, and patch reef) facies in the Dezful Embayment. In contrast, deep-marine (outer ramp and basin) facies formed 67% of this formation in the Fars Zone (i.e., Sarvestan Field, Well-X5; Figure 12B). In the eastern Persian Gulf, shallow-marine (inner ramp) facies become dominant, with a frequency of 55% (Figure 12B). Similar drastic changes in the frequency of facies associations and thickness of the Sarvak Fm. were previously reported from the Zagros Basin and the Persian Gulf (see [7]). They can be attributed to the effects of tectonic activities that were related to the onset of Neo-Tethys subduction beneath Central Iran and its consequent compressional regime, which reactivated the basement faults and salt diapirs [54–56].

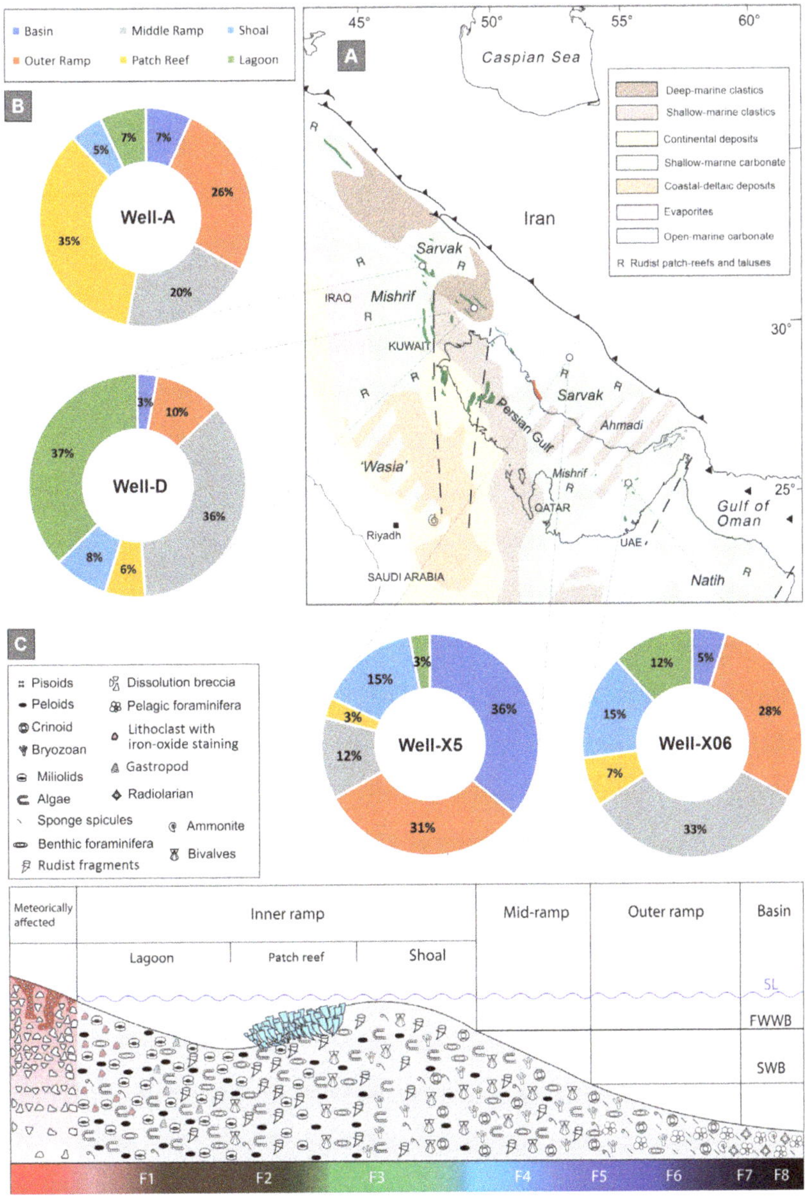

Figure 12. (**A**) Paleogeographic map of the Zagros Basin and Persian Gulf during the Cenomanian (modified after [28]). (**B**) Frequency diagrams of facies associations of the Sarvak Fm. in the Abteymour (Well-A), Rag-e-Sefid (Well-D), Sarvestan (Well-X5), and Sirri (Well-X06) fields. (**C**) Conceptual depositional model of the Sarvak Fm. in the Zagros Basin and the Persian Gulf.

5.2. Diagenetic History

A schematic diagenetic model and paragenetic sequence of the Sarvak Fm. are presented in Figures 13 and 14, respectively. The diagenetic model is specifically provided for rudist-dominated facies of this formation in the studied fields (Figure 13). The paragenetic sequence of diagenetic processes includes marine, meteoric, and shallow- to deep-burial diagenesis (Figure 14).

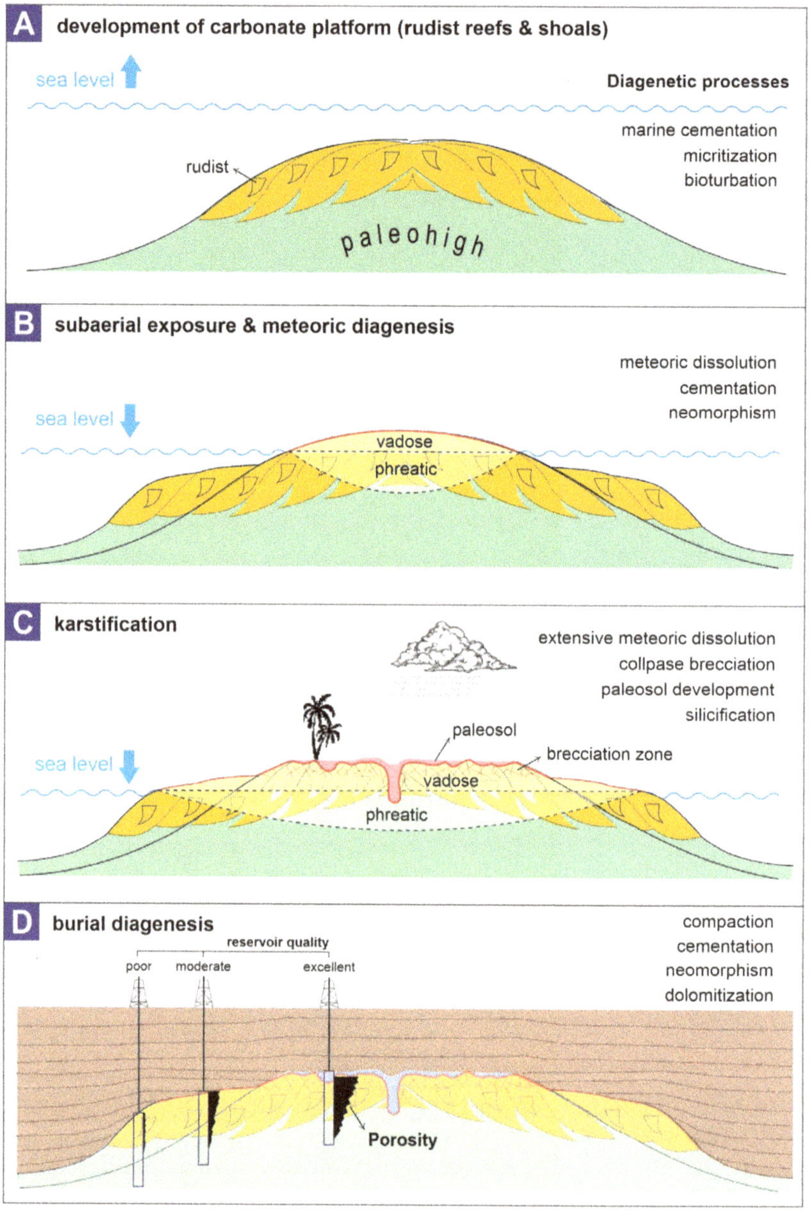

Figure 13. Schematic cartoons illustrating the diagenetic history of the Sarvak Fm. in the Zagros Basin and the Persian Gulf (modified after [47]). This model includes marine (**A**), meteoric (**B**,**C**), and burial (**D**) diagenetic realms and their associated diagenetic processes.

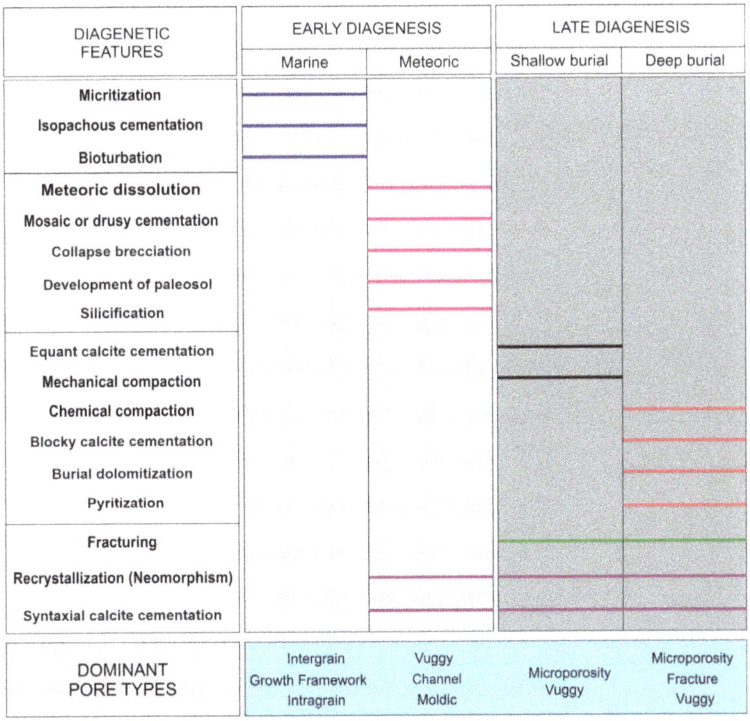

Figure 14. Paragenetic sequence of diagenetic features recorded in the Sarvak Formation.

The diagenetic history of the Sarvak Fm. is highly variable across the Zagros Basin and the Persian Gulf. Depending on the paleo-structural altitude, one, two, and sometimes three disconformable surfaces were formed following the subaerial exposure of carbonates in the mid-Cenomanian (mC-ES), Cenomanian–Turonian boundary (CT-ES), and middle Turonian (mT-ES) [1,57,58]. Consequently, meteoric diagenesis was the most important phase in the diagenetic evolution of the Sarvak Fm. [14,59,60].

Meteoric diagenetic features, including dissolution (karstification), brecciation, cementation, and paleosol formation, had remarkable control over the reservoir properties of this formation throughout the Zagros and Persian Gulf [21,58,61,62].

The number and intensity of meteoric phases were directly controlled by tectonics. On paleohighs (uplifted blocks), three paleoexposure events occurred along with karstified intervals with variable maturities. For example, in the Gachsaran and Rag-e-Sefid fields, the maximum development of karstified profiles is observed. In these fields, thick paleosol units and brecciated intervals point to an over-matured karst profile, especially beneath the mT-ES. In these localities, the diagenetic history of the Sarvak Fm. includes two phases of meteoric diagenesis: (a) eogenetic meteoric diagenesis at the time of exposure and (b) telogenetic meteoric diagenesis in the Cenomanian sequence at the mid-Turonian exposure event [1].

Similar phenomena were also recorded in the Hendijan and Bahregansar fields (NW Persian Gulf). However, in the Fars Zone (Sarvestan, Saadat Abad) and eastern Persian Gulf (Sirri Field), immature (young) karsts occur below the disconformable surfaces, which are associated with preserved dissolution features such as vugs, caves, and channels, especially beneath the CT-ES. Moreover, minor effects of marine diagenesis and intensive burial diagenetic features are distinguished within the Sarvak Fm. (Figure 14). Isopachous cementation, micritization, and bioturbation are marine diagenetic processes.

Burial diagenetic processes include blocky calcite cementation, mechanical and chemical compaction, stylolite-related dolomitization, pyritization, and fracturing. The mineral stabilization and cementation of meteorically affected carbonates have resulted in lesser effects of burial diagenesis on karstified intervals of the Sarvak Fm. Our previous studies proved that the dolomitized units of the Sarvak Fm. provide some productive (reservoir) zones of this formation in the Dezful Embayment, Abadan Plain, and the Persian Gulf [8,9,20]. This diagenetic process improved the reservoir quality of the Sarvak Fm. by increasing the permeability. The possible sources of Mg are assumed to be the dissolution of skeletal grains with high-Mg calcite (HMC) mineralogy and Mg^{2+} released during the transformation of clay minerals [20].

The diagenetic history of the Sarvak Fm. is the subject of several studies in south and southwest Iran [14,15,58,61,63] and other places in the Middle East [64]. In all of these studies, relatively similar diagenetic models are discussed, and meteoric diagenesis has been considered as the main phase of diagenesis in Cenomanian–Turonian carbonate rocks.

Paleoexposure-related diagenetic processes were the main controlling factors on reservoir properties of the Sarvak Fm. Extensive dissolution (karstification) formed high amounts (up to 35%) of vuggy, moldic, channel, and enlarged fracture pores within the neritic facies of this formation. The high-energy shoal and rudist-dominated facies of reef talus environments were potentially high-quality reservoir facies. They contain high amounts of primary pores such as interparticle, intraskeletal, and growth framework pores (Figure 14). This primary good reservoir potential was multiplied through meteoric dissolution.

Considering their control over reservoir properties, the diagenetic processes of the Sarvak Fm. can be classified into three main classes: (a) diagenetic processes with positive effects on reservoir quality, including dissolution, dolomitization, and fracturing; (b) diagenetic processes with negative effects on reservoir properties, such as compaction, brecciation, and cementation; and (c) diagenetic processes with minor effects on reservoir quality, including micritization, recrystallization, bioturbation, silicification, and pyritization.

5.3. Geochemistry

The complex depositional–diagenetic evolution of the Sarvak Fm. has resulted in remarkable changes in elemental and isotopic compositions and conspicuous geochemical trends within this formation across the study area (see Section 4.4). Most of these geochemical trends are related to the paleoexposure surfaces and their associated meteoric diagenetic alteration. Therefore, they provide very useful tools for delineating the disconformities that are considered as important sequence boundaries for the sequence stratigraphic analysis of the Sarvak Fm.

5.3.1. Trace Elements

Beneath the paleoexposure surfaces, an increase in Mg, Fe, and Rb concentrations and a decrease in Sr content all point to a major phase of meteoric diagenesis. Such variations in elemental concentrations are observed in the karstified intervals below the CT-ES and mT-ES (Figure 7). The amplitude of elemental changes is higher below the mT-ES (Figure 7). The meteoric waters contain lower Sr and higher Mn and Fe concentrations compared to the marine carbonates [42]. As a result, it is routinely expected that meteorically altered carbonates show a decrease in Sr and an increase in Mn and Fe contents [42]. The subaerial exposure of carbonates usually increases the aluminosilicate detritus input (i.e., clay minerals) and, consequently, increases the Rb content of altered carbonates [37,65,66]. Afterwards, this Rb content can alter the $^{87}Sr/^{86}Sr$ signature of the host rocks due to in situ decay [16,18,19,38,66,67]. Therefore, the lowest Sr/Rb ratios are measured from the weathered intervals below the paleoexposure surfaces.

The cross plots of Sr/Ca vs. Mn and $\delta^{18}O_{carb}$ vs. Mn also indicate an extensive effect of meteoric waters on the carbonates of the Sarvak Fm. in an open diagenetic system (Figure 8). The decrease in the Sr/Ca ratios and $\delta^{18}O_{carb}$ along with the increase in the

Mn concentration of the analyzed samples all point to a major meteoric diagenesis under the high water-to-rock ratio condition [42]. These samples are not suitable for absolute age calculation using the $^{87}Sr/^{86}Sr$ ratios. Samples taken far from the paleoexposure surfaces show lesser effects of meteoric diagenesis, which are reflected in their higher Sr/Ca ratios and $\delta^{18}O_{carb}$ values and lower Mn concentrations. These samples belong to the transgressive system tracts of the Sv_Tur and Sv_Cen sequences of the Sarvak Fm. and the total interval of the Ilam Fm. (IL_San sequence).

5.3.2. Carbon and Oxygen Isotopes

The $\delta^{13}C_{carb}$ and $\delta^{18}O_{carb}$ values provide useful tools for evaluating the diagenetic impact on carbonate rocks of the Sarvak Fm. Decreased values of $\delta^{13}C_{carb}$ and $\delta^{18}O_{carb}$ are recorded from the RSTs of the Cenomanian and Turonian sequences, below the CT-ES and mT-ES (see Figure 9). Such low values point to a severe meteoric diagenesis, including the dissolution of calcium carbonate and its reprecipitation in the form of LMC (low-magnesium calcite) cements containing ^{13}C and ^{18}O-depleted CO_3 [68–70]. The sharp inverted-J patterns on the $\delta^{13}C_{carb}$ vs. $\delta^{18}O_{carb}$ cross plots provide another strong clue for the meteoric diagenesis effect on carbonates of the Sarvak Fm. [71,72]. The samples with decreased values of $\delta^{13}C_{carb}$ and $\delta^{18}O_{carb}$ stratigraphically belong to the uppermost parts (RSTs) of the Cenomanian (Sv_Cen) and Turonian (Sv_Tur) sequences of the Sarvak Fm. (Figures 7 and 9). These meteorically altered samples mostly include drusy calcite cements and recrystallized micrite along with samples taken from paleosols and brecciated intervals, which were not appropriate for absolute dating using their $^{87}Sr/^{86}Sr$ ratios (Figure 15). Relatively uniform values of both $\delta^{13}C_{carb}$ and $\delta^{18}O_{carb}$ in the Ilam Fm. indicate lesser diagenetic effects on the carbonates of this formation. Similar geochemical trends were also reported from the Sarvak and Ilam formation by other researchers from various parts of the Zagros Basin in Iran [13–15,30,43] and other places in the Middle East [3,48].

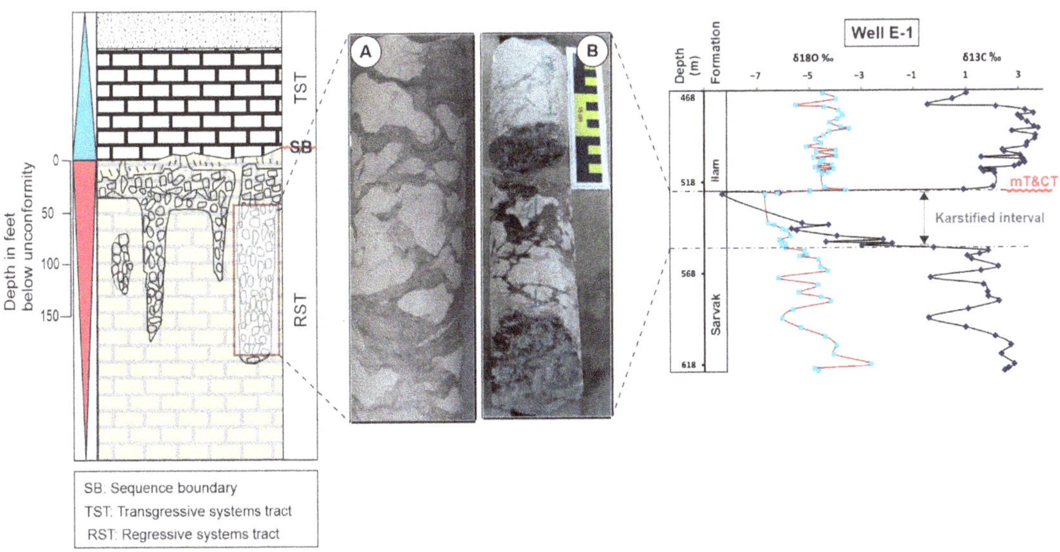

Figure 15. Stable isotopic signature of karstified and brecciated intervals of the Sarvak Fm. in the Gachsaran Field (Well-E1). As shown, a pronounced decrease in the $\delta^{13}C_{carb}$ and $\delta^{18}O_{carb}$ values provides a useful tool for delineating the paleoexposure surfaces and karstified units.

5.3.3. $^{87}Sr/^{86}Sr$ Ratios

The $^{87}Sr/^{86}Sr$ ratios of carbonates are used for (a) delineating the precise stratigraphic position of disconformable (paleoexposure) surfaces, (b) the absolute dating of studied sequences, (c) calculating the length of the time gap (hiatus duration) along the paleoexposures, and (d) the chemostratigraphic correlation and sequence stratigraphic analysis of Cenomanian–Santonian rocks throughout the Zagros Basin and the Persian Gulf.

An increase in the $^{87}Sr/^{86}Sr$ ratios may also be caused by the admixture of isotopically heavy strontium derived from the weathering of continental deposits [66]. A sharp increase in the $^{87}Sr/^{86}Sr$ ratios of carbonates just below the disconformities makes it easy to delineate the precise stratigraphic positions of CT-ES and mT-ES in the Sarvak Fm. (Figure 10). As mentioned, the in situ decay of Rb was responsible for the increase in the $^{87}Sr/^{86}Sr$ ratios of meteorically altered carbonates [66,67]. In all of the studied wells, such sharp spikes in the $^{87}Sr/^{86}Sr$ logs of the studied sequences are visible, which provided useful tools for stratigraphic correlation.

Absolute dating of the studied sequences revealed that the exposure duration at the CT-ES was generally lower than 1 Myr. This short-term exposure resulted in the development of dissolution features such as vugs and channels (immature/young karst) in the Cenomanian sequence (Sv_Cen) of the Sarvak Fm. This karstified interval, which commonly formed within the rudist-dominated facies of the Sarvak Fm., provides the best reservoir unit (productive zone) of this formation throughout the Zagros Basin [8,9,20,58]. On the other hand, the exposure duration was calculated as 2–7 Myr for mT-ES, which is much longer than the CT-ES. Such prolonged subaerial exposure of carbonates under the warm and humid (tropical) paleoclimatic conditions of the Arabian Plate during the Late Cretaceous resulted in the formation of an over-matured (senile) karst profile. Karst over-maturation is marked by the collapse and brecciation of dissolved carbonates and the development of paleosols [73]. Such over-matured karst profiles are typically recorded in oilfields located on the structural paleohighs such as the Hendijan, Bahregansar, Gachsaran, and Rag-e-Sefid fields in the Dezful Embayment and NW Persian Gulf. In the Fars Zone and eastern Persian Gulf, salt tectonics had the main control over the development and diagenetic evolution of Late Cretaceous platforms.

6. Conclusions

The integrated sedimentological and geochemical analysis of the Sarvak Fm. and lower parts of the Ilam Fm. in 13 subsurface sections from various parts of the Zagros Basin and the Persian Gulf has resulted in the following conclusions:

1. Microfacies analysis and palaeoenvironmental reconstruction indicate that the Sarvak Fm. is composed of eight facies types representing a general ramp-like depositional model. Rudists were the main carbonate producers in this platform. Tectonic and eustatic factors were responsible for highly variable depositional settings. Accordingly, the large variations in spatial facies distribution and thickness of this formation were associated across the studied sections. Close to the intra-platform basins, the general depositional model can be considered as an open (non-rimmed) shelf, where abrupt changes from inner to outer platform facies are associated with frequent debrites and tempestites (e.g., in the Fars Zone, close to the platform edge).
2. Paleo-structural features such as the paleohighs, troughs, intra-platform basins, and seaways had major control over facies characteristics and the diagenetic evolution of the Sarvak Fm. The maximum development of shallow-marine facies (lagoon, patch reefs, and shoals) is visible in some areas of the Dezful Embayment (e.g., Gachsaran and Rag-e-Sefid fields) and the NW Persian Gulf (Hendijan and Bahregansar fields). In these localities, well-known paleohighs were formed by the reactivated basement faults, during and after the deposition of the Sarvak Fm. Far from these paleohighs, the depositional–diagenetic evolution of this formation was completely different.
3. Diagenesis had a major impact on the carbonates of the Sarvak Fm. Among the marine, meteoric, and shallow- to deep-burial diagenetic realms, the main diagenetic phases occurred at the end of the Cenomanian and Turonian in rocks of this formation. Consequently, two karstified intervals are commonly distinguishable in the Sarvak Fm. within the RSTs of the Cenomanian and Turonian third-order sequences. Intensive

meteoric dissolution, cementation, brecciation, silicification, and development of paleosols are major paleoexposure-related diagenetic alterations in this formation. The frequency, type, and intensity of these alterations are remarkably different throughout the Zagros Basin and the Persian Gulf.

4. Conspicuous geochemical trends are observed in the studied sections, which were related to the disconformities. An increase in Mg, Fe, Mn, and Rb and a decrease in Ca and Sr together with decreased $\delta^{13}C_{carb}$ and $\delta^{18}O_{carb}$ values and sharp increases in the $^{87}Sr/^{86}Sr$ ratios are recorded from the meteorically altered intervals below the paleoexposure surfaces. Sr-isotope dating of the Sarvak Fm. provides a reliable chemostratigraphic framework for this formation across the study area.
5. Short-term subaerial exposures are calculated for the CT-ES with a duration of <1 Myr. In contrast, exposure duration was much higher for the mT-ES (2 to 7 Myr). Such remarkable differences in exposure durations have resulted in two different scenarios of karst formation and evolution below these disconformities. A short-term subaerial exposure is inferred for the Cenomanian–Turonian boundary event (CT-ES), which resulted in the development of dissolution features such as vugs and channels and their preservation (i.e., immature stage of karst). Below this disconformity, the reservoir properties of the Sarvak Fm. were improved, and the best productive zones of this formation were formed within the dissolved rudist-dominated facies. However, long-term exposure at mT-ES resulted in the collapse and brecciation of the karstified strata (senile stage) and, consequently, lower reservoir quality. Such differences are also reflected by geochemical trends and had major control over the reservoir properties of the Sarvak Fm. in its different depositional sequences.
6. There is a close correspondence between the development (spatial and temporal) of rudistid facies, the intensity of karstification, and, finally, the reservoir quality of the Sarvak Fm. Consequently, thicker rudistid zones commonly show higher degrees of meteoric dissolution (diagenesis) and better reservoir quality. The interplay between the depositional facies characteristics and diagenetic alteration resulted in the development of the second most important hydrocarbon reservoir in Iran (i.e., the Sarvak Fm.) in the Zagros Basin and the Persian Gulf.

Funding: This research received no external funding.

Data Availability Statement: Data will be available on request.

Acknowledgments: The University of Tehran provided the facilities for this research, for which the author is grateful. The National Iranian Oil Company is thanked for data support.

Conflicts of Interest: The author declares no conflict of interest.

References

1. Rahimpour-Bonab, H.; Mehrabi, H.; Enayati-Bidgoli, A.H.; Omidvar, M. Coupled imprints of tropical climate and recurring emergence on reservoir evolution of a mid-Cretaceous carbonate ramp, Zagros Basin, southwest Iran. *Cretac. Res.* **2012**, *37*, 15–34.
2. Rahimpour-Bonab, H.; Mehrabi, H.; Navidtalab, A.; Omidvar, M.; Enayati-Bidgoli, A.H.; Sonei, R.; Sajjadi, F.; Amiri-Bakhtyar, H.; Arzani, N.; Izadi-Mazidi, E. Palaeo-exposure surfaces in Cenomanian-Santonian carbonate reservoirs in the Dezful Embayment, SW Iran. *J. Pet. Geol.* **2013**, *36*, 335–362. [CrossRef]
3. Bromhead, A.D.; van Buchem, F.S.P.; Simmons, M.D.; Davies, R.B. Sequence stratigraphy, palaeogeography and petroleum plays of the Cenomanian–Turonian succession of the Arabian Plate: An updated synthesis. *J. Pet. Geol.* **2022**, *45*, 119–161. [CrossRef]
4. Ghabeishavi, A.; Vaziri-Moghaddam, H.; Taheri, A.; Taati, F. Microfacies and depositional environment of the Cenomanian of the Bangestan anticline, SW Iran. *J. Asian Earth Sci.* **2010**, *37*, 275–285. [CrossRef]
5. Mehrabi, H.; Rahimpour-Bonab, H. Paleoclimate and tectonic controls on the depositional and diagenetic history of the Cenomanian-early Turonian carbonate reservoirs, Dezful Embayment, SW Iran. *Facies* **2014**, *60*, 147–167. [CrossRef]
6. Mehrabi, H.; Rahimpour-Bonab, H.; Enayati-Bidgoli, A.H.; Navidtalab, A. Depositional environment and sequence stratigraphy of the Upper Cretaceous Ilam Formation in central and southern parts of the Dezful Embayment, SW Iran. *Carbonates Evaporites* **2014**, *29*, 263–278. [CrossRef]
7. Mehrabi, H.; Rahimpour-Bonab, H.; Hajikazemi, E.; Jamalian, A. Controls on depositional facies in upper cretaceous carbonate reservoirs in the Zagros area and the Persian Gulf, Iran. *Facies* **2015**, *61*, 23. [CrossRef]

8. Mehrabi, H.; Rahimpour-Bonab, H.; Enayati-Bidgoli, A.H.; Esrafili-Dizaji, B. Impact of contrasting paleoclimate on carbonate reservoir architecture: Cases from arid Permo-Triassic and humid Cretaceous platforms in the south and southwestern Iran. *J. Pet. Sci. Eng.* **2015**, *126*, 262–283. [CrossRef]
9. Mehrabi, H.; Bagherpour, B.; Honarmand, J. Reservoir quality and micrite textures of microporous intervals in the Upper Cretaceous successions in the Zagros area, SW Iran. *J. Pet. Sci. Eng.* **2020**, *192*, 107292. [CrossRef]
10. Mohseni, H.; Zeybaram Javanmard, R. New data on sequence stratigraphy of the Sarvak Formation in Malekshahi city, (Ilam province) Zagros basin, Iran. *Mar. Pet. Geol.* **2020**, *112*, 104035. [CrossRef]
11. Omidvar, M.; Mehrabi, H.; Sajjadi, F.; Bahramizadeh-Sajjadi, H.; Rahimpour-Bonab, H.; Ashrafzadeh, A. Revision of the foraminiferal biozonation scheme in Upper Cretaceous carbonates of the Dezful Embayment, Zagros, Iran: Integrated palaeontological, sedimentological and geochemical investigation. *Rev. Micropaléontol.* **2014**, *57*, 97–116. [CrossRef]
12. Bagherpour, B.; Mehrabi, H.; Faghih, A.; Vaziri-Moghaddam, H.; Omidvar, M. Tectono-eustatic controls on depositional setting and spatial facies distribution of Coniacian–Santonian sequences of the Zagros Basin in Fars area, S. *Iran. Mar. Pet. Geol.* **2021**, *129*, 105072. [CrossRef]
13. Bagherpour, B.; Nkemata, A.F.; Vaziri–Moghaddam, H.; Blattmann, T.M.; Mohtadi, M. Paleoenvironmental significance of the carbon isotope record across the Cenomanian–Turonian transition and the Oceanic Anoxic Event 2 (OAE2) in the southeastern Neotethys, Zagros, Iran. *Cretac. Res.* **2023**, *150*, 105574. [CrossRef]
14. Hajikazemi, E.; Al-Aasm, I.S.; Coniglio, M. Subaerial exposure and meteoric diagenesis of the Cenomanian–Turonian upper Sarvak Formation, southwestern Iran. *Geol. Soc. Lond. Spec. Publ.* **2010**, *330*, 253–272. [CrossRef]
15. Hajikazemi, E.; AL-Aasm, I.S.; Coniglio, M. Chemostratigraphy of Cenomanian–Turonian carbonates of the Sarvak Formation, Southern Iran. *J. Pet. Geol.* **2012**, *35*, 187–205. [CrossRef]
16. Navidtalab, A.; Rahimpour-Bonab, H.; Huck, S.; Heimhofer, U. Elemental geochemistry and strontium-isotope stratigraphy of Cenomanian to Santonian neritic carbonates in the Zagros Basin, Iran. *Sediment. Geol.* **2016**, *346*, 35–48. [CrossRef]
17. Navidtalab, A.; Heimhofer, U.; Huck, S.; Omidvar, M.; Rahimpour-Bonab, H.; Aharipour, R.; Shakeri, A. Biochemostratigraphy of an upper Albian–Turonian succession from the southeastern Neo-Tethys margin, SW Iran. *Palaeogeogr. Palaeoclimatol. Palaeoecol.* **2019**, *533*, 109255. [CrossRef]
18. Mehrabi, H.; Navidtalab, A.; Enayati, A.; Bagherpour, B. Age, duration, and geochemical signatures of paleo-exposure events in Cenomanian–Santonian sequences (Sarvak and Ilam formations) in SW Iran: Insights from carbon and strontium isotopes chemostratigraphy. *Sediment. Geol.* **2022**, *434*, 106136. [CrossRef]
19. Mehrabi, H.; Navidtalab, A.; Rahimpour-Bonab, H.; Heimhofer, U. Geochemical expression of sequence stratigraphic surfaces: A case from Upper Cretaceous shallow-water carbonates of southeastern Neo-Tethys margin, SW Iran. *Cretac. Res.* **2022**, *140*, 105329. [CrossRef]
20. Rahimpour-Bonab, H.; Mehrabi, H.; Navidtalab, A.; Izadi-Mazidi, E. Flow unit distribution and reservoir modelling in Cretaceous carbonates of the Sarvak Formation, Abteymour Oilfield, Dezful Embayment, SW Iran. *J. Pet. Geol.* **2012**, *35*, 213–236. [CrossRef]
21. Mehrabi, H.; Bagherpour, B. Scale, origin, and predictability of reservoir heterogeneities in shallow-marine carbonate sequences: A case from Cretaceous of Zagros, Iran. *J. Pet. Sci. Eng.* **2022**, *214*, 110571. [CrossRef]
22. Foroshani, J.S.; Mehrabi, H.; Rahimpour-Bonab, H. Reservoir heterogeneity of Upper Cretaceous Sarvak Formation in the Dezful Embayment, SW Iran: Implications of flow unit distribution, electrofacies analysis and geological-based reservoir zonation. *J. Afr. Earth Sci.* **2023**, *200*, 104882. [CrossRef]
23. Alavi, M. Structures of the Zagros fold-thrust belt in Iran. *Am. J. Sci.* **2007**, *307*, 1064–1095. [CrossRef]
24. Moghadam, H.S.; Stern, R.J.; Griffin, W.; Khedr, M.; Kirchenbaur, M.; Ottley, C.; Whattam, S.; Kimura, J.-I.; Ghorbani, G.; Gain, S. Subduction initiation and back-arc opening north of Neo-Tethys: Evidence from the Late Cretaceous Torbat-e-Heydarieh ophiolite of NE Iran. *GSA Bull.* **2020**, *132*, 1083–1105. [CrossRef]
25. Moghadam, H.S.; Stern, R.J. Subduction initiation causes broad upper plate extension: The Late Cretaceous Iran example. *Lithos* **2021**, *398–399*, 106296. [CrossRef]
26. Ahmadhadi, F.; Lacombe, O.; Daniel, J.-M. Early Reactivation of Basement Faults in Central Zagros (SW Iran): Evidence from Pre-folding Fracture Populations in Asmari Formation and Lower Tertiary Paleogeography. In *Thrust Belts and Foreland Basins*; Lacombe, O., Roure, F., Lavé, J., Vergés, J., Eds.; Frontiers in Earth Sciences; Springer: Berlin/Heidelberg, Germany, 2007. [CrossRef]
27. Sharland, P.R.; Archer, D.M.; Casey, R.B.; Davies, S.H.; Hall, A.P.; Heward, A.D.; Horbury, A.D.; Simmons, M.D. Arabian Plate Sequence Stratigraphy. *GeoArabia Spec. Publ.* **2001**, *2*, 371.
28. Ziegler, M.A. Late Permian to Holocene Paleofacies Evolution of the Arabian Plate and Its Hydrocarbon Occurrences. *GeoArabia* **2001**, *6*, 445–504.
29. Alsharhan, A.S. Petroleum systems in the Middle East. In *Tectonic Evolution of the Oman Mountains*; Rollinson, H.R., Searle, M.P., Abbasi, A.I., Al-Lazki, A.I., Al Kindi, M.H., Eds.; Geological Society, London, Special Publications: London, UK, 2014; Volume 392, pp. 361–408.
30. Asadi Mehmandosti, E.; Asadi, A.; Daneshian, J.; Woods, A.D.; Loyd, S.J. Evidence of Mid-Cretaceous carbon cycle perturbations and OAE2 recorded in Cenomanian to middle Campanian carbonates of the Zagros fold–thrust belt basin, Iran. *J. Asian Earth Sci.* **2021**, *218*, 104863. [CrossRef]

31. Esrafili-Dizaji, B.; Rahimpour-Bonab, H. Carbonate reservoir rocks at giant oil and gas fields in SW Iran and the adjacent offshore: A review of stratigraphic occurrence and poro-perm characteristics. *J. Pet. Geol.* **2019**, *42*, 343–370. [CrossRef]
32. Özkan, R.; Altiner, D. The Cretaceous Mardin Group carbonates in southeast Turkey: Lithostratigraphy, foraminiferal biostratigraphy, microfacies and sequence stratigraphic evolution. *Cretac. Res.* **2019**, *98*, 153–178. [CrossRef]
33. Simmons, M.D.; Miller, K.G.; Ray, D.C.; Davies, A.; van Buchem, F.S.P.; Gr'eselle, B. Phanerozoic Eustasy. In *Geologic Time Scale*; Gradstein, F.M., Ogg, J.G., Schmitz, M.D., Ogg, G.M., Eds.; Elsevier: Amsterdam, The Netherlands, 2020; pp. 357–400.
34. Dunham, R.J. Classification of Carbonate Rocks According to Depositional Texture. In *Classification of Carbonate Rocks*; Ham, W.E., Ed.; AAPG: Tulsa, OK, USA, 1962; pp. 108–121.
35. Embry, A.F.; Klovan, J.E. A late Devonian reef tract on northeastern Banks Island. *Bull. Can. Pet. Geol.* **1971**, *19*, 730–781.
36. Flügel, E. *Microfacies of Carbonate Rocks, Analysis, Interpretation and Application*; Springer: Berlin/Heidelberg, Germany, 2010; 976p.
37. McArthur, J.M. Recent trends in strontium isotope stratigraphy. *Terra Nova* **1994**, *6*, 331–358. [CrossRef]
38. Faure, G.; Mensing, T.M. *Isotopes: Principles and Applications*; John Wiley Sons: New York, NY, USA, 2004; 928p.
39. Scotese, C.R. An Atlas of Phanerozoic Paleogeographic Maps: The seas come in and the seas go out. *Annu. Rev. Earth Planet. Sci.* **2021**, *49*, 679–728. [CrossRef]
40. Ahmadnejad, F.; Mongelli, G. Geochemistry of Upper cretaceous bauxite deposits, Zagros Fold Thrust Belt, SW Iran: Paleoenvironment and provenance constraints. *Sediment. Geol.* **2023**, *454*, 106461. [CrossRef]
41. Wynd, J. *Biofacies of the Iranian Consortium Agreement Area*; No. 1082; Iranian Offshore Oil Company Report: Tehran, Iran, 1965; 213p.
42. Brand, U.; Veizer, J. Chemical diagenesis of a multicomponent carbonate system: 1. Trace elements. *J. Sediment. Petrol.* **1980**, *50*, 1219–1236.
43. Asadi Mehmandosti, E.; Adabi, M.H.; Woods, A.D. Microfacies and geochemistry of the Middle Cretaceous Sarvak Formation in Zagros Basin, Izeh Zone, SW Iran. *Sediment. Geol.* **2013**, *293*, 9–20. [CrossRef]
44. Kalanat, B.; Vaziri-Moghaddam, H.; Bijani, S. Depositional history of the uppermost Albian–Turonian Sarvak Formation in the Izeh Zone (SW Iran). *Int. J. Earth Sci.* **2021**, *110*, 305–330. [CrossRef]
45. Esfandyari, M.; Mohseni, H.; Heidari, M. Facies analysis, depositional sequences and platform evolution of the Sarvak Formation (late Albian-Turonian) in the Zagros Basin, West of Iran. *J. Afr. Earth Sci.* **2023**, *198*, 104811. [CrossRef]
46. Moeini, M.; Rahimpour-Bonab, H.; Tavakoli, V. Evidence for Isolated Platform Development in the Cenomanian on the Passive Margin of Neotethys, Southwest Iran. *Minerals* **2023**, *13*, 757. [CrossRef]
47. Esrafili-Dizaji, B.; Rahimpour-Bonab, H.; Mehrabi, H.; Afshin, S.; Kiani-Harchegani, F.; Shahverdi, N. Characterisation of rudist-dominated units as potential reservoirs in the middle Cretaceous Sarvak Formation, SW Iran. *Facies* **2015**, *61*, 14. [CrossRef]
48. Yu, Y.; Sun, L.; Song, X.; Guo, R.; Gao, X.; Lin, M.; Yi, L.; Han, H.; Li, F.; Liu, H. Sedimentary diagenesis of rudist shoal and its control on reservoirs: A case study of Cretaceous Mishrif Formation, H Oilfield, Iraq. *Pet. Explor. Dev.* **2018**, *45*, 1075–1087. [CrossRef]
49. Wang, H.; Shi, K.; Ma, Y.; Liu, B.; Song, X.; Ge, Y.; Liu, H.; Hoffmann, R.; Immenhauser, A. Control of depositional and diagenetic processes on the reservoir properties of the Mishrif Formation in the AD oilfield, Central Mesopotamian Basin, Iraq. *Mar. Pet. Geol.* **2021**, *132*, 105202. [CrossRef]
50. Al-Zaabi, M.; Taher, A.; Azzam, I.; Witte, J. Geological Overview of the Middle Cretaceous Mishrif Formation in Abu Dhabi. In Proceedings of the International Petroleum Exhibition and Conference, Abu Dhabi, United Arab Emirates, 1–4 November 2010. [CrossRef]
51. Sadooni, F.N. Impact of the demise mechanisms of the Cretaceous rudist buildups in the Arabian Plate on their reservoir characteristics. *Carbonates Evaporites* **2018**, *33*, 465–476. [CrossRef]
52. Philip, J.; Borgomano, J.; Al-Maskiry, S. Cenomanian-Early Turonian carbonate platform of northern Oman: Stratigraphy and palaeo-environments. *Palaeogeogr. Palaeoclimatol. Palaeoecol.* **1995**, *119*, 77–92. [CrossRef]
53. Droste, H.J. *Stratigraphic Framework of the Natih Formation in the Sultanate of Oman*; European Association of Geoscientists & Engineers: Utrecht, The Netherlands, 2010. [CrossRef]
54. Piryaei, A.; Reijmer, J.J.G.; Van Buchem, F.S.P.; Yazdi-Moghadam, M.; Sadouni, J.; Danelian, T. The influence of Late Cretaceous tectonic processes on sedimentation patterns along the northeastern Arabian plate margin (Fars Province, SW Iran). *Geol. Soc. Lond. Spec. Publ.* **2010**, *330*, 211–251. [CrossRef]
55. Jahani, S.; Hassanpour, J.; Mohammadi-Firouz, S.; Letouzey, J.; Frizon de Lamotte, D.; Alavi, S.A.; Soleimany, B. Salt tectonics and tear faulting in the central part of the Zagros Fold-Thrust Belt, Iran. *Mar. Pet. Geol.* **2017**, *86*, 426–446. [CrossRef]
56. Mohammadrezaei, H.; Alavi, S.A.; Cardozo, N.; Ghassemi, M.R. Deciphering the relationship between basement faulting and two-phase folding in the Hendijan anticline, northwest Persian Gulf, Iran. *Mar. Pet. Geol.* **2020**, *122*, 104626. [CrossRef]
57. Hajikazemi, E.; Al-Aasm, I.S.; Coniglio, M. Diagenetic history and reservoir properties of the Cenomanian-Turonian carbonates in southwestern Iran and the Persian Gulf. *Mar. Pet. Geol.* **2017**, *88*, 845–857. [CrossRef]
58. Mehrabi, H.; Yahyaei, E.; Navidtalab, A.; Rahimpour-Bonab, H.; Abbasi, R.; Omidvar, M.; Assadi, A.; Honarmand, J. Depositional and diagenetic controls on reservoir properties along the shallow-marine carbonates of the Sarvak Formation, Zagros Basin: Petrographic, petrophysical, and geochemical evidence. *Sediment. Geol.* **2023**, *454*, 106457. [CrossRef]

59. Assadi, A.; Honarmand, J.; Moallemi, S.A.; Abdollahie-Fard, I. Impacts of Depositional Facies and Diagenesis on Reservoir Quality: A Case Study from the Rudist-bearing Sarvak Formation, Abadan Plain, SW Iran. *Acta Geol. Sin. Engl. Ed.* **2023**, *97*, 190–206. [CrossRef]
60. Mehdipour, V.; Rabbani, A.R.; Kadkhodaie, A. Geological modeling of diagenetic logs of the Sarvak reservoir in Dezful Embayment, southwestern Iran: Implications for geostatistical simulation and reservoir quality assessment. *J. Pet. Explor. Prod. Technol.* **2023**, *13*, 2083–2107. [CrossRef]
61. Taghavi, A.A.; Mørk, A.; Emadi, M.A. Sequence stratigraphically controlled diagenesis governs reservoir quality in the carbonate Dehluran Field, southwest Iran. *Pet. Geosci.* **2006**, *12*, 115–126. [CrossRef]
62. Malekzadeh, H.; Daraei, M.; Bayet-Goll, A. Field-scale reservoir zonation of the Albian–Turonian Sarvak Formation within the regional-scale geologic framework: A case from the Dezful Embayment, SW Iran. *Mar. Pet. Geol.* **2020**, *121*, 104586. [CrossRef]
63. Sabouhi, M.; Moussavi-Harami, R.; Kadkhodaie, A.; Rezaee, P.; Jalali, M. A qualitative-quantitative approach for studying the impact of facies and diagenesis control on the rudist biostrome of the Sarvak formation, Abadan plain, SW Iran. *J. Pet. Sci. Eng.* **2022**, *212*, 110245. [CrossRef]
64. Cantrell, D.L.; Shah, R.A.; Ou, J.; Xu, C.; Phillips, C.; Li, X.L.; Hu, T.M. Depositional and diagenetic controls on reservoir quality: Example from the upper Cretaceous Mishrif Formation of Iraq. *Mar. Pet. Geol.* **2020**, *118*, 104415. [CrossRef]
65. Brand, U. Strontium isotope diagenesis of biogenic aragonite and low-Mg calcite. *Geochim. Cosmochim. Acta* **1991**, *55*, 505–513. [CrossRef]
66. McArthur, J.M.; Howarth, R.J.; Shields, G.A. Strontium isotope stratigraphy. In *The Geologic Time Scale*; Gradstein, F., Ogg, J., Schmitz, M., Ogg, G., Eds.; Elsevier: Amsterdam, The Netherlands, 2012; pp. 127–144.
67. McArthur, J.M.; Howarth, R.J. Strontium isotope stratigraphy. In *A Geological Time Scale*; Gradstein, F., Ogg, J., Smith, A., Eds.; Cambridge University Press: Cambridge, UK, 2004; pp. 96–105.
68. Allan, J.R.; Matthews, R.K. Isotope signatures associated with early meteoric diagenesis. *Sedimentology* **1982**, *29*, 797–817. [CrossRef]
69. Swart, P.K.; Kennedy, M. Does the global stratigraphic reproducibility of δ13C in Neoproterozoic carbonates require a marine origin? A Pliocene–Pleistocene comparison. *Geology* **2012**, *40*, 87–90. [CrossRef]
70. Dyer, B.; Maloof, A.C.; Higgins, J.A. Glacioeustasy, meteoric diagenesis, and the carbon cycle during the Middle Carboniferous. *Geochem. Geophys. Geosyst.* **2015**, *16*, 3383–3399. [CrossRef]
71. Meyers, W.J.; Lohmann, K.C. Isotope geochemistry of regionally extensive calcite cement zones and marine components in Mississippian limestones, New Mexico. In *Society of Economists Paleontologists and Mineralogists Special Publications*; Schneidennann, N., Harris, P.M., Eds.; Carbonate Cements; Society of Economists Paleontologists and Mineralogists: Tulsa, OK, USA, 1985; Volume 36, pp. 223–239.
72. Lohmann, K.C. Geochemical Patterns of Meteoric Diagenetic Systems and Their Application to Studies of Paleokarst. In *Paleokarst*; James, N.P., Choquette, P.W., Eds.; Springer: New York, NY, USA, 1988; pp. 58–80. [CrossRef]
73. Fritz, R.D.; Wilson, J.L.; Yurewicz, D.A. *Paleokarst Related Hydrocarbon Reservoirs*; SEPM (Society for Sedimentary Geology): Tulsa, OK, USA, 1993. [CrossRef]

minerals MDPI

Article

Sediment Texture, Geochemical Variation, and Ecological Risk Assessment of Major Elements and Trace Metals in the Sediments of the Northeast Persian Gulf

Alireza Vaezi and Razyeh Lak *

Research Institute for Earth Sciences, Geological Survey and Mineral Exploration of Iran, Tehran 1387835841, Iran; al.vaezi@yahoo.com
* Correspondence: dr.lak@gsi.ir

Abstract: This study presents a comprehensive analysis of sediment texture along with geochemical variation, potential sources, and ecological risk assessment of major elements and trace metals in the bottom sediments of Iranian territorial waters situated in the northeast Persian Gulf. To achieve this, diverse laboratory methods, GIS techniques, statistical analyses, and index analysis approaches were employed. Over 50% of the sediment particles are mud, while one-third are sand-sized particles, primarily composed of skeletal fragments. The sediment's elements concentrations were ranked in descending order as Ca > Al > Fe > Sr > Mn > Ba > Cr > Ni > V > Zn > Cu > Pb > As > Co > Mo > Cd. Cluster analysis revealed strong correlations among Al-Ni-Cr-V, Cd-Cu-Zn-Pb, Ba-Fe, silt-clay, and Ca-Sr. Calcium and strontium showed extremely severe enrichment due to high content of carbonate matter. Arsenic and Mo were significantly enriched, while Ba, V, Co, Zn, and Cu demonstrated moderate enrichment. Nevertheless, all the sampling stations were classified as having zero to very low levels of contamination, indicating a low potential ecological risk. Arsenic emerged as the primary contributor to the ecological risk index. Notably, no strong correlation was found between As, Mo, and other elements, indicating that As and Mo likely originate from distinct sources.

Keywords: carbonate sediments; heavy metals; Persian Gulf; pollution; sediment

Citation: Vaezi, A.; Lak, R. Sediment Texture, Geochemical Variation, and Ecological Risk Assessment of Major Elements and Trace Metals in the Sediments of the Northeast Persian Gulf. *Minerals* **2023**, *13*, 850. https://doi.org/10.3390/min13070850

Academic Editor: Maria Economou-Eliopoulos

Received: 28 May 2023
Revised: 14 June 2023
Accepted: 19 June 2023
Published: 23 June 2023

1. Introduction

The Persian Gulf is a foreland basin and a marginal sea known for its rich biodiversity and fishery resources, despite its extreme salinity and high temperature [1]. The ecological importance of this semi-enclosed and shallow basin has made it one of the most remarkable marine ecosystems in the world [2]. However, the region is subjected to severe threats from pollutants, including potentially toxic elements (PTEs), due to having about two-thirds of the world's oil sources [3–5].

The northeast Persian Gulf is exposed to diverse sources of pollution by PTEs, which pose risks to the health of both humans and ecosystems [6]. Due to limited circulation, shallow depth, and high temperature and salinity [7,8], regional environmental and ecological contamination can be considerable. Industrial activities such as oil and gas production, petrochemical plants, power plants, and shipyards are among the primary sources of PTEs in this region [9,10]. These industries may release PTEs such as Cd, Cr, Cu, Ni, Pb, V, and Zn into surrounding water bodies via direct effluent discharge or atmospheric deposition [11]. Furthermore, crude oil and natural gas deposits may contain PTEs such as Pb, Cd, and Cu, which can be mobilized during drilling, transportation, and refining processes [12]. The discharge of wastewater from petroleum production facilities can also contribute to PTEs pollution in nearby marine environments [13]. Agriculture runoff from neighboring farms and untreated sewage discharge from populated areas are additional contributors to PTEs pollution in the region [14]. In addition, natural geological processes, including erosion and weathering of rock formations, can lead to the release of PTEs into nearby water bodies [15].

PTEs are contaminants found in both terrestrial and aquatic environments [16–18] that have gained widespread attention due to their toxic, persistent, and bioaccumulative nature in ecosystems [19–21]. These contaminants can enter the environment from natural and anthropogenic sources [12,22] and can be adsorbed into particles or re-mineralized via various biogeochemical cycles [20,23]. PTEs accumulate in the bottom sediments of aquatic environments and can disrupt the balance of aquatic ecosystems [24,25]. Therefore, aquatic sediments have been recognized as sensitive indicators for monitoring environmental contamination [23,26,27].

Studies on the geochemical variation and ecological risk assessment of major elements and trace metals have primarily been confined to the northwestern region of the Persian Gulf [2,28–32]. These studies are limited in their scope due to their reliance on a small number of sampling stations and a restricted range of elements [33]. As such, they offer only a narrow understanding of the spatial distribution, enrichment, and accumulation of major elements and trace metals in addition to pollution levels [33,34]. Moreover, these studies fail to provide an adequate overview of the northeastern segment of this valuable water body.

This study investigates the distribution, ecological risk assessment, and potential sources of major elements and trace metals in conjunction with sediment texture and particle size analysis in the bottom sediments of Iranian territorial waters. Specifically, this study analyzed 140 sediment samples using inductively coupled plasma optical emission spectroscopy to determine the concentrations of 16 elements, including Al, As, Ba, Ca, Cd, Co, Cr, Cu, Fe, Mn, Mo, Ni, Pb, Sr V, and Zn. The data was then utilized to generate interpolated maps using GIS techniques, followed by statistical analyses such as cluster analysis, Pearson coefficient, and correlation coefficient to identify similarities among various elements and particle size fractions in surface sediments. An index analysis approach employing the enrichment factor, modified degree of contamination, and potential ecological risk index (RI) was employed to classify the degree of contamination at the tested stations.

2. Geographical Setting

The Persian Gulf is a crescent-shaped marginal sea located in a foreland basin, hemmed in by the Arabian Peninsula to its south and west and the Zagros Mountains to its north and east [35]. It is connected to the Oman Sea and the Indian Ocean via the Strait of Hormuz [36]. The average water depth of the Persian Gulf stands at 35 m, with its maximum depth reaching up to 100 m near the Strait of Hormuz in the northeast [37].

The sedimentary deposits in the region consist of detrital riverine, biogenic, as well as loess materials [36]. Sediments from riverine sources are sourced from the Shatt-al-Arab and small rivers along the northern coasts, while aeolian sediments mainly originate from southwestern winds [38,39]. These dust storms bring fine-grain detrital materials from coastal plains, northern deserts, Iraq, and the Arabian Desert to the Persian Gulf [38,39]. The presence of rivers such as Kol and Shoor (Minab) in the northern part of the Persian Gulf contributes to a lower production of carbonate sediments [40] compared to the southern part of the Gulf due to their detrital components [37].

Surface sediments in the northern Persian Gulf contain high organic-biogenic components such as Gastropoda, Ostracoda, echinoderm, bryozoan, and benthonic-planktonic foraminifera, while detrital components including quartz, feldspar, mica, clay minerals, and rock fragments are derived from the Anatolia–Zagros mountain belt via inlet rivers that flow into the northern part of the Persian Gulf [37]. The entrapment of sediments in the delta results in Arvand flooding conditions, which have a slight impact on sediment supply to the Persian Gulf. Only about 10% of material reaches the Gulf due to this process [33].

Near Bandar-e-Abbas in shallow waters, the average sedimentation rate was found to be 1.04 mm per year, with a maximum sedimentation rate of 1.2 mm/year [34]. Studies conducted in the region have reported an average bottom-dissolved oxygen level of 6.5 ppm in shallow waters up to 25 m in depth and 3.5 ppm in deeper areas [37,41].

Tidal currents are a dominant factor in the Persian Gulf's hydrodynamics, typically flowing parallel to the catchment. Maximum tidal current velocities near the Strait of Hormuz have been estimated at around 90 cm/s, while the central part of the Khoran Strait between the ports of Pohl (Bandar-e-Abbas) and Laft (Qeshm Island) may experience maximum tide speeds of up to 137 cm/s [37,42]. High tide heights in the Persian Gulf vary from 1 to 5 m [37].

3. Materials and Methods

3.1. Sampling

The Geological Survey of Iran conducted a comprehensive study on physical, chemical, and biological parameters in the northern Persian Gulf. Sampling was carried out during the MG-2008-PG cruise, where 140 surface samples were collected from the bottom sediments of the marine boundary of Iranian coastal territory in the northeastern Persian Gulf using a Van Veen Grab Sampler (Figure 1). The sampling points' location was recorded using the Universal Transverse Mercator coordinate system (UTM) and later converted to decimal degrees using the World Geodetic System 1984 (WGS84). The top 10 cm of sediment was subsampled using plastic syringes and stored at a temperature of 4 °C. To ensure homogeneity, mechanical shaking was applied to the samples before being separated for various analyses such as grain size and chemical composition.

A

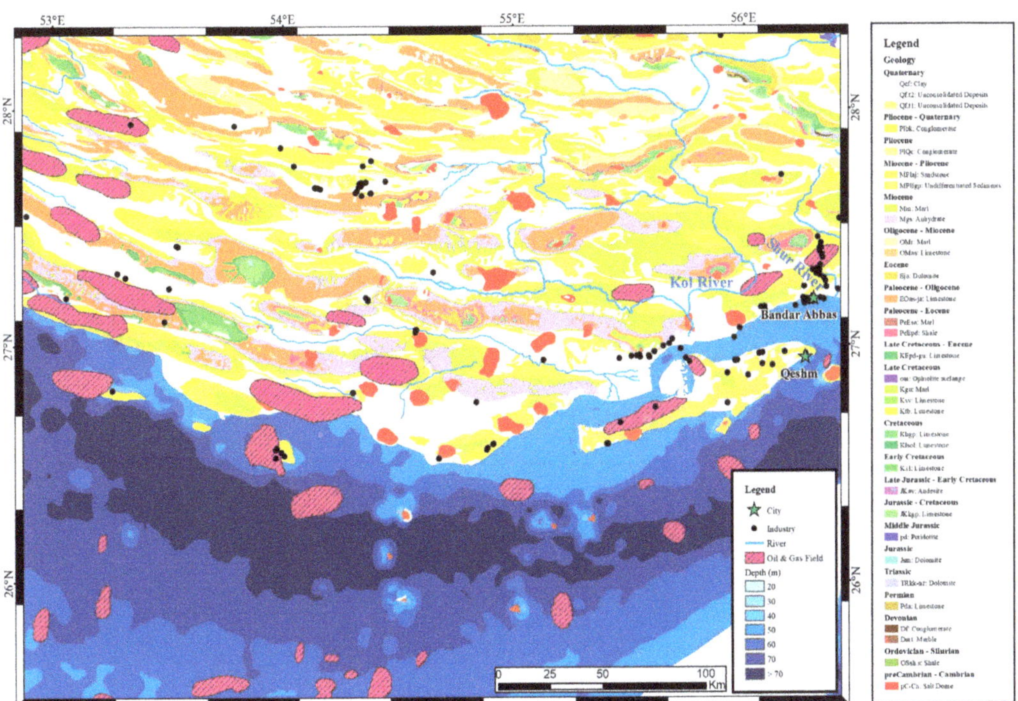

Figure 1. *Cont.*

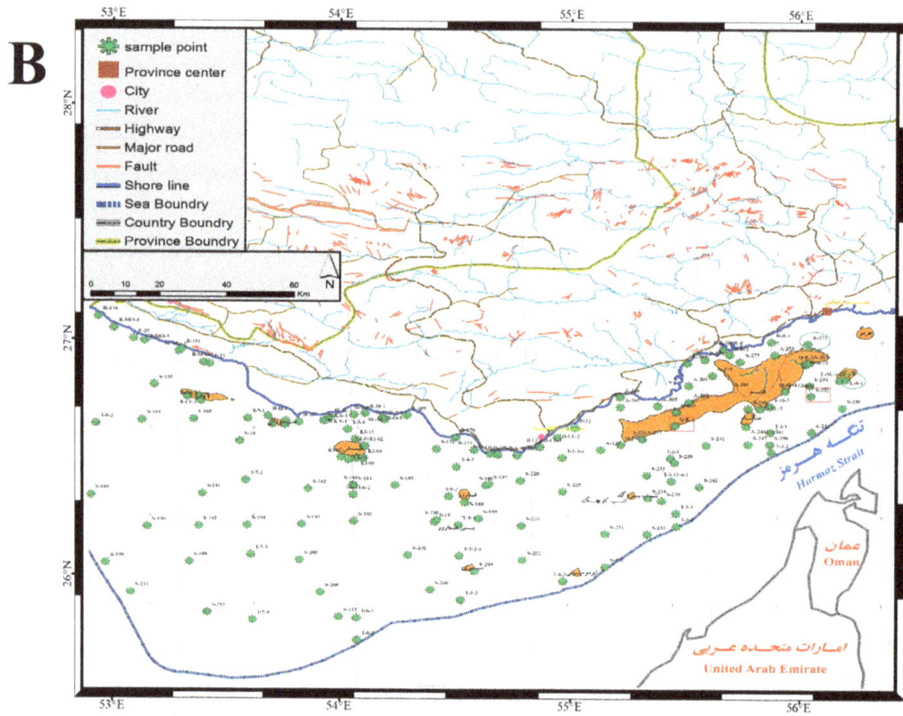

Figure 1. Location of the study area and sampling points in the northeastern Persian Gulf. The upper panel (**A**) displays the bathymetry and geological features of the region, as well as major river networks, industries, oil and gas fields. Panel (**B**) displays sampling points. The green ovals represent the examined samples in Figure 2. The two red rectangles represent the two most polluted sampling stations based on the mcd index.

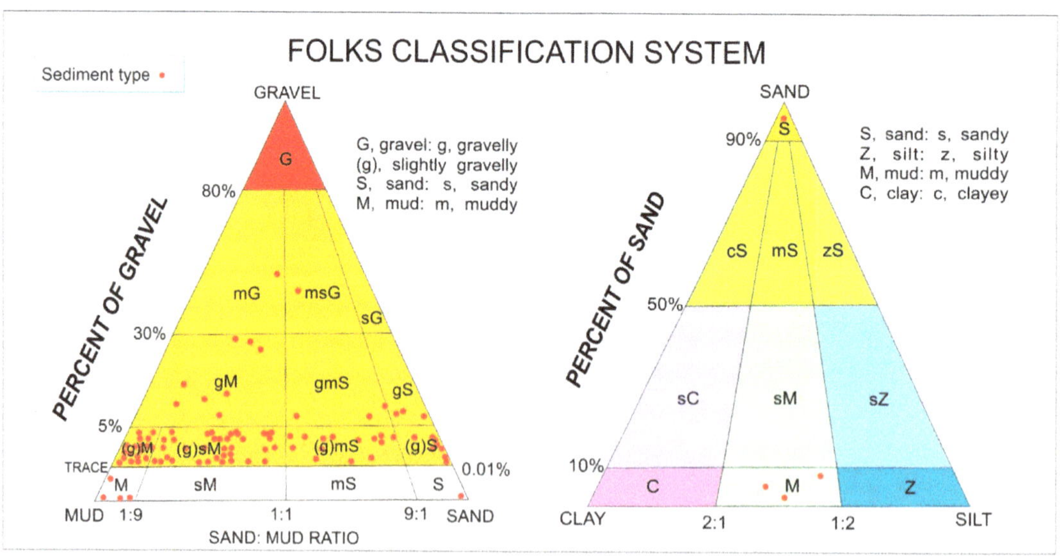

Figure 2. Ternary diagram showing particle size distribution of sediment.

3.2. Grain-Size Analysis

Prior to granulometry and chemical analysis, the sediments were dried at 75 °C for 24 h. The wet sieving method (Fritch model) was used for grain size analysis up to 63 μm, while the laser grain size test (Analysette 22 compact) was used for sizes less than 63 μm. To assess the precision of the method, several samples were subjected to repeated measurements, which yielded a relative standard deviation of <5%. Sediment description was carried out using the Folk classification method [43]. According to this system, clay particles are smaller than 2 microns; silt particles range from 2 to 63 microns; sand particles range from 63 microns to 2 mm; and gravel particles are larger than 2 mm.

3.3. Sediment Constituent Analysis

After drying, samples from different fractions, including 2 mm, 1 mm, 500 μ, 250 μ, and 125 μ, were selected for analysis of mineralogy, morphometry, and morphoscopy. The sediment particles were examined using an Binocular microscope (SMZ 1500 Nikon, USA) to evaluate their shape, roundness, and granularity within each sediment class. Specific and essential cases were photographed using a DS-Fi1 Nikon camera after studying the sediment particles under the microscope.

3.4. Elemental Analyses

To determine the composition of the sediment sample, 0.5 g of bulk sediments were weighed and placed in Teflon vessels. Next, 7 mL of aqua-regia solution (1:3 HCl: HNO_3) was added to each vessel [44,45]. Digestion was carried out using a Milestone Ethos1 microwave digester, with the samples being heated to 185 °C for 30 min and then held for 15 min. Subsequently, the samples were diluted with deionized distilled water up to 50 mL and analyzed for major and trace elements using an ICP-MS (Agilent 7700x, USA) device at the Iranian Geological Survey and mineral exploration.

It is important to note that the analysis focused on the "pseudo-total composition" rather than the "total composition" due to aqua-regia's limitations in fully digesting the sample. As such, the analysis only reflects the amount of element that has been extracted which is ideal for environmental purposes.

To ensure measurement accuracy and precision, procedural blanks, duplicates, and a standard sample (MESS-1) were used. The accuracy was found to be within ±5% of the suggested values in the standard.

3.5. Estimation of Sediment Contamination

The assessment and evaluation of sediment contamination have been addressed through the utilization of several reliable pollution indices [2,46]. Specifically, this study employed the enrichment factor (EF), modified degree of contamination (mC_d), and potential ecological risk index (RI) as the chosen indices for pollution assessment and evaluation.

3.5.1. Enrichment Factor

The Enrichment Factor (EF) is a helpful indicator that reflects the status of environmental contaminants, particularly heavy elements, in marine environments [47,48]. The following equation expresses the EF Equation (1) [40,41].

$$EF = ([(M_c)/(M_r)]_s)/([(M_c)/(M_r)]_b) \quad (1)$$

Here, M_c represents the concentration of metals; M_r represents the concentration of the reference element, which in this study is aluminum (Al); s denotes the studied sample; and b indicates the background used in this study, which was the element concentration in the Upper Continental Crust (UCC; [49]). Six contamination classes were defined based on the EF values (Table 1; [43]).

Table 1. Categories of sediment pollution based on the calculated indices.

EF	State of Pollution	mC_d	State of Pollution	Eri	State of Pollution	RI	Ecological Risk
<1	No enrichment	<1.5	Zero to the very low degree of contamination	<40	Low	<150	Low
1–2	Deficiency to minimal enrichment	1.5–2	Low degree of contamination	40–80	Moderate	150–300	Moderate
2–5	Moderate enrichment	2–4	Moderate degree of contamination	80–160	High	300–600	High
5–20	Significant enrichment	4–8	High degree of contamination	160–320	Serious	>600	Very high
20–40	Very high enrichment	8–16	Very high degree of contamination	>320	Severe		
>40	Extremely high enrichment	16–32	Extremely high degree of contamination				
		≥32	Ultra-high degree of contamination				

3.5.2. Degree of Contamination

The contamination factor (C_f) defined by [50] is proposed by dividing the concentration of each element in the sediment sample by the background concentration (UCC). The modified degree of contamination (mC_d) was proposed by [51] and consists of seven classes (Table 1) for an overall assessment of sediment pollution.

$$C_f = \frac{C_{\text{metal}}}{C_{\text{background}}} \quad (2)$$

$$mC_d = \frac{\sum_{i=1}^{n} C_f^i}{n} \quad (3)$$

3.5.3. Potential Ecological Risk Index

The potential ecological risk index (RI) is a measure of contamination levels based on the adverse ecological impact of each potentially toxic element [52]. It is calculated using Equations (4)–(6) as follows:

$$Cf^i = \frac{C^i_{\text{sample}}}{C^i_{\text{background}}} \quad (4)$$

$$Er^i = Tr^i \times Cf^i \quad (5)$$

$$RI = \sum_{i=1}^{n} Er^i \quad (6)$$

In these equations, the contamination factor (Cf^i) represents the concentration of each element in the sample relative to the background level. The toxicity coefficient (Tr^i) of a particular element indicates the relative harm it poses, with Cd, As, Pb, Ni, Cu, Cr, and Zn having coefficients of 30, 10, 5, 5, 5, 2, and 1, respectively [52]. Higher toxicity coefficients indicate greater harm caused by the potentially toxic element due to its higher toxicity, persistence, and bioaccumulation in ecosystems [53]. Er^i represents the potential ecological risk of each PTE as determined by [50]. Five categories are defined for Er^i, and four categories are defined for RI, as shown in Table 1.

3.5.4. Cluster Analysis

Cluster analysis (CA) is a statistical method commonly used in identifying groups of samples that exhibit similar behavior [54,55]. To determine the degree of similarity between various elements and particle size fractions present in the surface sediments of the northeastern part of the Persian Gulf, we utilized the multivariate statistical package analytical software (MVSP 3.1). In our study, we employed the weighted pair group method and Pearson's correlation coefficient since these widely used methods are effective in identifying clustering tendencies [27,55,56]. By utilizing these methods, we created a cluster tree by joining alternatives of the same weight to form larger clusters.

4. Results and Discussion

4.1. Sediments Texture

The surface sediments in the study area are soft and have not undergone diagenesis, likely due to their young age [40]. These sediments appear gray in color and were deposited in water with a pH greater than 7, as determined by physicochemical analysis. Table 2 presents the sand content, ranging from 0.3% to 99.9%, with an average value of 41.9%. Based on Folk's classification [43], more than 50% of the sediments consist of mud (silt and clay) particles, while sand-sized particles, primarily composed of skeletal fragments (such as benthonic and planktonic shells of carbonate composition), constitute one-third of the total bed sediments. The analysis categorized 75% of bed sediments into four major types, namely slightly gravelly sandy mud, slightly gravelly mud, slightly gravelly muddy sand, and slightly gravelly sand (Figure 2). Sedimentological studies reveal that gravel-sized sediments are mainly composed of skeletal and rock fragments. In order to provide a comprehensive understanding of the textural properties of the bottom sediments in the northeastern Persian Gulf, GIS techniques were employed to generate several interpolated maps. Figure 3 illustrates the accumulation of coarse materials such as gravel and sand in the east and west regions of the study area. This observation may be attributed to the prevailing current velocities in those areas.

Table 2. Descriptive statistics for major and trace elements, textural properties, mean enrichment factors (EF), mean contamination factor (Cf), and potential ecological risk associated with these elements and metals in bottom sediments located in the northeastern Persian Gulf. The contents of these elements and metals in Upper Continental Crust and Earth's shale are also included for comparative purposes.

	Min	Max	Mean	SD	C.V. (%)	UCC [1]	Shale [2]	EF	CF	Er
Ca (%)	0.24	31.09	18.17	7.45	41	3.00	2.2	91.2	8.3	
Al (%)	0.11	5.79	2.52	1.59	63	8.04	8.0			
Fe (%)	0.09	21.14	2.18	2.30	106	3.50	4.7	4.3	0.6	
Sr (mg/kg)	71	3550	1288	755	59	350	300	43.5	3.7	
Mn (mg/kg)	13	3122	387	374	97	600	850	2.4	0.5	
Ba (mg/kg)	2	3436	142	327	230	550	580	1.8	0.3	
Cr (mg/kg)	2	228	70	45	65	85	90	3.0	0.8	1.7
Ni (mg/kg)	2	130	50	34	68	50	68	3.2	1.0	5.0
V (mg/kg)	1	142	41	26	64	110	90	1.8	0.5	
Zn (mg/kg)	1	437	34	39	115	71	95	1.4	0.4	0.4
Cu (mg/kg)	1	278	13	24	178	25	45	1.1	0.3	1.5
Pb (mg/kg)	1	145	9	14	146	16	20	3.7	0.5	2.3
Co (mg/kg)	1	18	8	4	53	17	19	1.5	0.5	
As (mg/kg)	1	14	4	2	54	2	13	14.6	2.7	27.3
Mo (mg/kg)	0.2	12.5	0.9	1	138	1.5	2.6	7.5	0.6	
Cd (mg/kg)	0.06	1.53	0.20	0.15	75	0.10	0.30	3.7	0.7	20.4
Gravel (%)	0.0	51.7	3.9	8.8	222					
Sand (%)	0.3	99.9	41.9	31.5	75					
Silt (%)	0.0	69.9	32.1	21.7	67					
Clay (%)	0.0	69.6	21.9	17.4	79					

[1] Upper Continental Crust (UCC; [42]). [2] Average contents in the Earth's shale [57].

Sample A-271, which represents group A in Figure 4, is located in the northeast of Qeshm Island. The sediment sample's granulometric curve indicates a dominance of sand particles. The rock fragments present in the sedimentary basin vary between 8% to 22% in different fractions. Quartz is the most abundant mineral in this sediment sample, indicating that marine currents carried these particles from the Gulf of Oman and beaches around rivers such as Kol and Shur Rivers and deposited them in the study area. Therefore, it can be concluded that the main source of sediment in this group is detrital in nature. Calcite is also relatively abundant, suggesting that the sediment has both detrital and chemical origins.

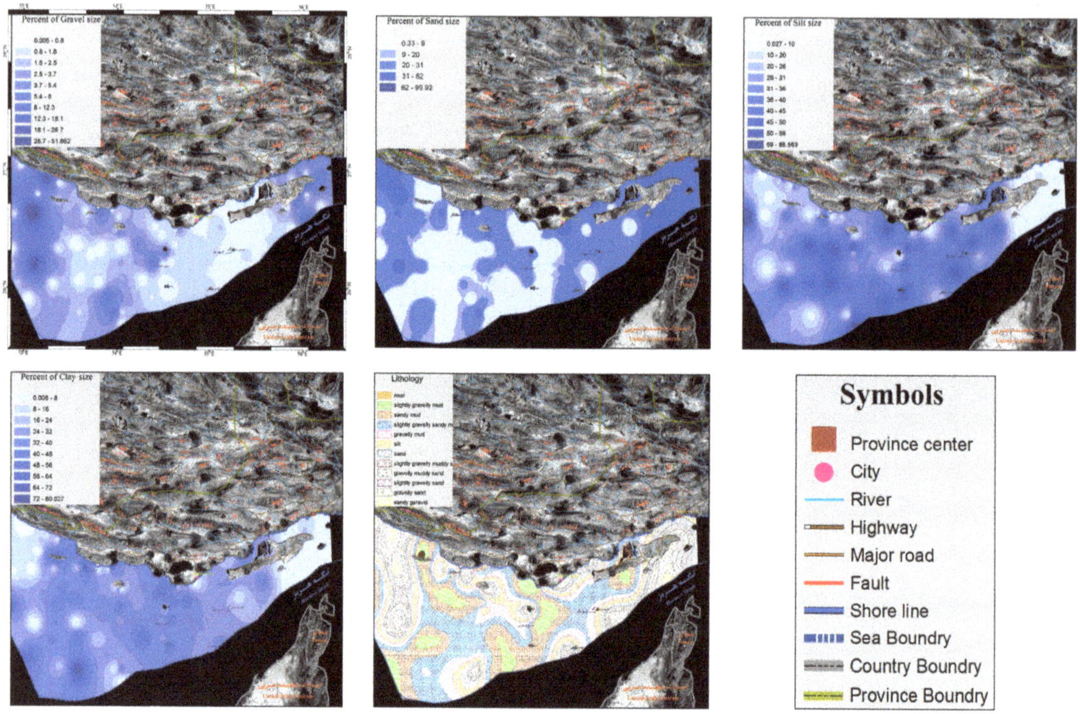

Figure 3. Interpolated maps depicting the distribution of gravel, sand, silt, and clay-sized particles in conjunction with sedimentology data, across the surface sediments of the northeastern region of the Persian Gulf.

Sample L-SL-1, which represents group B in Figure 4, is located near Larak Island. The sediment sample's granulometric curve also indicates a dominance of sand particles. The rock fragments constitute only 3% of the 125 μ fractions. Calcite is the most abundant mineral in this sediment sample and can be related to chemical and biogenic sources. Quartz is relatively abundant after calcite, and the presence of quartz, feldspar, and clay minerals supports the idea that the part of sediment has a detrital origin.

Based on the granulometric curve, statistical parameters, and components of sediments analyzed under a binocular microscope, two main groups of sediments can be identified in the study area. The first group (A) consists of coarse-grained sediments of biogenic origin with tiny detrital particles, composed mainly of shell fragments. The large size of these sediments in the Persian Gulf belongs to shells with biogenic sources. The second group (B) belongs to mixed fine and coarse-grain, bimodal or trimodal sediments with more than one source, including detrital, chemical, biogenic, and wind origins. These sediments contain siliciclastic minerals such as quartz, feldspar, rock fragments, and clay minerals. Additionally, chemical carbonate precipitated from seawater and biogenic sediments are present.

4.2. Variations of Elemental Composition

Table 2 presents the descriptive statistics for the concentrations of major elements and trace metals, as well as the textural properties of bottom sediments in the northeastern Persian Gulf. The statistical parameters reported in Table 2 include the mean, maximum, minimum, standard deviation, and average contents in the Earth's shale, as reported by [57].

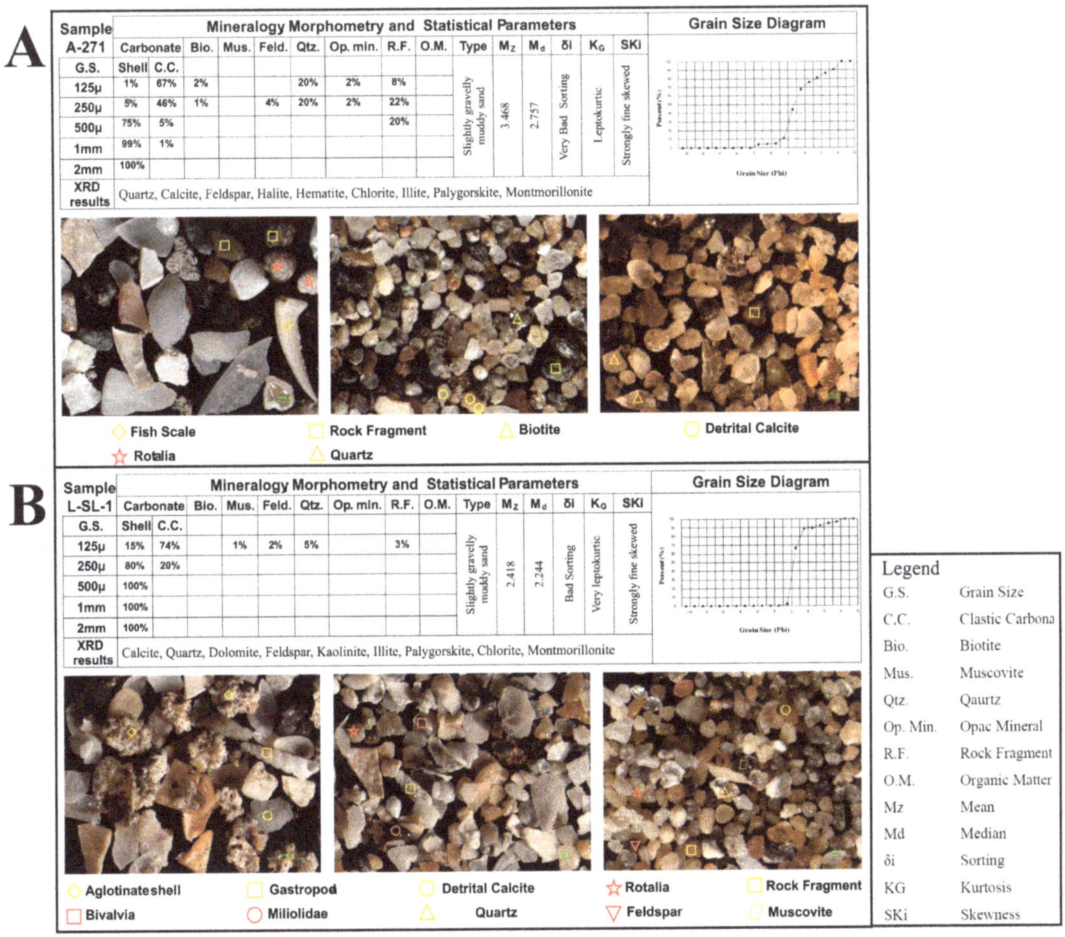

Figure 4. The granulometric curves, statistical parameters, and sediment component analysis are presented for two sediment samples that are representative of the two main groups of sediment (**A**,**B**) found in the northeastern Persian Gulf. The graph provides detailed information about the particle size distribution in the samples, including the proportion of sand, silt, and clay particles. Additionally, statistical parameters such as mean grain size and sorting coefficient are presented. Sediment component analysis reveals the composition of each sample, including the mineralogy and organic matter content.

The major and trace elements present in the sediments (Table S1) were ranked in descending order as follows: Ca > Al > Fe > Sr > Mn > Ba > Cr > Ni > V > Zn > Cu > Pb > As > Co > Mo > Cd. The average concentrations of Ca, Sr, As, and Cd exceeded the upper continental crust (UCC) values. These results indicate that calcium (Ca) is the most abundant element present in the bottom sediments of the northeastern Persian Gulf, with a concentration range from 0.24 to 31.09 percent mg/kg and a mean value of 18.17 percent mg/kg. Aluminum (Al) is the second most abundant element, but significant differences were observed between the mean concentration of Al in the marine sediment of the study area and the corresponding mean values for shale and crust. The concentration of strontium (Sr) varied from 71 to 3550 mg/kg, with a mean value of 1288 mg/kg. Notably, there is a remarkable difference between the mean concentration of Ba in the sediments of the study area and the corresponding UCC values (Ba is remarkably lower).

The coefficient of variation (C.V.) is the ratio of the standard deviation to the mean. The higher the coefficient of variation, the greater the level of dispersion around the mean. It is generally expressed as a percentage. Fe, Ba, Zn, Cu, Pb, and Mo had C.V. greater than 100%, indicating considerable variability [58] in the concentrations of these elements. The lowest coefficient of variation (C.V.) belongs to Ca, indicating less variability in the concentrations of calcium in the study area.

To better visualize the concentrations of major elements and trace metals, GIS techniques were employed to generate a series of interpolated maps (Figure 5). The distribution maps of calcium and aluminum indicate that the islands of Kish, Lavan, Faro, Beni Faro, Hengam, and Hondurabi in the Persian Gulf are dominated by carbonate sediments. This inference is supported by the high levels of calcium and low levels of aluminum in these areas.

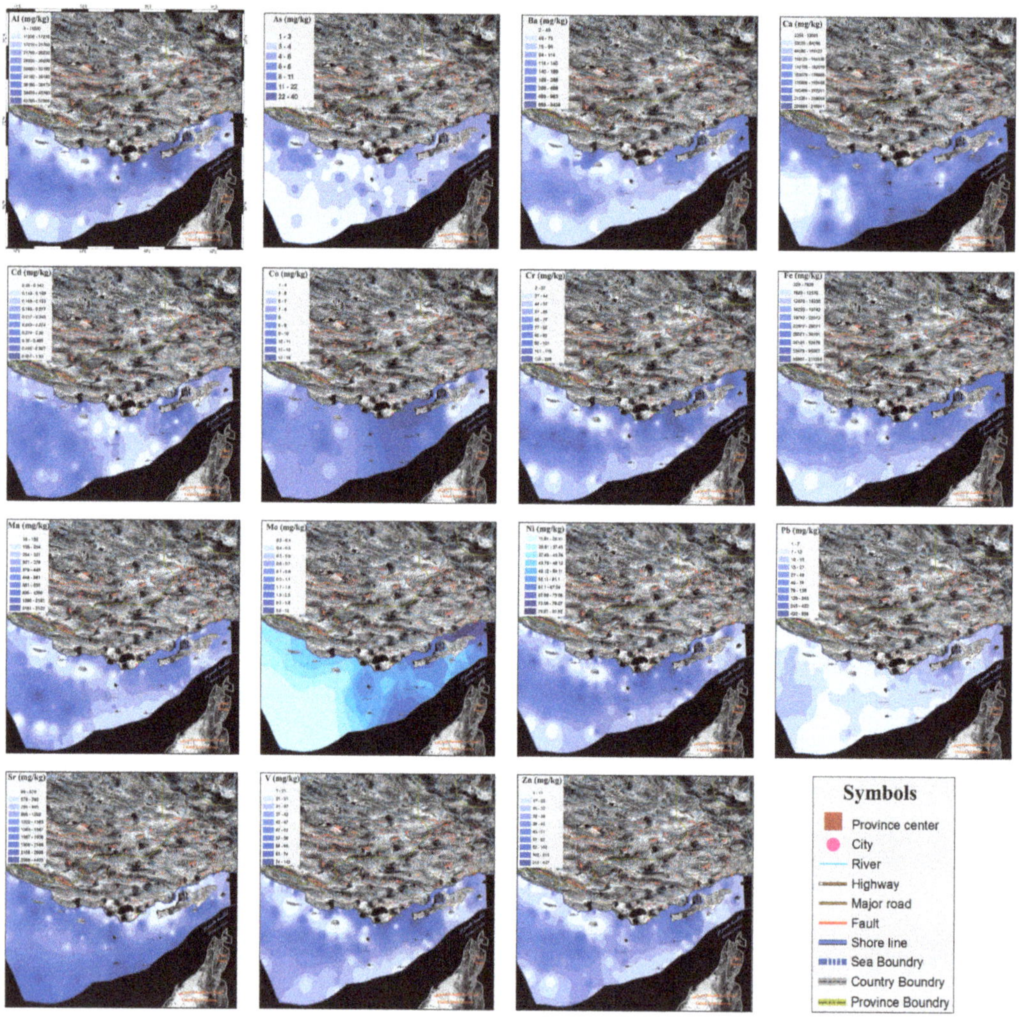

Figure 5. Interpolated concentration maps for each chemical element detected in the surface sediments of the northeastern part of the Persian Gulf.

4.3. Assessment of Sediment Contamination

4.3.1. Enrichment Factor

The results indicated that Ca and Sr had the highest mean EF values, suggesting that they belong to the extremely high enrichment class (Table 2). However, it should be noted that the presence of these elements in high concentrations does not necessarily indicate anthropogenic pollution, as they can naturally occur in sediments as a result of geologic and biochemical processes. Extremely severe enrichment was observed for Ca and Sr, while As and Mo exhibited significant enrichment based on their mean EF values. Fe, Pb, Cd, Ni, Cr, and Mn were found to belong to the moderate enrichment class. Finally, Ba, V, Co, Zn, and Cu displayed EF values less than two, indicating deficiency to minimal enrichment.

4.3.2. Modified Degree of Contamination

The results of the modified degree of contamination (mC_d) values indicate that, except for three stations, all other sites fall under the category of zero to very low degree of contamination, based on the classification proposed by [50]. The remaining three stations belong to the Low degree of contamination category. Among the sampled stations, the highest mC_d values were recorded at Stations Q-R-1 (2.62) and S-252 (1.52).

4.3.3. Potential Ecological Risk Index

The potential ecological risk index (RI) of bottom sediments in the northeastern Persian Gulf was assessed based on the presence of studied potentially toxic elements. Results showed that the potential ecological risk (Er^i) exhibited a decreasing order as follows: As > Cd > Ni > Pb > Cr > Cu > Zn, with As being the main contributor to the RI. Specifically, As and Cd accounted for 47% and 35%, respectively, of the Eri index in bottom sediments of the northeastern Persian Gulf. The average potential ecological risk values for all elements were lower than 40 (as shown in Table 2), indicating a low potential ecological risk. The samples had an average potential ecological risk index (RI) value of 46, suggesting that they can be characterized as having a low ecological risk.

4.4. Statistical Analyses

The resulting dendrogram (Figure 6) provided us with detailed information on the similarities between these clusters, enabling us to better interpret data. The highest correlation coefficients were observed between Zn-Cu (0.95), Al-Ni (0.92), Cr-V (0.89), and Zn-Pb (0.88). The UPGMA analysis of the dendrogram revealed several significant clusters with strong correlations: (A) Al-Ni-Cr-V; (B) Cd-Cu-Zn-Pb; (C) Ba-Fe; (D) silt-clay; (E) Ca-Sr.

The concentration of Cr, Ni, and V in the area is partly from geogenic sources, as confirmed by their high correlation with aluminum. Furthermore, the low levels of these elements in the sediments around the islands can be attributed to a relatively lower amount of detrital sediment input. This leads us to believe that the origin of these particular elements in the sediments of the Persian Gulf is predominantly geogenic in nature. However, there are also some amounts of these elements that have a secondary anthropogenic origin. These results are consistent with previous research by [2] who found high similarity coefficients between Ni, Cr, and Total Petroleum Hydrocarbons in sediment from Mahshahr Bay in the northwest Persian Gulf. Other studies suggest that some of these elements in the mangroves located in the northwest Qeshm and Asalouye region may have a partly anthropogenic origin [59].

Cu, Zn, Pb, and Cd also showed a relatively high correlation coefficient indicative of a common origin. The next cluster, consisting of Ca-Sr-Sand, showed a relatively low correlation compared to the previous ones. Thus, Ca and Sr that accumulated in coarse sediments most likely have biogenic sources. As mentioned above, calcite is the most abundant sediment mineral related to chemical, biogenic, and detrital sources. Lastly, a strong negative correlation coefficient between Al and Ca showed that Al has a non-carbonate source.

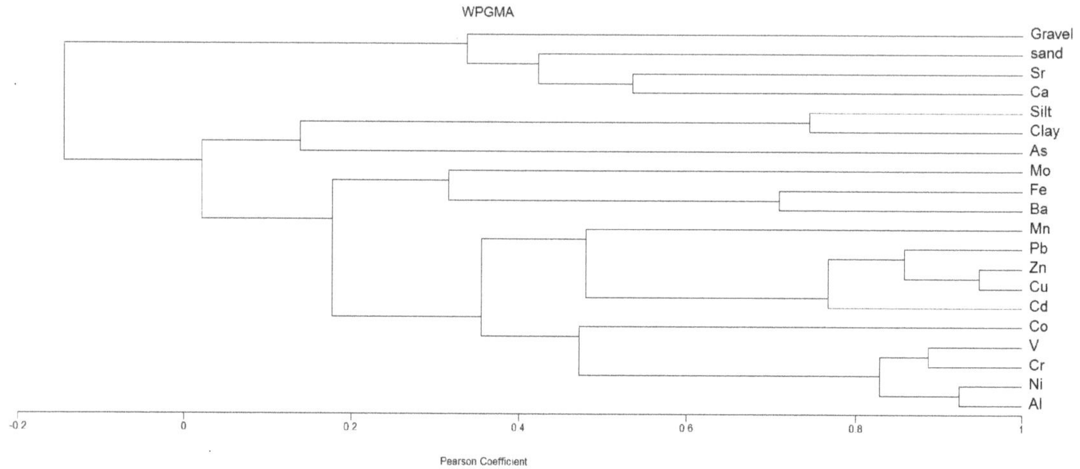

Figure 6. Dendrogram showing the clustering patterns of major elements, trace metals, and particle size fractions.

Notably, no strong correlation was found between As and other elements. The significant enrichment of As suggests that it likely originated from a different source. The same argument can be done for Mo. The distribution map of arsenic and molybdenum clearly indicates that the elevated concentrations of these elements in the northeast section of the study area are closely linked to the presence of numerous industries in Bandar-e-Abbas and Qeshm cities, as shown in Figure 1. Previous studies have recorded high amounts of arsenic in dust particles in Iran [60]. Therefore, it is possible that some of the high concentration of arsenic may have originated from wind.

5. Conclusions

The Persian Gulf is a globally significant body of water that plays a vital role in maintaining the ecological balance of the region. In recent years, there has been growing concern about the impact of human activities on the Gulf, and particularly on the bottom sediments in Iranian territorial waters located in the northeast region. To address this concern, this study presented a comprehensive analysis of sediment texture, geochemical variation, potential sources, and ecological risk assessment of major elements and trace metals in the bottom sediments of Iranian territorial waters situated in the northeast Persian Gulf region.

The samples underwent an analysis utilizing granulometric curve, statistical parameters, and binocular microscope components, which resulted in the identification of two distinct sediment groups. The first group was composed of coarse-grained sediments of biogenic origin, primarily consisting of shell fragments with small detrital particles. The second group included mixed fine and coarse-grain, bimodal or trimodal sediments originating from multiple sources, such as detrital, chemical, biogenic, and wind origins. These sediments contained siliceous minerals like rock fragments, quartz, feldspar, and clay minerals. Additionally, they featured chemical carbonate precipitated from seawater and biogenic sediments.

Additionally, calcium is the most abundant element present in the bottom sediments, followed by aluminum and iron. Arsenic and molybdenum exhibit extremely high enrichment, while iron, lead, cadmium, nickel, chromium, and manganese belong to the moderate enrichment class. Other elements were classified as deficient to minimal enrichment (Ba > V > Co > Zn > Cu). Stations were classified as having a low degree of contamination based on modified degree of contamination values.

The distribution maps of calcium and aluminum in the Persian Gulf suggest that carbonate sediments dominate around several islands in the region. The low levels of Cr, Ni, and V around these islands imply that the origin of these elements in the sediments is predominantly geogenic. On the other hand, the distribution map of arsenic and molybdenum shows high concentrations of these elements in the northeast section of the study area, which can be attributed to industrial activities in Bandar-e-Abbas and Qeshm cities. As previous studies have recorded high amounts of arsenic in dust particles in Iran, it is possible that some of the elevated concentration of arsenic may have originated from wind transport. The findings of this study provide valuable information for continued research in this ecologically sensitive region, particularly regarding the potential sources of pollution and their impact on the marine ecosystem.

Supplementary Materials: The following supporting information can be downloaded at: https://www.mdpi.com/article/10.3390/min13070850/s1, Table S1: Analysis of textural properties and major/trace element composition in bottom sediments from the northeastern Persian Gulf.

Author Contributions: Authors have made substantial contributions. They have contributed to interpreting the data and writing the manuscript. They have approved the final version of the paper. Conceptualization: A.V. and R.L.; data curation: A.V. and R.L.; formal analysis: A.V. and R.L.; funding acquisition: R.L.; writing—original draft: A.V.; writing—review and editing: A.V. and R.L. All authors have read and agreed to the published version of the manuscript.

Funding: This research was funded by the Geological Survey and Mineral Exploration of Iran (GSI) to Razyeh Lak, grant number MG-2008-PG.

Data Availability Statement: The authors confirm that the data supporting the findings of this study are available within the article. Raw data from this study are also available from the corresponding author on request.

Acknowledgments: This research was funded by the Geological Survey and Mineral Exploration of Iran (GSI) to Razyeh Lak. We are very grateful for the support that the Geological Survey and Mineral Exploration of Iran (GSI) extended for sampling operations during a sea cruise (MG-2008-PG Cruise), various logistic activities, and relevant experiments.

Conflicts of Interest: The authors declare no conflict of interest. The funders had no role in the design of the study; in the collection, analyses, or interpretation of data; in the writing of the manuscript; or in the decision to publish the results.

References

1. Hume, B.; D'angelo, C.; Burt, J.; Baker, A.C.; Riegl, B.; Wiedenmann, J. Corals from the Persian/Arabian Gulf as models for thermotolerant reef-builders: Prevalence of clade C3 Symbiodinium, host fluorescence and ex situ temperature tolerance. *Mar. Pollut. Bull.* **2013**, *72*, 313–322. [CrossRef]
2. Vaezi, A.R.; Karbassi, A.R.; Fakhraee, M. Assessing the trace metal pollution in the sediments of Mahshahr Bay, Persian Gulf, via a novel pollution index. *Environ. Monit. Assess.* **2015**, *187*, 613. [CrossRef]
3. Al-Awadhi, F.M.A. The Year of the Ocean and its crucial importance to the Gulf. *Desalination* **1999**, *123*, 127–133. [CrossRef]
4. Nadim, F.; Bagtzoglou, A.C.; Iranmahboob, J. Coastal management in the Persian Gulf region within the framework of the ROPME programme of action. *Ocean Coast. Manag.* **2008**, *51*, 556–565. [CrossRef]
5. Price, A.R.G. Impact of the 1991 Gulf War on the coastal environment and ecosystems: Current status and future prospects. *Environ. Int.* **1998**, *24*, 91–96. [CrossRef]
6. Soltani, N.; Moore, F.; Keshavarzi, B.; Sorooshian, A.; Javid, R. Potentially toxic elements (PTEs) and polycyclic aromatic hydrocarbons (PAHs) in fish and prawn in the Persian Gulf, Iran. *Ecotoxicol. Environ. Saf.* **2019**, *173*, 251–265. [CrossRef]
7. Pourang, N.; Nikouyan, A.; Dennis, J.H. Trace Element Concentrations in Fish, Surficial Sediments and Water from Northern Part of the Persian Gulf. *Environ. Monit. Assess.* **2005**, *109*, 293–316. [CrossRef]
8. Kazemi, A.; Bakhtiari, A.R.; Kheirabadi, N.; Barani, H.; Haidari, B. Distribution Patterns of Metals Contamination in Sediments Based on Type Regional Development on the Intertidal Coastal Zones of the Persian Gulf, Iran. *Bull. Environ. Contam. Toxicol.* **2012**, *88*, 100–103. [CrossRef] [PubMed]
9. Allami, H.; Afzali, A.; Mirzaei, R. Determination and investigation of heavy metal concentrations in sediments of the Persian Gulf coasts and evaluation of their potential environmental risk. *Anal. Methods Environ. Chem. J.* **2020**, *3*, 60–71. [CrossRef]

10. Neyestani, M.R.; Bastami, K.D.; Esmaeilzadeh, M.; Shemirani, F.; Khazaali, A.; Molamohyeddin, N.; Afkhami, M.; Nourbakhsh, S.; Dehghani, M.; Aghaei, S.; et al. Geochemical speciation and ecological risk assessment of selected metals in the surface sediments of the northern Persian Gulf. *Mar. Pollut. Bull.* **2016**, *109*, 603–611. [CrossRef]
11. Saadatmand, M.; Dadolahi-Sohrab, A.; Tavani, M.B.; Khazaei, S.-H.; Saadatmand, F. Monitoring heavy metal contamination on the Iranian coasts of the Persian Gulf using biological indicators: Risk assessment for the consumers. *Environ. Monit. Assess.* **2022**, *194*, 83. [CrossRef]
12. Akhbarizadeh, R.; Moore, F.; Keshavarzi, B.; Moeinpour, A. Microplastics and potentially toxic elements in coastal sediments of Iran's main oil terminal (Khark Island). *Environ. Pollut.* **2017**, *220*, 720–731. [CrossRef]
13. Lahijanzadeh, A.R.; Rouzbahani, M.M.; Sabzalipour, S.; Nabavi, S.M.B. Ecological risk of potentially toxic elements (PTEs) in sediments, seawater, wastewater, and benthic macroinvertebrates, Persian Gulf. *Mar. Pollut. Bull.* **2019**, *145*, 377–389. [CrossRef]
14. Diagomanolin, V.; Farhang, M.; Ghazi-Khansari, M.; Jafarzadeh, N. Heavy metals (Ni, Cr, Cu) in the Karoon waterway river, Iran. *Toxicol. Lett.* **2004**, *151*, 63–67. [CrossRef]
15. Haghnazar, H.; Sabbagh, K.; Johannesson, K.H.; Pourakbar, M.; Aghayani, E. Phytoremediation capability of Typha latifolia L. to uptake sediment toxic elements in the largest coastal wetland of the Persian Gulf. *Mar. Pollut. Bull.* **2023**, *188*, 114699. [CrossRef]
16. Samani, A.R.V.; Karbassi, A.R.; Fakhraee, M.; Heidari, M.; Vaezi, A.R.; Valikhani, Z. Effect of dissolved organic carbon and salinity on flocculation process of heavy metals during mixing of the Navrud River water with Caspian Seawater. *Desalin. Water Treat.* **2015**, *55*, 926–934. [CrossRef]
17. Wagh, V.M.; Panaskar, D.B.; Mukate, S.V.; Gaikwad, S.K.; Muley, A.A.; Varade, A.M. Health risk assessment of heavy metal contamination in groundwater of Kadava River Basin, Nashik, India. *Model. Earth Syst. Environ.* **2018**, *4*, 969–980. [CrossRef]
18. Vaezi, A.R.; Karbassi, A.R.; Valavi, S.; Ganjali, M.R. Ecological risk assessment of metals contamination in the sediment of the Bamdezh wetland, Iran. *Int. J. Environ. Sci. Technol.* **2015**, *12*, 951–958. [CrossRef]
19. Bacha, M.E.; Nedjimi, B.; Ararem, A.; Beladel, B. Seasonal variation of selected trace elements in rare endemic species Thuriferous Juniper growing in the Aurès Mountains of Algeria. *J. For. Res.* **2017**, *28*, 249–254. [CrossRef]
20. Jafarabadi, A.R.; Bakhtiari, A.R.; Spanò, N.; Cappello, T. First report of geochemical fractionation distribution, bioavailability and risk assessment of potentially toxic inorganic elements in sediments of coral reef Islands of the Persian Gulf, Iran. *Mar. Pollut. Bull.* **2018**, *137*, 185–197. [CrossRef]
21. Karbassi, A.R.; Heidari, M.; Vaezi, A.R.; Samani, A.R.V.; Fakhraee, M.; Heidari, F. Effect of pH and salinity on flocculation process of heavy metals during mixing of Aras River water with Caspian Sea water. *Environ. Earth Sci.* **2014**, *72*, 457–465. [CrossRef]
22. Bastami, K.D.; Neyestani, M.R.; Esmaeilzadeh, M.; Haghparast, S.; Alavi, C.; Fathi, S.; Nourbakhsh, S.; Shirzadi, E.A.; Parhizgar, R. Geochemical speciation, bioavailability and source identification of selected metals in surface sediments of the Southern Caspian Sea. *Mar. Pollut. Bull.* **2017**, *114*, 1014–1023. [CrossRef] [PubMed]
23. Karbassi, A.R.; Fakhraee, M.; Heidari, M.; Vaezi, A.R.; Samani, A.R.V. Dissolved and particulate trace metal geochemistry during mixing of Karganrud River with Caspian Sea water. *Arab. J. Geosci.* **2015**, *8*, 2143–2151. [CrossRef]
24. Elsagh, A.; Jalilian, H.; Aslshabestari, M.G. Evaluation of heavy metal pollution in coastal sediments of Bandar Abbas, the Persian Gulf, Iran: Mercury pollution and environmental geochemical indices. *Mar. Pollut. Bull.* **2021**, *167*, 112314. [CrossRef]
25. Venugopal, T.; Giridharan, L.; Jayaprakash, M. Characterization and risk assessment studies of bed sediments of river adyar-an application of speciation study. *Int. J. Environ. Res.* **2009**, *3*, 581–598.
26. Seshan, B.R.R.; Natesan, U.; Deepthi, K. Geochemical and statistical approach for evaluation of heavy metal pollution in core sediments in southeast coast of India. *Int. J. Environ. Sci. Technol.* **2010**, *7*, 291–306. [CrossRef]
27. Vaezi, A.R.; Karbassi, A.R.; Habibzadeh, S.K.; Heidari, M.; Samani, A.R.V. Heavy metal contamination and risk assessment in the riverine sediment. *Indian J. Geo-Marine Sci.* **2016**, *45*, 1017–1023.
28. Karbassi, A.R.; Bayati, G.R.N.-B.I. Environmental geochemistry of heavy metals in a sediment core off bushehr, persian gulf. *J. Environ. Health Sci. Eng.* **2005**, *2*, 255–260.
29. Bastami, K.D.; Afkhami, M.; Mohammadizadeh, M.; Ehsanpour, M.; Chambari, S.; Aghaei, S.; Esmaeilzadeh, M.; Neyestani, M.R.; Lagzaee, F.; Baniamam, M. Bioaccumulation and ecological risk assessment of heavy metals in the sediments and mullet Liza klunzingeri in the northern part of the Persian Gulf. *Mar. Pollut. Bull.* **2015**, *94*, 329–334. [CrossRef] [PubMed]
30. Delshab, H.; Farshchi, P.; Keshavarzi, B. Geochemical distribution, fractionation and contamination assessment of heavy metals in marine sediments of the Asaluyeh port, Persian Gulf. *Mar. Pollut. Bull.* **2017**, *115*, 401–411. [CrossRef] [PubMed]
31. El-Taher, A.; Alshahri, F.; Elsaman, R. Environmental impacts of heavy metals, rare earth elements and natural radionuclides in marine sediment from Ras Tanura, Saudi Arabia along the Arabian Gulf. *Appl. Radiat. Isot.* **2018**, *132*, 95–104. [CrossRef]
32. Sharifinia, M.; Taherizadeh, M.; Namin, J.I.; Kamrani, E. Ecological risk assessment of trace metals in the surface sediments of the Persian Gulf and Gulf of Oman: Evidence from subtropical estuaries of the Iranian coastal waters. *Chemosphere* **2018**, *191*, 485–493. [CrossRef] [PubMed]
33. Rezaei, M.; Mehdinia, A.; Saleh, A.; Modabberi, S.; Daneshvar, M.R.M. Environmental assessment of heavy metal concentration and pollution in the Persian Gulf. *Model. Earth Syst. Environ.* **2021**, *7*, 983–1003. [CrossRef]
34. Maghrebi, M.; Karbassi, A.; Lak, R.; Noori, R.; Sadrinasab, M. Temporal metal concentration in coastal sediment at the north region of Persian Gulf. *Mar. Pollut. Bull.* **2018**, *135*, 880–888. [CrossRef] [PubMed]
35. Orang, K.; Motamedi, H.; Azadikhah, A.; Royatvand, M. Structural framework and tectono-stratigraphic evolution of the eastern Persian Gulf, offshore Iran. *Mar. Pet. Geol.* **2018**, *91*, 89–107. [CrossRef]

36. Mokhtari, S.; Hosseini, S.M.; Danehkar, A.; Azad, M.T.; Kadlec, J.; Jolma, A.; Naimi, B. Inferring spatial distribution of oil spill risks from proxies: Case study in the north of the Persian Gulf. *Ocean Coast. Manag.* **2015**, *116*, 504–511. [CrossRef]
37. Caouette, C. *The Persian Gulf: Oceanography and Wildlife, Oceanography and Ocean Engineering Series*; Nova Science Publishers: Hauppauge, NY, USA, 2020; ISBN 9781536183047.
38. Sugden, W. Some Aspects of Sedimentation in the Persian Gulf. *SEPM J. Sediment. Res.* **1963**, *33*, 355–364. [CrossRef]
39. Al-Bakri, D.; El-Sayed, M. Mineralogy and provenance of the clastic deposits of the modern intertidal environment of the northern Arabian Gulf. *Mar. Geol.* **1991**, *97*, 121–135. [CrossRef]
40. Ghazban, F. *Petroleum Geology of Persian Gulf*; Tehran University; National Iranian Oil Company: Tehran, Iran, 2007.
41. Behbahani, R.; Hosseinyar, G.; Lak, R. The controlling parameters on organic matter preservation within the bottom sediments of the northern part of the Persian Gulf. *Neues Jahrb. fur Geol. und Palaontol.—Abh.* **2015**, *276*, 267–283. [CrossRef]
42. Hosseinyar, G.; Moussavi-Harami, R.; Behbahani, R. Shallow gas accumulations and seepage in the sediments of the Northeast Persian Gulf. *Acta Geophys.* **2014**, *62*, 1373–1386. [CrossRef]
43. Folk, R.L. *Petrology of Sedimentary Rocks*; Hemphill Publishing Company: Austin, TX, USA, 1980; ISBN 0914696149.
44. Karbassi, A.R.; Monavari, S.M.; Bidhendi, G.N.; Nouri, J.; Nematpour, K. Metal pollution assessment of sediment and water in the Shur River. *Environ. Monit. Assess.* **2008**, *147*, 107–116. [CrossRef] [PubMed]
45. Loring, D.H.; Rantala, R.T.T. Manual for the geochemical analyses of marine sediments and suspended particulate matter. *Earth-Sci. Rev.* **1992**, *32*, 235–283. [CrossRef]
46. Pejman, A.; Bidhendi, G.N.; Ardestani, M.; Saeedi, M.; Baghvand, A. A new index for assessing heavy metals contamination in sediments: A case study. *Ecol. Indic.* **2015**, *58*, 365–373. [CrossRef]
47. Soltani, N.; Keshavarzi, B.; Moore, F.; Tavakol, T.; Lahijanzadeh, A.R.; Jaafarzadeh, N.; Kermani, M. Ecological and human health hazards of heavy metals and polycyclic aromatic hydrocarbons (PAHs) in road dust of Isfahan metropolis, Iran. *Sci. Total. Environ.* **2015**, *505*, 712–723. [CrossRef]
48. Yuen, J.Q.; Olin, P.H.; Lim, H.S.; Benner, S.G.; Sutherland, R.A.; Ziegler, A.D. Accumulation of potentially toxic elements in road deposited sediments in residential and light industrial neighborhoods of Singapore. *J. Environ. Manage* **2012**, *101*, 151–163. [CrossRef]
49. Taylor, S.R.; McLennan, S.M. Chemical Composition and Element Distribution in the Earth's Crust. *Encycl. Phys. Sci. Technol.* **2003**, *312*, 697–719. [CrossRef]
50. Håkanson, L. An ecological risk index for aquatic pollution control. A sedimentological approach. *Water Res.* **1980**, *14*, 975–1001. [CrossRef]
51. Abrahim, G.M.S.; Parker, R.J. Assessment of heavy metal enrichment factors and the degree of contamination in marine sediments from Tamaki Estuary, Auckland, New Zealand. *Environ. Monit. Assess.* **2008**, *136*, 227–238. [CrossRef]
52. Mihankhah, T.; Saeedi, M.; Karbassi, A. A comparative study of elemental pollution and health risk assessment in urban dust of different land-uses in Tehran's urban area. *Chemosphere* **2020**, *241*, 124984. [CrossRef] [PubMed]
53. Li, F.; Zeng, X.-Y.; Wu, C.-H.; Duan, Z.-P.; Wen, Y.-M.; Huang, G.-R.; Long, X.-L.; Li, M.-J.; Li, M.-J.; Xu, J.-Y. Ecological Risks Assessment and Pollution Source Identification of Trace Elements in Contaminated Sediments from the Pearl River Delta, China. *Biol. Trace Elem. Res.* **2013**, *155*, 301–313. [CrossRef]
54. Vaezi, A.; Karbassi, A.; Fakhraee, M.; Samani, A.V.; Heidari, M. Assessment of sources and concentration of metal contaminants in marine sediments of Musa Estuary, Persian Gulf. *J. Environ. Stud.* **2014**, *40*, 345–360.
55. Reyment, R.A.; Davis, J.C. Statistics and Data Analysis in Geology. *Biometrics* **1988**, *44*, 918. [CrossRef]
56. Sharmad, T.; Bidhendi, G.N.; Karbassi, A.; Moatar, F.; Adabi, M.H. Historical changes in distribution and partitioning of natural and anthropogenic shares of heavy metals in sediment core from the southern Caspian Sea. *Environ. Earth Sci.* **2012**, *67*, 799–811. [CrossRef]
57. Turekian, K.K.; Wedepohl, K.H. Distribution of the Elements in Some Major Units of the Earth's Crust. *Bull. Geol. Soc. Am.* **1961**, *72*, 175–192. [CrossRef]
58. Madadi, R.; Mohamadi, S.; Rastegari, M.; Karbassi, A.; Rakib, R.J.; Khandaker, M.U.; Faruque, M.R.I.; Idris, A.M. Health risk assessment and source apportionment of potentially toxic metal(loid)s in windowsill dust of a rapidly growing urban settlement, Iran. *Sci. Rep.* **2022**, *12*, 19736. [CrossRef]
59. Zarezadeh, R.; Rezaee, P.; Lak, R.; Masoodi, M.; Ghorbani, M. Distribution and accumulation of heavy metals in sediments of the northern part of mangrove in Hara Biosphere Reserve, Qeshm Island (Persian Gulf). *Soil Water Res.* **2017**, *12*, 86–95. [CrossRef]
60. Behrooz, R.D.; Kaskaoutis, D.G.; Grivas, G.; Mihalopoulos, N. Human health risk assessment for toxic elements in the extreme ambient dust conditions observed in Sistan, Iran. *Chemosphere* **2021**, *262*, 127835. [CrossRef] [PubMed]

minerals MDPI

Article

Facies, Depositional Environment and Reservoir Quality of an Early Cambrian Carbonate Ramp in the Tarim Basin, NW China

Yongjin Zhu [1,2], Jianfeng Zheng [1,2], Jiankun Zhang [3,4,*], Xinsheng Luo [5], Guang Yu [1], Jun Li [6], Fangjie Hu [5] and Guo Yang [5]

1 PetroChina Hangzhou Institute of Geology, Hangzhou 310023, China; zhuyj_hz@petrochina.com.cn (Y.Z.); zhengjf_hz@petrochina.com.cn (J.Z.); yug_hz@petrochina.com.cn (G.Y.)
2 PetroChina Key Laboratory of Carbonate Reservoir, Hangzhou 310023, China
3 State Key Laboratory of Nuclear Resources and Environment, East China University of Technology, Nanchang 330013, China
4 Jiangxi Engineering Technology Research Center of Nuclear Geoscience Data Science and System, East China University of Technology, Nanchang 330013, China
5 PetroChina Tarim Oilfield Company, Korla 841000, China; luoxs-tlm@petrochina.com.cn (X.L.); hufangjie100@163.com (F.H.); yangguo-tlm@petrochina.com.cn (G.Y.)
6 PetroChina Research Institute of Petroleum Exploration & Development, PetroChina, Beijing 100083, China; lijun69@petrochina.com.cn
* Correspondence: sparkler1984@126.com

Abstract: The Xiaoerbulake Formation in the Tarim Basin is considered one of the most important deep to ultradeep hydrocarbon reservoirs in the world. The objective of the present study is to analyze the facies, depositional environment and reservoir quality of the Early Cambrian Xiaoerbulake Formation in the Tarim Basin through integrated analysis of 120 m of cores, 3240 m of well cuttings, wireline logs and over 1100 thin sections from 17 exploration wells. Early Cambrian Xiaoerbulake Formation deposits in the Tarim Basin were deposited on a ramp setting. The ramp was occupied by seven facies associations and ten facies types ranging from the inner ramp to the outer ramp. These facies associations include tidal flat, lagoon, high-energy shoal, sabkha, inter-shoal, low-energy shoal and open shelf. Their distribution was controlled by paleogeographic patterns, sea level changes and the paleoclimate. Reservoir quality is considerably affected by facies together with diagenetic processes, including dolomitization and anhydrite cementation. High-quality reservoirs are found in the dolograinstone of high-energy shoal environments, which are favorable exploration facies in the carbonate ramp. Their pore space types consist of interparticle pores, intraparticle pores, intercrystalline pores, vuggy pores and moldic pores, with a porosity of 1.4%–7.5%. This study will help with our understanding of the stratigraphic framework, sedimentary-facies evolution and high-quality reservoir distribution of the Early Cambrian carbonate ramp in the Tarim Basin, facilitating exploration and the production of hydrocarbons from the Xiaoerbulake Formation.

Keywords: facies; depositional environments; reservoir quality; Xiaoerbulake Formation; Early Cambrian carbonate ramp; Tarim Basin

Citation: Zhu, Y.; Zheng, J.; Zhang, J.; Luo, X.; Yu, G.; Li, J.; Hu, F.; Yang, G. Facies, Depositional Environment and Reservoir Quality of an Early Cambrian Carbonate Ramp in the Tarim Basin, NW China. *Minerals* **2023**, *13*, 791. https://doi.org/10.3390/min13060791

Academic Editors: Hamzeh Mehrabi, Vahid Tavakoli and Thomas Gentzis

Received: 9 May 2023
Revised: 4 June 2023
Accepted: 8 June 2023
Published: 10 June 2023

1. Introduction

The concept of carbonate ramps was established to characterize a very low-gradient sloping depositional surface (with a gradient < 1°) that passes gradually from a shallow, high-energy environment of offshore and downslope into a progressively deeper, lower-energy environment and ultimately into basinal facies [1,2]. Compared to the shelf model, carbonate ramps generally lack slope break belts that separate the shallow water areas from the basin [3]. They also differ from rimmed shelves as there are usually no continuous reef trends; high-energy lime sands situated close to the shoreline and clasts of shallow shelf-edge facies are generally not observed in deeper water breccias (if present) [2–4].

Based on the study of various types of carbonate platforms, ramps were subdivided into homoclinal and distally steepened ramps by [4]. Ramp deposits comprise the foundation phases for many large-scale carbonate platforms and occur as major basin fills in some structural and depositional settings. They also host a considerable number of petroleum and mineral deposits. Recently, with advancing carbonate hydrocarbon exploration and development, important oil and gas fields in connection with carbonate ramp deposits have been detected successively, such as the Early Carboniferous carbonate ramp of Brabant in Wales [5], the Lower Triassic Kangan Formation carbonate ramp of the South Pars Field in the Persian Gulf Superbasin [6], the carbonate ramp of the Early Cambrian Longwangmiao Formation in the Sichuan Basin [7], the Ordovician carbonate ramp of the Gucheng area and the Early Cambrian Xiaoerbulake Formation carbonate ramp in the Tazhong Uplift of the Tarim Basin [8]. Ramp-type carbonate platforms have attracted increasing attention [9,10] and have become favorable exploration targets. Despite their abundance and undoubted economic importance, the facies characteristics of Cambrian carbonate ramps and the relationships between facies and high-quality reservoirs are poorly understood.

The Middle–Early Cambrian sedimentary succession (540–513 Ma) of the Tarim Basin is configured with high-quality source rocks in the Early Cambrian Yuertusi Formation, large-scale dolostone reservoir rocks in the Early Cambrian Xiaoerbulake Formation and the Early Cambrian Wusongeer Formation and widespread seals of gypsum caprocks in the Middle Cambrian Shayilike Formation and the Awatage Formation. These favorable conditions for hydrocarbon accumulation enable this interval to be an important strategic replacement of deep to ultradeep oil and gas exploration [11]. The recent breakthroughs of Wells ZS1, LT1 and JNKT1 in the Cambrian subsalt dolostone have rendered the strata increasingly prominent for oil and gas exploration. The Xiaoerbulake Formation has become a key hydrocarbon exploration target in the Tarim Basin as the main horizon of the Early Cambrian subsalt dolostone [12]. However, deep to ultradeep carbonate rocks in China are characterized by old age, great burial depth (4600–8500 m) and long-term tectonic evolution [13]. The prototype basins are poorly preserved, and the distribution of high-energy facies zones is limited; therefore, exploration and development are impeded by huge technical and commercial risks. For example, Wells YL6, XH1, CT1 and HT2 failed successively (Figure 1a), calling into question the exploration direction and potential of deep to ultradeep carbonate rocks to a certain extent. The major theoretical and technical bottleneck is identifying whether to detect and finely describe the deep to ultradeep large-scale high-quality reservoirs [11,14,15]. Global case studies show that the distribution of large-scale carbonate reservoirs is mostly facies-controlled [6,10,16,17]. Therefore, an accurate determination of facies types and the distribution of facies is the key to predict high-quality reservoirs in the deep to ultradeep Xiaoerbulake Formation in the Tarim Basin. Economic interest has driven increased research into the sedimentary facies of the Early Cambrian Xiaoerbulake Formation in the Tarim Basin. Researchers have proposed that the Xiaoerbulake Formation belongs to the sedimentary facies zone of ramp, ramp–tidal flat, restricted platform, semi-evaporative platform, open platform margin or slope–basin [14,18–21]. However, previous studies mainly focused on limited well logging and cutting data and low-resolution 2D and 3D seismic data, but did not thoroughly address with the facies and facies associations of the Early Cambrian Xiaoerbulake Formation carbonate ramp. There is no fine characterization of the facies. To date, intensive research has not been conducted on the types and characteristics of the ramp facies and their impacts on favorable reservoirs. Thus, this will restrict the further exploration of the Early Cambrian subsalt dolostone in the Tarim Basin.

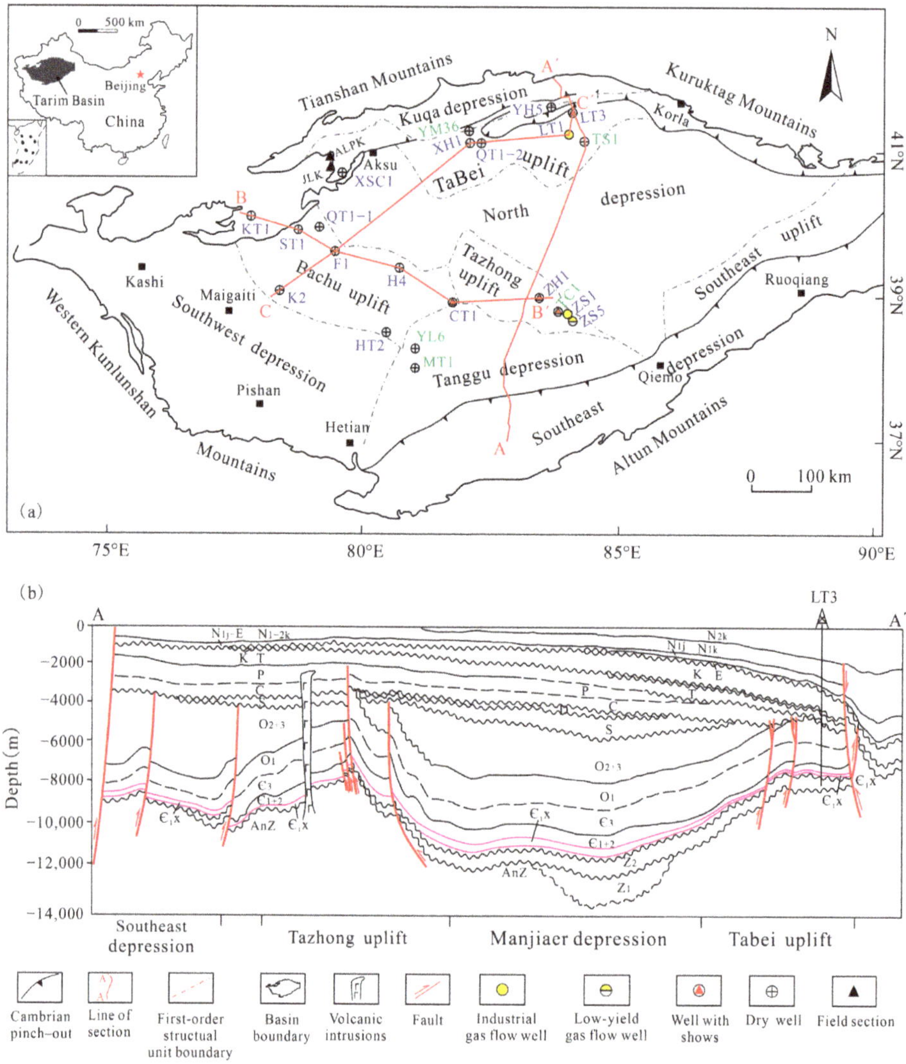

Figure 1. (**a**) Schematic tectonic provinces in the Tarim Basin (adapted from [22]); (**b**) Structural–stratal configuration section in the north–south direction (A–A′) (adapted from [23]. AnZ: Ediacaran; Z_2q: Qigebulake Formation; $Є_1$y: Yuertusi Formation; $Є_1$x: Xiaoerbulake Formation; $Є1x_1{}^1$: the first member of the Lower Xiaoerbulake Formation; $Є_1$w: Wusongeer Formation; $Є_2$s: Shayilike Formation; $Є_2$a: Awatage Formation; O_1: Lower Ordovician; O_{2+3}: Mid–upper Ordovician; S: Silurian; C: Carboniferous; P: Permian; K, T: Jurassic and Triassic; E: Eocene; N: Neogene. Exploration wells (well names in blue) of the Lower Cambrian: KT1: Well Ketan1; ST1: Well Shutan1; QT1−1: Well Qiaotan1; K2: Well Kang2; F1: Well Fang1; HT2: Well Hetian2; H4: Well He4; CT1: Well Chutan1; ZH1: Well Zhonghan1; ZS1: Well Zhongshen1; ZS5: Well Zhongshen5; XSC1: Well Xinsucan1; XH1: Well Xinhe1; QT1−2: Well Qitan1; YH5: Well Yaha5; LT1: Well Luntan1; LT3: Well Luntan3. Exploration wells (well names in green) of the Early Cambrian Sayilike Formation and Awatage Formation: YL6: Well Yulong6; MT1: Matan1; TC1: Well Tacan1; YM36: Well Yingmai36; TS1: Well Tashen1. ALPK: Aolinpike section; JLK: Jinlinkuang section. Thickness data of field sections from [14].

Based on sedimentological methods, this study systematically analyzes the facies, depositional environment and reservoir quality of an Early Cambrian carbonate ramp in the Tarim Basin using drilling, core, cutting and well logging data. The aim of this paper is to (1) reveal facies and facies association characteristics; (2) establish a facies model; (3) discuss the controlling factors of facies and facies associations; (4) clarify high-quality reservoirs and (5) explore the relationship between facies and high-quality reservoirs in an Early Cambrian carbonate ramp in the Tarim Basin.

2. Geological Setting

The Tarim Basin is located in the south of the Xinjiang Uygur Autonomous Region of Northwest (NW) China. With an area of 56×10^4 km^2, the Tarim Basin is the largest inland petroliferous superimposed basin developed on the Archean–Early Mesoproterozoic metamorphic crystalline basement and metamorphic fold basement. The Tarim Basin has experienced several periods of major structural changes during its formation and evolution, which gave rise to the present tectonic pattern of "four uplifts and five depressions" surrounded by mountains of the basin margin. The mountains include the Tianshan Mountains to the northwest, the Kuruktag Mountains to the southeast, the Kunlun Mountains to the southwest and the Altun Mountains to the southeast. The "four uplifts" are the Tabei Uplift, Bachu Uplift, Tazhong Uplift and Southeast Uplift. The "five depressions" are the Kuqa Depression, North Depression, Tanggu Depression, Southeast Depression and Southwest Depression (Figure 1a).

The Tarim Basin has undergone multiple periods of subsidence and uplift [24]. From the Early Cryogenian to the Early Paleozoic, influenced by aggregation and breakup events on the Rodinia supercontinent and the Gondwana continent, the Tarim Basin experienced strong regional extensional tectonic movements, which led to the formation of a unified paleo–Tarim plate. In the late Ediacaran, the Tarim Basin entered the rift–depression transition stage of the ancient rift system from the rift stage owing to mantle bulge, crustal attenuation or extension (Figure 1b). In addition to local synsedimentary faults, the basin was dominated by thermal subsidence and demonstrated differential characteristics, namely "uplift in the south and platform in the north, high in the south and low in the north" [14,25]. At the end of Ediacaran, the Cryogenian–Ediacaran Formations in the whole basin were denuded to varying degrees, while a regional large-scale unconformity between Ediacaran Formations and Cambrian Formations was generated owing to the Keping movement [26]. From the Early Cambrian to the Middle Cambrian, the post-rift subsidence stage and construction stage of the craton occurred under an extensional background in the Tarim Basin (Figure 2). In addition to the late Ediacaran differentiation characteristics of "north–south differentiation, structural high in the south and structural low in the north", the basin exhibited a new characteristic of "structural high in the west and structural low in the east" [27].

The Cryogenian–Middle Cambrian sedimentary succession of the Tarim Basin has been extensively studied, mostly through drilling data for hydrocarbon exploration [12,14,19,25,28]. The Cryogenian–Ediacaran sedimentary succession in the Tarim Basin was deposited with extremely thick coarse clastic deposits interbedded with mudstone and carbonate rocks, multiple sets of volcanic extrusive rocks, intrusive rocks and four sets of lodgement till deposits. These deposits recorded the tectono–sedimentary response from the initial rifting stage to the main rifting stage. The Early–Middle Cambrian sedimentary succession includes the Early Cambrian Yuertusi, Xiaoerbulake and Wusonggeer Formations and the Middle Cambrian Shayilike and Awatage Formations, which were revealed in the drilling results from 22 exploration wells (Figures 1a and 2). The sediments in the Tarim Basin were restricted by the paleogeographic framework with three paleohighs (Uqia, Tanan and Keping–Wensu) and two depressions (southwestern Tarim Depression and North Depression) in the Early Cambrian. The Yurtusi Formation was clearly controlled by the Precambrian tectonic framework of the alternated uplifts and depressions, and has a sediment thickness of 8–70 m. A set of diamictite rocks can be observed in the Bachu Uplift, and the Tabei Uplift is dominated by black bioclastic mudstone, siliceous mudstone and muddy limestone interbedded with thin-bedded dolostone, which are favorable source rocks. The Xiaoerbulake Formation comprises a set of carbonate ramp deposits with a thickness ranging from 38 m to 372 m. According to the drilling results of Wells ZS1, ZH1, CT1, QT1–2 and KT1, in the southeast of the Tazhong Uplift, the Bachu Uplift and the northern margin of the Tanggu Depression, the formation is mainly characterized by terrigenous diamictite sediment, peloidal algal dolostone and dolostone with a set of diamictite and muddy dolostone in the middle part and anhydrite–salt rock and local volcanic rock intrusions. The upper part evolves into a set of argillaceous limestone and marl deposits in the QT1–1 well block of the Bachu Uplift. In the Tabei Uplift, represented by Wells XH1, LT1, LT3 and QT1–2, the lower part of the Xiaoerbulake Formation mainly comprises medium–thick-bedded dark gray (muddy) dolomitic or dolostone-bearing limestone, algal limestone and peloidal micritic limestone, while the upper part is mainly composed of (mud-bearing) peloidal or fine-grained dolomitic limestone and bioclastic (dolomitic) limestone. The lithologic associations of thick-bedded dolostone interbedded with thin-bedded algal dolostone are well developed in the local well blocks such as the YH5 well block of the Tabei Uplift. The Wusonggeer Formation has a thickness ranging from 32 m to 423 m. Argillaceous or terrigenous clastic dolostones interbedded with algal dolostone, dolomitic mudstone, gypsiferous rocks and local intrusive rocks were recorded in wells QT1–1, K2, HT2 and ZS1 in the Bachu Uplift, Tanggu Depression and Tazhong Depression. In the Tabei Uplift, the lower part of the Wusonggeer Formation in Well XH1 and Well LT1 mainly comprises a set of thick-bedded (dolostone-bearing) peloidal or fine-grained limestone with thin-bedded marl and (mud-bearing) dolomitic limestone. The middle–upper part in Wells LT3, TS1 and LT1 consists of thick-bedded peloidal dolostone, (mud-bearing) calcareous dolostone or dolomitic limestone. This set of formations is missing in the QT1–2 well block owing to the presence of a submarine uplift in the Tabei Uplift. The Shayilike and Awatage Formations are characterized by extensive thick anhydrite–salt rock and anhydritic dolostone with thicknesses of 48–459 m and 158–494 m, reflecting the sedimentary environment of strongly rimmed platforms dominated by evaporative lagoons under the prolonged hot arid paleoclimate.

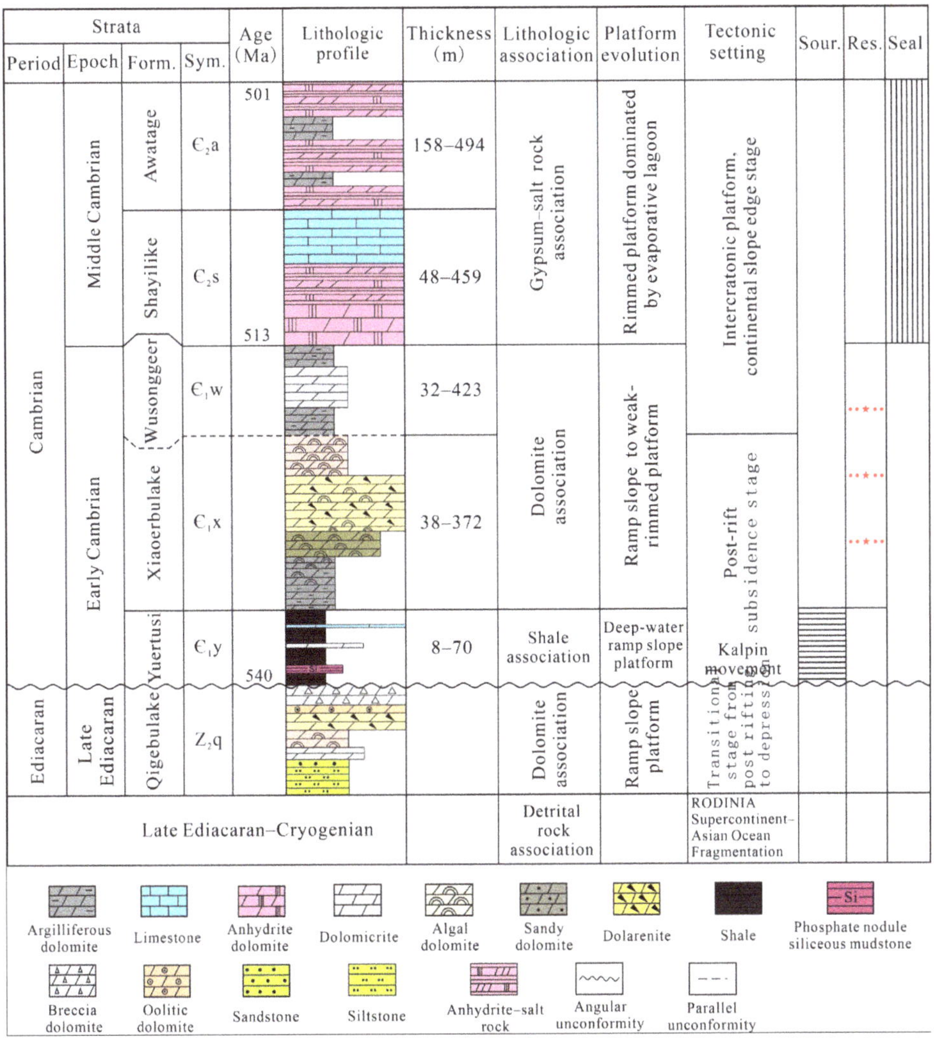

Figure 2. Generalized stratigraphic column (adapted from [11]). Age data from [29,30]. The thickness data are derived from the 17 exploration wells shown in Figure 1a.

3. Materials and Methods

A combination of 120 m of cores, 3240 m of cuttings and wireline logs from 17 wells were used to study the facies and to present a sedimentary model of the Early Cambrian Xiaoerbulake Formation in the Tarim Basin (Table 1). Over 1100 thin sections were prepared from the core and cutting samples collected at intervals of 2–5 m and at each lithological change in each well. The cores, cuttings and drilling data were provided by the Exploration and Development Research Institute of PetroChina Tarim Oilfield Company in Korla, Bayingolin Mongolian Autonomous Prefecture, Xinjiang Uyghur Autonomous Region, Northwest China. These thin sections were stained with alizarin red S and observed and photographed using a Leica DM2500 transmitted and reflected polarizing microscope. Over 12,000 photographs of the representative lithologic characteristics were obtained. The facies characteristics, such as the lithology (studied in terms of bedding, color, grain size, proportions of grains and amount of mud), thickness, grain types, fossils, and depositional textures, were studied in detail.

Table 1. Core and cuttings information, thickness and depth of the Early Cambrian Xiaoerbulake Formation from 17 wells.

Well	Thickness (m)	Depth (m)	Sampling Depth of Cuttings (m)	Length of Cuttings (m)	Coring Depth (m)	Core Length (m)	Core Recovery (%)
ZH1	109	7381–7490	7381–7490	109	7356.9–7369.8 7387.5–7391.5 7400–7413 7473.69–7480.85	12.74 2.68 8.66 7.16	98.8 67 70.6 100
XH1	372	7420–7792	7420–7792	372	7473.69–7480.85	7.16	100
ZS1	49.85	6760.65–6810.5	6760.65–6810.5	49.85	6804–6807.36	2.62	78
YH5	273.8	6023.48–6297.28	6023.48–6297.28	273.8	6023.48–6025 6122.2–6127.2 6240.86–6246.26	1.52 5 5.4	100 100 100
LT3	160.45	8343–8503.45	8343–8503.45	160.45	8370–8378	7.99	99.9
F1	204.85	4409.15–4614	4409.15–4614	204.85	4513.1–4520.25 4579.63–4582.53	7.15 2.9	100 100
QT1−1	210	5860–6070	5860–6070	210	6006–6013	7	100
HT2	74.5	6460.5–6535	6460.5–6535	74.5	6492.12–6500.12	8	100
K2	195.65	5438.85–5634.5	5438.85–5634.5	195.65	5490.2–5497.2 5630.61–5634.5	6.6 3	94.3 77.1
ST1	188.2	1871.8–2000	1871.8–2000	188.2	1883–1889.3 1915–1919.43 1988.67–1996	5.98 4.3 7.29	94.9 97.1 99.5
KT1	212	4692–4904	4692–4904	212	4895.48–4903.48	7	87.5
XSC1	221.5	4852.5–5074	4852.5–5074	221.5		No coring	
CT1	71.15	7725.85–7797	7725.85–7797	71.15			
H4	140.8	5669.2–5810	5669.2–5810	140.8			
QT1−2	356.05	5600.3–5956.35	5600.3–5956.35	356.05			
ZS5	37.87	6749.5–6787.37	6749.5–6787.37	37.87			
LT1	363	8260–8623	8260–8623	363			

To interpret the paleoenvironment, we first created a lithologic description presenting different lithotypes, including color, thickness and sedimentary structures such as lamination and bioturbation and vertical and lateral distribution of individual lithofacies types and cycles. Subsequently, we determined the arrangement of the lithofacies types and conducted facies and facies association analyses. The classifications of carbonate rocks and mixed siliciclastic–carbonate rocks were conducted according to the nomenclature of [31–33]. Classification and descriptive terms for crystallization textures and crystallization fabrics of dolostone were provided by [33–35]. Carbonate platforms and carbonate ramp depositional systems developed by [1,4,33,36] were referred to for the sedimentary models of the Early Cambrian Xiaoerbulake Formation.

To investigate the main factors controlling reservoir quality, we first characterized the pore system such as space types and physical properties using thin sections and the porosity of log interpretation. Thin-section samples were impregnated with blue or pink epoxy to facilitate the recognition of porosity. Then, we investigated the diagenesis associated with the pore system of hydrocarbon reservoirs through thin sections and casting thin sections of well cuttings and drill cores. The porosity was computed using acoustic, neutron and density logs. The pore type classification developed here is based on the classification systems of [37].

4. Results

4.1. Facies Analysis

Based on the depositional attributes and textural constituents, ten types of facies were recognized in the Early Cambrian Xiaoerbulake Formation in the Tarim Basin. They were further grouped into seven facies associations (FA1–FA7; Table 2) and twenty-six main types of facies associations were present (TFA1–TFA26; Table 2).

Table 2. Facies, facies association and depositional environment of an Early Cambrian carbonate ramp in the Tarim Basin, NW China.

Facies Association (FA)	Type of Facies Association (TFA)	Facies Type	Well	Depositional Environment
Tidal flat facies association (FA1)	Thick-bedded sandy dolostone (TFA1).	Mixed siliciclastic–carbonate rock (F1)	ZS1 and ZH1	
	Dolomitic sandstone with thin interlayer of argillaceous rock (TFA2).		KT1	
	Sandstone and dolomitic sandstone with thin interlayer of mudrock (TFA3).		ZH1	
	Medium- to thick-bedded dolostone (TFA4).	Crystalline dolostone (F2)	YH5, ST1, QT1–1, ZS1, ZH1, K2 and H4	
	Medium-bedded dolostone with thin interlayers of argillaceous dolostone (TFA5).		ZS1, ZS5	
	Dolostone with interbedded thin layers of oolitic dolostone and anhydrite-bearing dolostone (TFA6).		CT1	
	Dolomitized stromatolite interbedded with medium–thin interlayers of argillaceous dolostone (TFA7).	Dolobindstone (F3)	YH5	
	Medium- to thick-bedded dolomitized stromatolite (TFA8).		YH5, ST1 and K2	Inner ramp
Lagoon facies association (FA2)	Argillaceous dolostone and dolomicrite with medium–thin interlayers of dolostone (TFA9).	Dolomudstone (F4)	F1 and ZH1	
	Mud-bearing limestone interbedded with dolomicrite and argillaceous limestone (TFA10).	Argillaceous dolostone (F5)	QT1–1	
	Dolostone, limy dolostone and dolomite-bearing limestone with thin interlayers of mudrock and silt-bearing argillaceous dolostone (TFA11).		ST1 and QT1–1	
High-energy shoal facies association (FA3)	Sand-bearing dolarenite and oolitic dolostone with thin interlayer of dolostone (TFA12).	Dolograinstone (F6)	CT1	
	Dolarenite and oolitic dolostone interbedded with dolostone (TFA13).		CT1, ZH1, ZS1 and H4	
	Thrombolites and algal lumps (TFA14).		KT1 and ST1	
High-energy shoal facies association (FA3)	Dolarenite with thin interlayers of algal dolostone (TFA15).	Dolograinstone (F6)	F1	
	Dolarenite and oolitic dolostone with thin interlayers of algal dolostone (TFA16).		CT1	
	Algal dolostone interbedded with oolitic dolostone, dolarenite and grain dolostone (TFA17).		K2	Inner ramp
Sabkha facies association (FA4)	Anhydrite–salt rock with thin–medium interlayers of anhydritic dolostone (TFA18).	Anhydrite–salt rock (F7)	K2	
	Anhydritic dolostone with thin interlayers of anhydrite–salt rock (TFA19).		H4 and K2	
Inter-shoal facies association (FA5)	Medium- to thick-bedded dolomite-bearing limestone with thin to medium interlayers of mud-bearing limestone (TFA20).	Microcrystalline limestone (F8)	QT1–2, XSC1 and LT3	
	Thick-bedded dolomite-bearing limestone (TFA21).		YH5, XSC1 and LT1	
	Dolostone, limestone and dolomite-bearing limestone with argillaceous limestone interbed (TFA22).		QT1–2	Mid-ramp
Low-energy shoal facies association(FA6)	Dolomite-bearing limestone, dolomitic limestone and limestone interbedded with thin-bedded peloidal dolomite-bearing limestone (TFA23).	Peloidal grainstone (F9)	XH1 and LT1	
	Medium–thick-bedded peloidal dolomite-bearing limestone (with peloidal dolomitic limestone and peloidal liny dolostone) (TFA24).		QT1–2, LT1 and LT3	
Open-shelf facies association (FA7)	Thick-bedded mud-bearing limestone with thin interlayers of argillaceous limestone and dolomite-bearing limestone (TFA25).	Argillaceous limestone (F10)	QT1–2 and LT3	Outer ramp
	Dolomite-bearing limestone and mud-bearing limestone with thin interlayers of argillaceous limestone (TFA26).		XH1	

Note that the terms of thick-bedded, medium-bedded and thin-bedded are determined by the relative thickness of lithotypes in TFA.

4.1.1. Mixed Siliciclastic–Carbonate Rock (F1)

Description: F1 mainly comprises sandy dolostone (Figure 3A,B), dolomitic sandstone (Figure 3C) and sandstone (Figure 3D). Sandy dolostone is generally gray in color and 6.6 m to over 19 m thick. Microscopically, it dominantly comprises dolomite and quartz, with a small amount of bitumen (Figure 3A) or argillaceous materials filling intercrystalline pores. The dolomite crystals are silt-sized to finely crystalline (<0.25 mm) and are characterized by inequigranular texture and hypidiotopic and idiotopic fabric. The concentration of quartz grains that are embedded in dolomite crystals accounts for 25%–35% of the total composition. The grains are moderately sorted, predominantly angular and mostly silt-sized to medium sand (0.04–0.3 mm in size). Dolomitic sandstone is gray to dark gray in color and 3–9 m thick. Microscopically, this lithotype mainly consists of silicified quartz and dolomite. Quartz grains are poorly sorted and angular. Furthermore, their size is fine to medium sand (0.1–0.3 mm), exhibiting monocrystalline and polycrystalline structures. Most quartz grain margins were dissolved (Figure 3C). Dolomite is microcrystalline in size. Quartz sandstone is commonly light gray to gray in color and ~8 m thick. Microscopically, this lithotype includes quartz, bituminous and dolomicritic matrix and glauconite (Figure 3D). Quartz grains are moderately sorted, and most of them are smaller than 0.1 mm, while medium or coarse-sized grains > 0.4 mm can be observed sporadically. Silt-sized quartz grains are mainly angular, and sand-sized quartz grains are rounded, mainly elliptical in shape. Siliciclastic sediments in the study area mainly occurred in the lower part of the Xiaoerbulake Formation adjoining the Southern Tarim Paleohigh and Wuqia Paleohigh [25].

F1 is present in TFA1–TFA3 (Table 2; Figure 4A–C). TFA1 occurs in the middle part of the Xiaoerbulake Formation in the ZS1 well block and the upper part of the Xiaoerbulake Formation in the ZH1 well block in the Tazhong Uplift. TFA2 mainly occurs at the base of the Xiaoerbulake Formation in the KT1 well block in the Keping Uplift, with a thickness of 18 m. It comprises dolomitic fine sandstone, dolomitic medium sandstone, dolomitic siltstone and argillaceous siltstone with thin interlayers of argillaceous rock. TFA3 is found at the base of the Xiaoerbulake Formation in the ZH1 well block in the Tazhong Uplift, with a thickness of up to 40 m. It is made up of thin interlayers of fine sandstone, siltstone, dolomitic siltstone, argillaceous dolomitic siltstone, argillaceous siltstone and mudrock. Vertically, fine sandstone, siltstone, dolomitic siltstone, argillaceous dolomitic siltstone and argillaceous siltstone form rhythmic superposition. TFA2 and TFA3 are packaged in upward-fining transgressive systems, tracting and draping over underlying Precambrian metamorphic basement rock of the paleohighs with a regional angle unconformity surface [25].

Interpretation: Mixed siliciclastic–carbonate sedimentation is related to paleo-environ mental conditions, such as paleoclimate and eustatic sea level change [38]. According to paleoclimate data reported by [39], siliciclastic sedimentation likely prevailed during the relatively hot and humid period, then palaeoclimatic conditions became more arid, favoring the deposition of carbonate [40] and evaporitic sediments [41]. This finding indicates that the early sedimentary period of the Xiaoerbulake Formation could have been a hot and humid climate in which siliciclastic sedimentation predominated. The middle sedimentary period of the Xiaoerbulake Formation might have evolved into an arid climate in which thick-bedded anhydrite–salt rock was formed in the middle part of the Xiaoerbulake Formation, as can be observed in some areas such as the K2 and H4 well blocks. The poorly sorted and angular siliciclastic grains in F1 suggest limited abrasion during transport, possibly reflecting a short transport to the tidal flat environments of the carbonate inner ramp from paleohighs by rivers [33,42]. The participation of airborne transport, however, cannot be excluded [42,43].

4.1.2. Crystalline Dolostone (F2)

Description: F2 is generally gray to dark gray in color and medium-bedded or thick-bedded (5 m to nearly 100 m thick). It mainly comprises silt-sized crystalline dolostone (Figure 3E), finely crystalline dolostone (Figure 3F,G) and medium crystalline dolostone (Figure 3H). Silt-sized crystalline dolostone and medium crystalline dolostone commonly occur in the Early Cambrian Xiaoerbulake Formation in the Tarim Basin, while medium crystalline dolostone is rarely observed. Microscopically, F2 is characterized by tightly packed inequigranular, mosaic, nonrhombic, anhedral and subhedral dolomite crystals exhibiting lobate and straight boundaries. The grain sizes of crystals vary from 0.04 mm to 0.3 mm. Intercrystalline pores are usually infilled with anhydrite (Figure 3F), with amounts ranging from 10% to 20%.

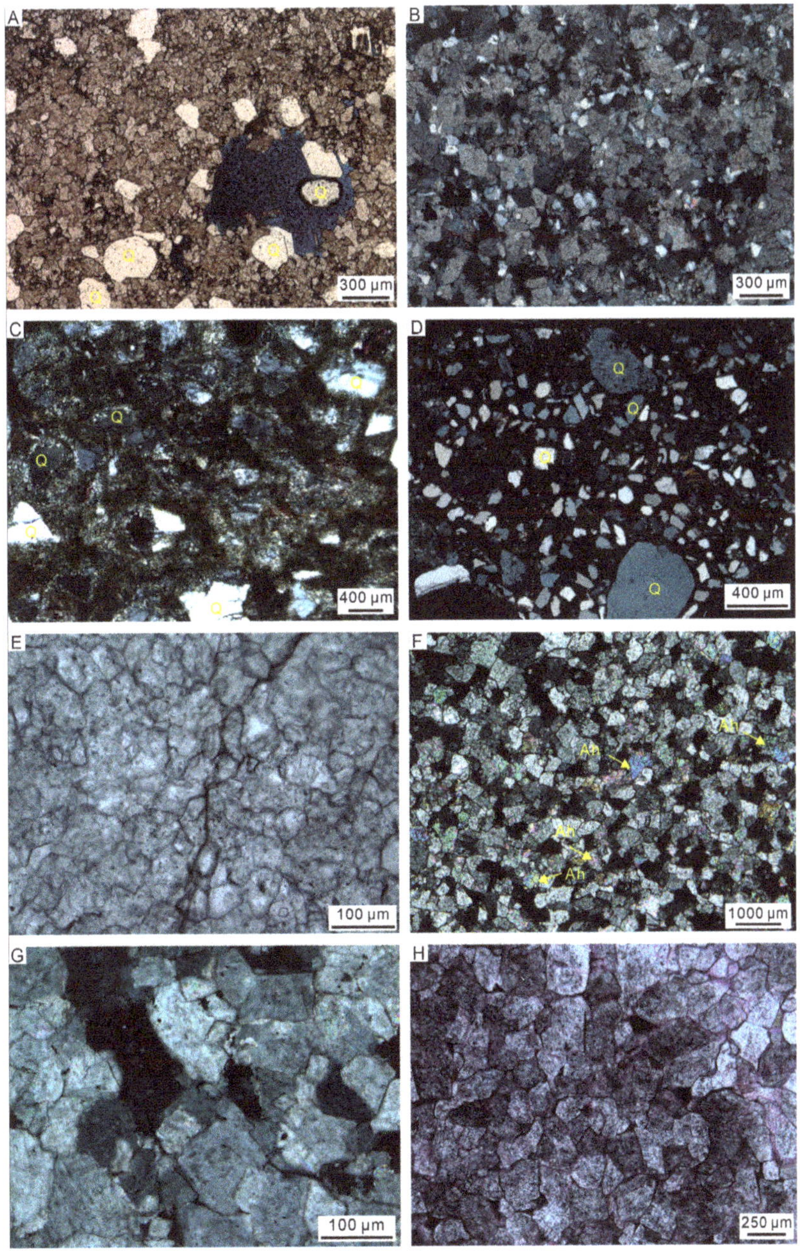

Figure 3. Photomicrographs of FA1 of the Early Cambrian Xiaoerbulake Formation in the Tarim Basin, Northwest China. (**A**) Sandy dolostone with subangular quartz grains, Well ZH1, depth 7387.5 m. (**B**) Silty dolostone, Well ZH1, depth 7479 m. (**C**) Dolomitic sandstone, Well KT1, 4900 m. (**D**) Bituminous sandstone, Well ZH1, 7484 m. (**E**) Silt-sized crystalline dolostone, Well CT1, depth 7794 m. (**F**) Finely crystalline dolostone with anhydrite, Well HT2, depth 6525 m. (**G**) Finely crystalline dolostone, Well HT2, depth 6525 m. (**H**) Medium crystalline dolostone, Well ST1, 1990.7 m. Q: grain; Ah: anhydrite.

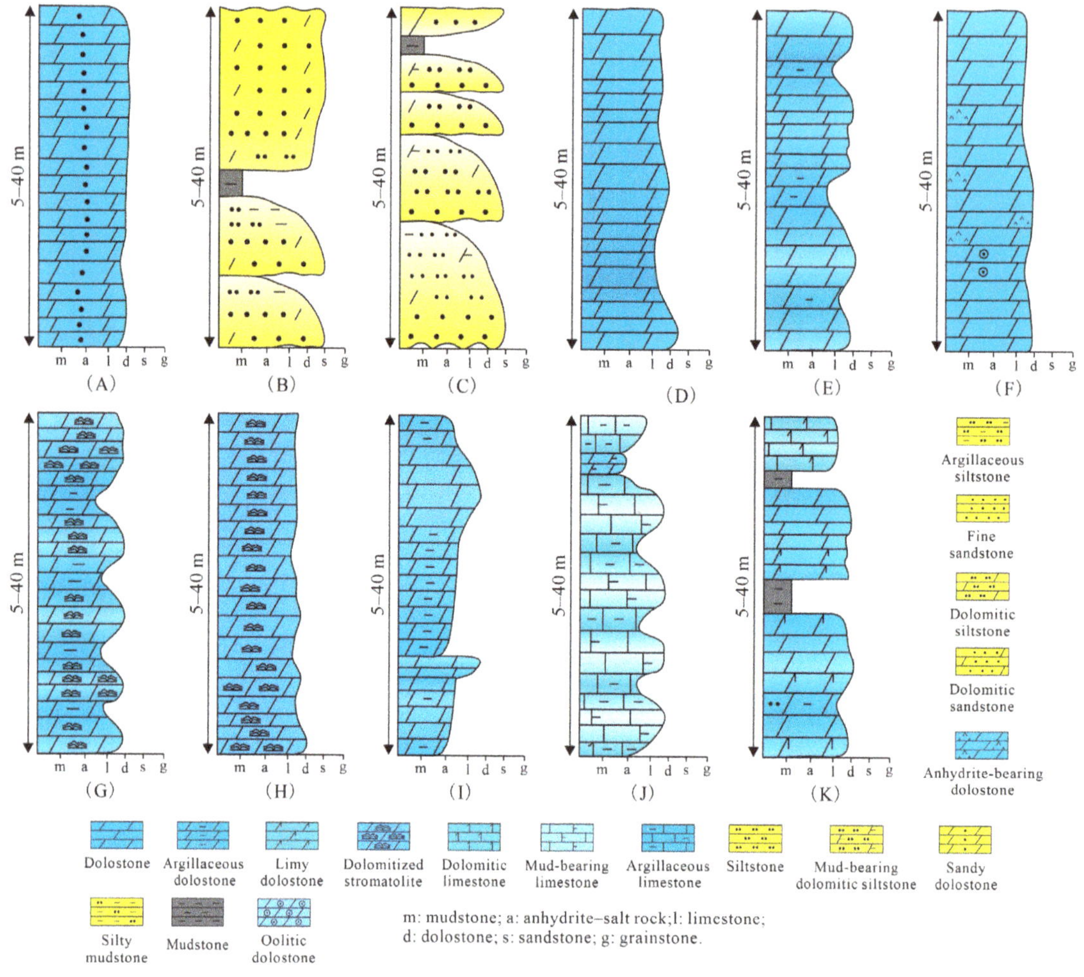

Figure 4. TFA1-10 of FA1 and FA2 of the Early Cambrian Xiaoerbulake Formation in the Tarim Basin, Northwest China. (**A**) TFA1; (**B**) TFA2; (**C**) TFA3; (**D**) TFA4; (**E**) TFA5; (**F**) TFA6; (**G**) TFA7; (**H**) TFA8; (**I**) TFA9; (**J**) TFA10; (**K**) TFA11.

This facies is present in TFA4, TFA5 and TFA6 (Table 2; Figure 4D–F). TFA4 is 5–100 m thick and better developed in the middle and upper part of the Xiaoerbulake Formation of the Keping Uplift, Bachu Uplift and Tazhong Uplift and the north of the Tabei Uplift, represented by Wells KT1, ST1, H4, QT1–1, K2, F1, ZS1, ZH1 and YH5. TFA5 is 5–10 m thick and is characterized by one to three sets of thin interlayers (0.9–2 m thick). This type of facies association is mainly found in the ST1, YH5 and ZH5 well blocks. TFA6 is developed in the top of the Xiaoerbulake Formation in the CT1 well block with a thickness of ~10 m and comprises dolostone and interbedded thin layers of oolitic dolostone and anhydrite-bearing dolostone (Figure 4F).

Interpretation: The occurrence of crystallization textures and crystallization fabrics in F2 (Figure 3E–H) has been extensively documented and is deemed to stem from dolomitization and recrystallization [33,44]. Anhydrite was created through gypsum dehydration when buried at depths below the reaction isotherm, provided that the fluids could escape from the dehydrating system [45]. This transformation could even occur at shallow depths owing to increased heat flows related to the emplacement of local crustal magmatic bodies. The gypsum formation can be described by two processes: (a) under near-surface conditions, gypsum in the overlying evaporite strata was dissolved by fresh water, and Ca and S infiltrated into the dolostone pores, replacing dolostone with gypsum [46]; (b) under the hot and arid paleoclimate conditions, the seawater evaporated continuously, the salinity increased and the sulfate in the water gradually crystallized and precipitated to form gypsum [47]. The medium-bedded and thick-bedded dolostone with anhydrite suggests that F2 was likely deposited in a tidal flat environment, as reported by other researchers [6,33,36].

4.1.3. Dolobindstone (F3)

Description: F3 is gray to dark gray in color and mainly comprises dolomitized stromatolites (Figure 5A,B), with a thickness exceeding 10 m. It is found mostly in the ST1 well block of the Bachu Uplift and the upper part of the Xiaoerbulake Formation in the YH5 well block in the Tabei Uplift. The stromatolite has a bright and dark layer structure, exhibiting wavy lamination (Figure 5A) and even lamination (Figure 5C) under the microscope. Birdseyes are common in the stromatolite of the ST1 well block. The birdseyes of wavy stromatolite exhibit plate and stripe shapes, incompletely filled with dolosparite, with a length of tens of centimeters (Figure 5A). Birdseyes in even stromatolite present isolated, spotted and irregular voids filled with dolosparite. The void walls of birdseyes in stromatolite have a pectinate or agate ring edge and lack geopetal structure.

F3 is present in TFA7 and TFA8 (Table 2; Figure 4G,H). TFA7 is ~30 m thick and is only developed in the YH5 well block. The dolomitized stromatolite in TFA7 is 1.5–5.3 m thick, and the thin-bedded argillaceous dolostone is 1–2.5 m thick. Dolomitized stromatolite and argillaceous dolostone occur in thin interbedded layers. TFA8 mainly comprises dolomitized stromatolites with a thickness of 5–17 m and is found in the YH5, ST1 and K2 well blocks.

Interpretation: The origin of the birdseye structures in stromatolites may be described by two main genetic models: bubbles and dry shrinkage after organic matter decomposition [48]. Dolosparite in birdseyes developed because of recrystallization owing to dolomitization [49]. Stromatolites are laminated organosedimentary deposits that have accreted due to a benthic microbial community dominated by cyanobacteria after having trapped and bound detrital sediment and/or formed the locus of mineral precipitation [33,50]. Growth forms of stromatolites are deemed to be predominantly controlled by water depths and the degree of water turbulence [33,51]. Even stromatolites are common in supratidal and intertidal calm water environments, while wavy stromatolites are common in intertidal and shallow subtidal environments with relatively strong hydrodynamic conditions. Therefore, the dominance of wavy stromatolites and stratiform stromatolites with birdseyes indicates that F3 was deposited in the tidal flat environment under hot and arid climatic conditions [52].

4.1.4. Dolomudstone (F4)

Description: F4 is generally dark gray to black in color and thin- to medium-bedded (1.4–9 m thick). It mainly comprises dolomicrite and is found in the upper beds of the Xiaoerbulake Formation in the QT1–1 well block, the lower part of the Xiaoerbulake Formation in the ZH1 well block and the middle part of the Xiaoerbulake Formation in the F1 well block. Microscopically, F4 mainly contains black organic-rich or bituminous dolomitic mud and dolomite (Figure 5D) without sedimentary structures and grain structures. Some dolomite crystals are dolomitized or recrystallized to microspar.

F4 is mainly present in TFA9 (Table 2; Figure 4I). TFA9 consists of lower dolomicrite (F4), middle–lower argillaceous dolostone, upper crystalline dolostone and thin-bedded argillaceous dolostone. It is found in the middle–lower part of the Xiaoerbulake Formation of the F1 and ZH1 well blocks.

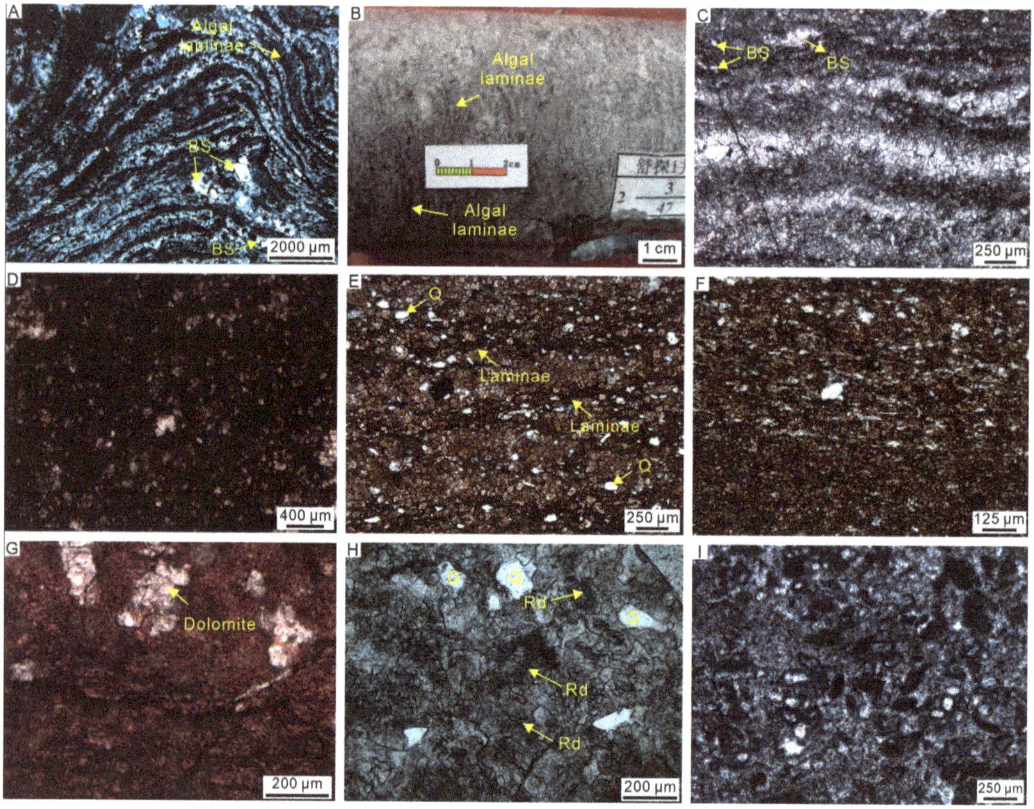

Figure 5. Photomicrographs and core images of FA1, FA2 and FA3 of the Early Cambrian Xiaoerbulake Formation in the Tarim Basin, Northwest China. (**A**) Wavy stromatolites with birdseyes, Well ST1, depth 1883.4 m. (**B**) Stromatolites, Well ST1, depth 1883.38 m. (**C**) Even stromatolites, Well ST1, 1886.22 m. (**D**) Dolomicrite, Well QT1–1, depth 5953 m. (**E**) Laminated argillaceous dolostone with silty quartz grains (Q), Well ST1, depth 1994.98 m. (**F**) Silt-bearing argillaceous dolomicrite, Well ST1, depth 1995.4 m. (**G**) Mud-bearing limestone, Well QT1–1, depth 5878 m. (**H**) Residual doloarenite (Rd) with fine-grained scattered quartz grains (Q), Well CT1, depth 7786 m. (**I**) Doloarenite, Well ST1, depth 2058.96 m. BS: birdseyes; Q: grain; Rd: residual doloarenite.

Interpretation: The absence of sedimentary and grain structures and the presence of black-colored dolomicrite in F4 suggest a protected lagoon marine setting with anoxic conditions [53].

4.1.5. Argillaceous Dolostone (F5)

Description: F5 is brown and dark gray in color and thin- to thick-bedded (0.8–10 m thick). It mainly comprises silt-bearing argillaceous dolostone (Figure 5E,F) and is found in the middle beds of the Xiaoerbulake Formation in the ZS1, F1 (Fang1) and ZS5 well blocks, in the lower part of the Xiaoerbulake Formation in the YH5 and ZH1 well blocks and in the upper part of the Xiaoerbulake Formation in the QT1–1 well block. Microscopically, F5 comprises silt-sized dolomite (Figure 5E) or mud-sized dolomite (Figure 5F), mud matrix and silty quartz grains with silty and argillaceous laminae. Crystallized textures of silt-sized dolomite are characterized by equigranular, tightly packed anhedral and subhedral crystals. Silty quartz grains are angular and scattered in the dolomite crystals and mud matrix (Figure 5E).

F5 is mainly present in TFA10 and TFA11 (Table 2; Figure 4J,K) and also observed in TFA5 and TFA7 (Figure 4E,G). TFA10 is ~40 m thick and is found in the upper part of the Xiaoerbulake Formation of the QT1–1 well block. In this type of facies association, F4 occurs as a thin interbedded layer in the upper part with a thickness of 3.7 m. TFA10 is ~25 m thick in the middle part of the Xiaoerbulake Formation in the ST1 well block and is up to 40 m thick in the middle–upper part of the Xiaoerbulake Formation in the QT1–1 well block. F5 is usually intercalated in limy dolostone, mud-bearing limestone (Figure 5G) and dolomicrite, as described in TFA10 and TFA11 (Figure 4J,K), and in dolostone and stromatolites, as described in TFA5 and TFA7 (Figure 4E,G).

Interpretation: The dominance of mud and the presence of scattered angular silty quartz grains with silty and argillaceous laminae point to a lower-energy depositional setting adjacent to the paleohighs [54]. The occurrence of silt-bearing argillaceous dolostone with silty and argillaceous laminae as interbedded thin layers in TFA5, TFA6 and TFA10 indicates that F5 was deposited in a low-energy, tidal flat environment [55] or restricted lagoonal environment [56].

4.1.6. Dolograinstone (F6)

Description: F6 comprises six lithotypes, sand-bearing dolarenite (Figure 5H), dolarenite (Figures 5I and 6A,B), oolitic dolopackstone (Figure 6C), oolitic dolograinstone (Figure 6D,E), thrombolites (Figure 6F) and algal lumpy dolograinstone (Figure 6G), which widely occur in the well blocks of the Bachu Uplift and the Tazzhong Uplift and in the KT1 well block, except for the oolitic dolopackstone. Sand-bearing dolarenite is gray in color with a bed thickness of ~9 m and is found at the base of the Xiaoerbulake Formation in the CT1 well block. Microscopically, fine-grained quartz grains are mainly angular and are scattered among grains of doloarenite, with amounts of 5%–10%. Dolarenite is brown-gray to gray in color and thin–thick-bedded (1.5–38 m thick). Microscopically, the residual texture of dolarenite is distinctly observed, and the grain size varies from 0.1 mm to 0.9 mm, with crystallized textures of inequigranular and anhedral crystals. Oolitic dolograinstone is yellowish white to gray in color and thin–medium-bedded (3–14 m thick). Microscopically, the ooids are 0.07–0.9 mm in diameter and are oval and spherical in shape. They are characterized by high concentrations, spheroidal and ellipsoidal shapes, and moderately to well-sorted and grain-supported texture with sparite dolomite cements. Dolomite crystals of ooids are equigranular, tightly packed anhedral, silt-sized and finely crystalline (Figure 6C,D). Although ooids' nuclei and concentric laminae are hardly observed (Figure 6D,E), some ooids exhibit traces of their concentric microstructure (Figure 6D). The ooids ranging from 0.05 mm to 0.12 mm in size are observed in the top of the Xiaoerbulake Formation in the CT1 well block with a thickness of ~2 m. According to the ooid structure and preservation, this lithotype can be classified into micrite, radial and recrystallized ooids (Figure 6C). Micrite ooids often occur together with radial ooids with anhydrite and silica (Figure 5B), displaying neither concentric nor radial microfabrics. Radial ooids vary from 0.08 mm to 0.12 mm in diameter. The radial–fibrous crystals build the inner part of the ooids and occur within the concentrically arranged outer laminae (cortices), with micritic fabric filling the ooid nuclei. Recrystallized ooids are 0.07–0.09 mm in size, and the crystal texture is characterized by a single dolomite crystal (Figure 6B). Thrombolites and algal lumps are usually observed in the same beds (Figure 6F,G). They are gray to dark gray in color and thin- to thick-bedded (from 1.5 m to more than 100 m thick). Under the microscope, clots and lumps exhibit irregular (Figure 4I), broken lobate (Figure 6F) and oval shapes (Figure 6G), with sizes of 0.15–1.5 mm. The components among them are silt-sized and finely crystalline dolosparites, and calcite found in low concentrations.

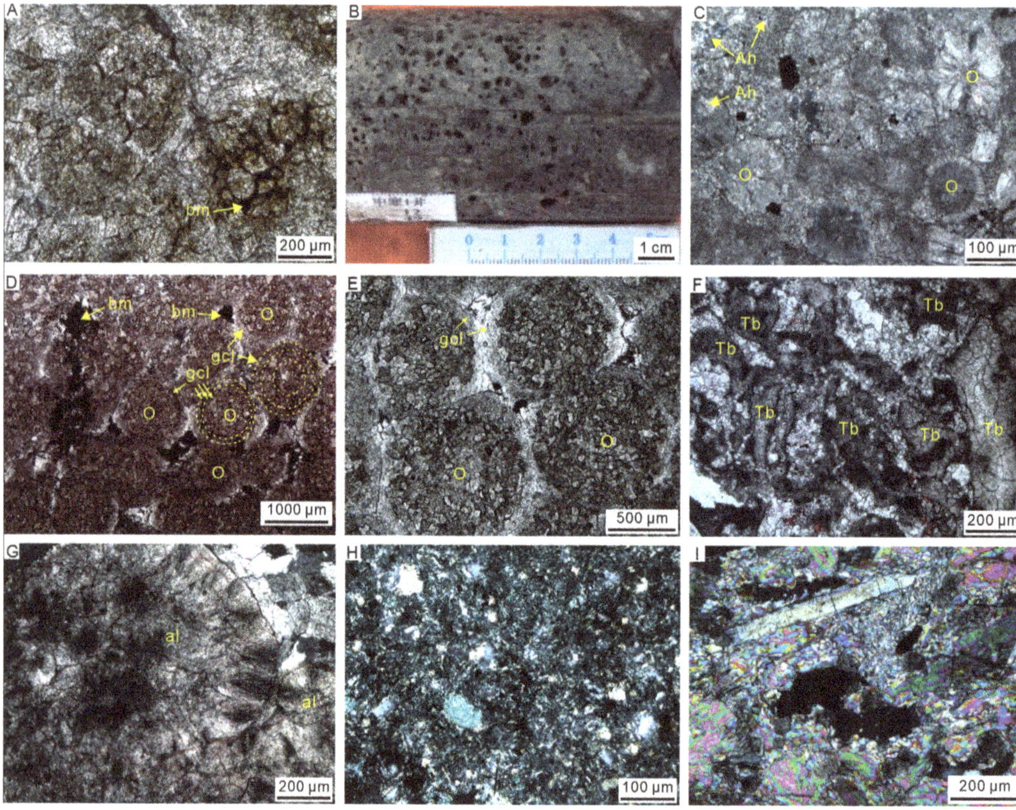

Figure 6. Photomicrographs and core images of FA3, intrusive rock and FA4 of the Early Cambrian Xiaoerbulake Formation in the Tarim Basin, Northwest China. (**A**) Doloarenite, Well ZS1, depth 6766 m. (**B**) Doloarenite with vuggy pores, Well ZH1, depth 7408.1 m. (**C**) Oolite with anhydrite, Well CT1, depth 7728 m. (**D**) Dolomitized ooids showing ghosts of concentric laminae and intergranular pores filled with bitumen (bm), Well CT1, depth 7767.93 m. (**E**) Ooids with complete replacement dolomitization, Well CT1, depth 7766.67 m. (**F**) Thrombolites, Well KT1, depth 4830 m. (**G**) Algal lump (al), Well KT1, depth 4835 m. (**H**) Diabase, Well KT1, depth 4820 m. (**I**) Anhydrite–salt rock, Well K2, depth 5548 m. gcl: ghosts of concentric laminae; bm: bitumen; Ah: anhydrite; Q: grain; Tb: thrombolite; O: ooid; al: algal lump.

F6 is mainly present in TFA12–17 (Table 2; Figure 7A–F). TFA12 comprises sand-bearing dolarenite, an interbedded thin layer of dolostone and oolitic dolostone from base to top (Figure 7A). It has a thickness of up to 30 m and occurs in the lower part of the Xiaoerbulake Formation in the CT1 well block. TFA13 demonstrates a vertical stacking pattern of dolarenite, oolitic dolostone and dolostone interlayers (Figure 7B), with a thickness of 20–60 m. It is mainly developed in the middle–upper part of the Xiaoerbulake Formation in the ZH1 well block and in the upper part of the Xiaoerbulake Formation in the H4 and CT1 well blocks. TFA14 consists of thrombolites and algal lumps (Figure 7C), with a thickness of 17–70 m, and is mainly found in the lower part of the Xiaoerbulake Formation in the ST1 and KT1 well blocks. TFA15 is ~40 m thick and mainly distributed in the upper part of the Xiaoerbulake Formation in the F1 well block. It is made up of dolarenite (30 m thick) and interbedded thin layers of algal dolostone (2–4 m thick) (Figure 7D). TFA16 is characterized by a vertical superposition of dolarenite and oolite with thin-bedded algal dolostone (Figure 7E). It is mainly found in the upper part of the Xiaoerbulake Formation in the CT1 well block with a thickness of ~20 m. TFA17 comprises oolitic, dolarenite and algal dolostone superimposed vertically (Figure 7F). It is mainly developed in the middle–lower part of the Xiaoerbulake Formation in the K2 well block, with a thickness of ~40 m. The overlying formation is anhydrite rock of the Wusonggeer Formation, while the underlying formation is TFA16.

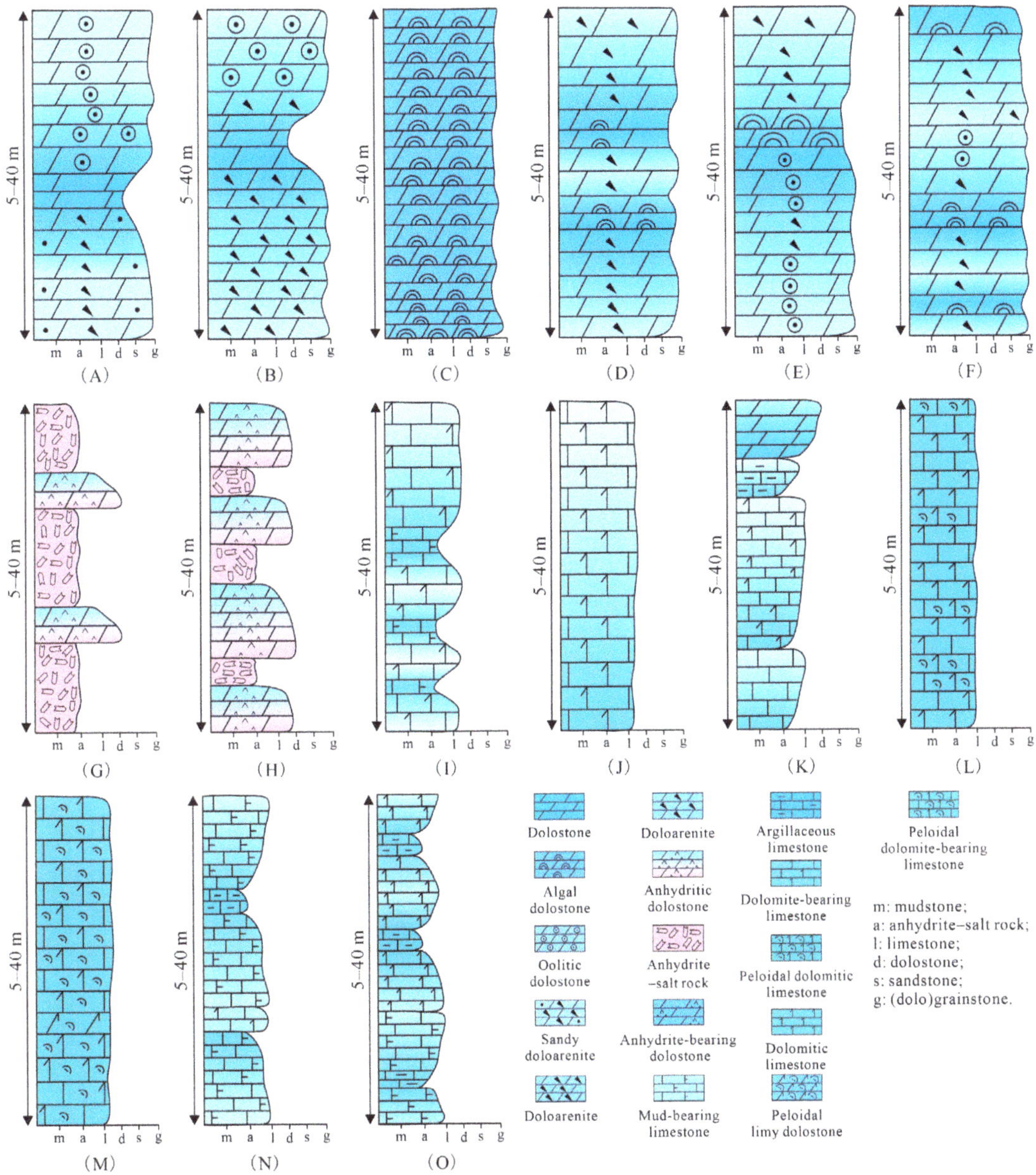

Figure 7. TFA12–26 of FA3–7 of the Early Cambrian Xiaoerbulake Formation in the Tarim Basin, Northwest China. (**A**) TFA12; (**B**) TFA13; (**C**) TFA14; (**D**) TFA15; (**E**) TFA16; (**F**) TFA17; (**G**) TFA18; (**H**) TFA19; (**I**) TFA20; (**J**) TFA21; (**K**) TFA22; (**L**) TFA23; (**M**) TFA24; (**N**) TFA25; (**O**) TFA26.

Interpretation: The angular shape of the quartz grains possibly suggests that the grains were transported to the carbonate ramp by rivers and currents for a short distance [57] or their transport was related to wind blowing [42,58]. The absence of ooid nuclei and some ghost features of concentric laminae imply the destruction of original structures [59]. Micritic fabric and silt-sized to sand-sized dolomite grains commonly filling ooid cortices, concentric laminae and nuclei are likely the result of diagenetic alteration (burial dolomitization) of primary minerals [59,60], which started with an initial precipitation along the grain periphery and was subsequently followed by dolomite precipitation along the porous cortical laminae [33,61]. The common presence of varied medium to coarse sand-sized grainstones and the vertical superpositions of oolites with high amounts of ooids, dolarenite, thrombolites and algal lumps indicate that F6 was deposited in a high-energy shoal environment [33,42,54,56,62]. Notably, the presence of all the ooids is not indicative of a high-energy sedimentary environment. Well-sorted and fine sand-sized oolites with micrite, radial and recrystallized ooids (Figure 8B) were likely formed in a relatively calm or low-energy lagoon. Alternatively, they were perhaps reworked and transported into a low-energy setting [33,63].

4.1.7. Anhydrite–Salt Rock (F7)

Description: F7 is light brown to brown in color and thin–medium-bedded (1.5–7 m thick). Under the microscope, the evaporite mainly comprises anhydrite and a small amount of dolostone (Figure 3H). The anhydrite crystals exhibit a lath-like shape, and their concentration exceeds 90%, while the dolostone concentration is <10% (Figure 6I).

F7 is present in TFA18 and TFA19 (Table 2; Figure 7G,H). TFA18 comprises anhydrite–salt rock and thin–medium interlayers of anhydritic dolostone and is found in the middle part and top of the Xiaoerbulake Formation in the K2 well block, with a thickness of ~40 m. The underlying and overlying stratum are TFA7 and TFA4, respectively. TFA19 is 8–24 m thick and composed of anhydritic dolostone and interbedded thin layers of anhydrite–salt rock. It mainly occurs in the upper part of the Xiaoerbulake Formation in the K2 well block and in the lower–middle part of the Xiaoerbulake Formation in the H4 well block. The underlying and overlying stratum of TFA19 is TFA4.

Interpretation: The anhydrite developed from the initial gypsum crystals that had grown and later dehydrated to form anhydrite in the buried environment [6,64]. The dominant occurrence of anhydrite in F7 points to a hot and arid, hypersaline sabkha setting, similar to that of the modern gypsum deposits along the western coast of the Persian Gulf [64].

4.1.8. Microcrystalline Limestone (F8)

Description: F8 is gray to dark gray in color and medium- to thick-bedded (18 m to over 100 m thick). It comprises dolomite-bearing and dolomitic microcrystalline (microsparitic) limestone. Microscopically, it is mainly made up of calcite, dolomite, bioclast and a small amount of pyrite (Figure 8A,B). Calcite was completely recrystallized and partially dolomitized. Calcite and dolomite are characterized by microspar textures, with crystal sizes < 0.03 mm in diameter. Wavy and horizontal laminae are locally observed.

F8 is extensively distributed in the Xiaoerbulake Formation in the Tabei Uplift and XSC1 well blocks and is present in TFA20–22 (Table 2; Figure 7I–K). TFA20 occurs within the Xiaoerbulake Formation in the QT1–2, XSC1 and LT3 well blocks; TFA21 is found in the Xiaoerbulake Formation in the QT1–2, XSC1 and LT3 well blocks and TFA22 is mainly developed in the top of the Xiaoerbulake Formation in the QT1–2 well block. TFA20, TFA21 and TFA22 coexist with F9 and F10 (Figure 7I,K,L,N,O).

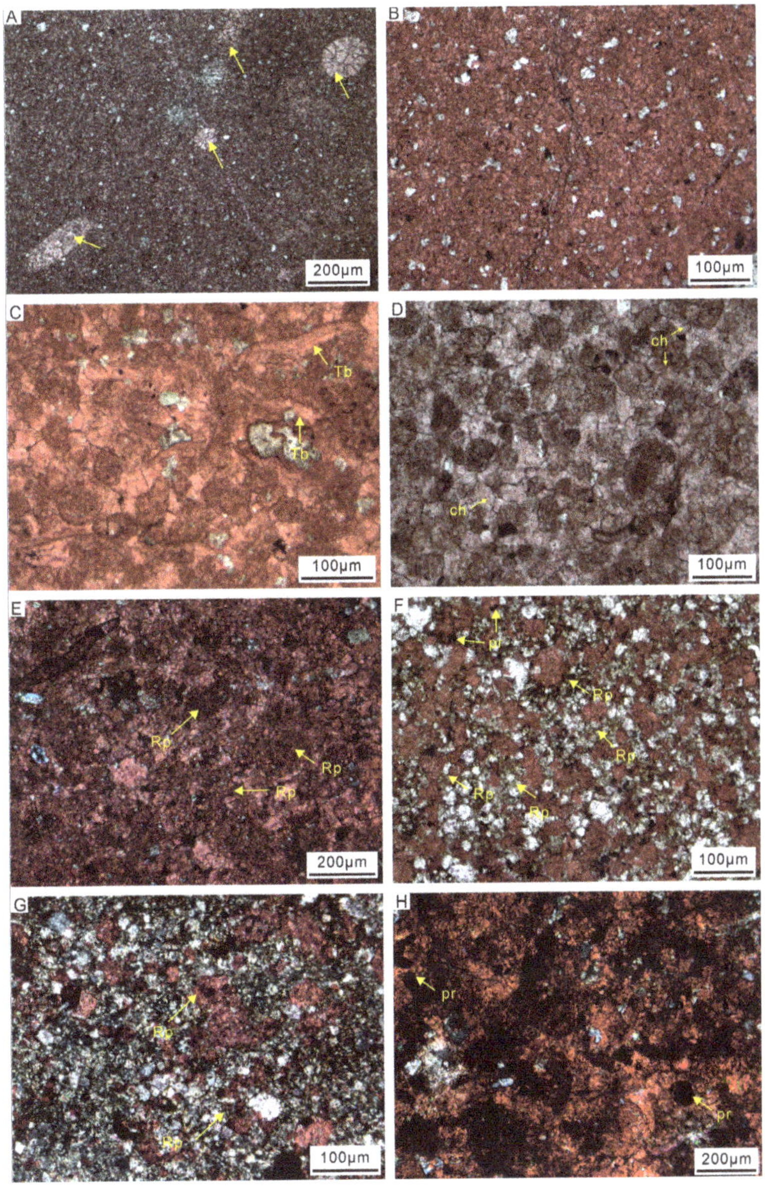

Figure 8. Photomicrographs of FA5, FA6 and FA7 of the Early Cambrian Xiaoerbulake Formation in the Tarim Basin, Northwest China. (**A**) Dolomitic limestone with bioclastics (arrows), Well XH1, depth 7510 m. (**B**) Dolomite-bearing limestone, Well LT1, depth 8400 m. (**C**) Peloidal dolomite-bearing limestone with trilobites, Well LT1, depth 8560 m. (**D**) Peloidal dolomite-bearing limestone with calthrops, Well LT3, depth 8369.01 m. (**E**) Peloidal dolomite-bearing limestone, Well XH1, depth 7565 m. (**F**) Peloidal dolomitic limestone with residual peloids and pyrite, Well LT3, depth 8365.38 m. (**G**) Peloidal limy dolostone, Well LT3, depth 8362.10 m. (**H**) Argillaceous limestone with a small amount of silicated dolostone and pyrite, Well LT3, depth 8495 m. Tb: trilobite; bm: bitumen; ch: calthrop; Rp: residual peloid; pr: pyrite.

Interpretation: Microspar limestone likely originated from lime mudstone or silt-sized carbonate grains (peloidal grainstone) [65]. The original textures of the grains were destroyed due to strong recrystallization during the diagenetic process. The presence of medium- to thick-bedded microcrystalline limestone with trilobite bioclasts and the vertical stratal relationship with F8 and F10 indicate that F9 was deposited in an inter-shoal environment of the mid-ramp [66].

4.1.9. Peloidal Grainstone (F9)

Description: Three lithotypes are recognized in F9 according to the dolomite and calcite concentrations. These are peloidal dolomite-bearing limestone (Figure 8C–E), peloidal dolomitic limestone (Figure 8F) and peloidal limy dolostone (Figure 8G). Dolomite-bearing limestone is gray to dark gray in color and commonly occurs in the Xiaoerbulake Formation of the Tabei Uplift, with a thickness ranging from 5 m to 80 m. Microscopically, it is dominated by a large amount of small (from 0.03 mm to 0.2 mm in size), poorly to moderately sorted, subrounded and subangular peloids with trilobite and monaxon fragments. Peloids comprise mud-sized and very finely crystalline calcites and have no internal structures. They are cemented by very finely crystalline calcite and small amounts of very finely crystalline dolomite. Peloidal dolomitic limestone and peloidal limy dolostone are gray in color. Peloidal dolomitic limestone is found in the upper part of the Xiaoerbulake Formation in the XH1 well block and in the top of the Xiaoerbulake Formation in the LT3 well block with a thickness of 4–10 m. Peloidal limy dolostone is only observed in the top of the Xiaoerbulake Formation in the LT3 well block with a thickness of 3.5 m. Though these two lithotypes were extensively dolomitized, a respectable amount of calcitized peloids have preserved grain structure and can be distinguished from dolomite. Microscopically, the peloid components also comprise very finely crystalline calcite as dolomite-bearing peloidal limestone is embedded in recrystallized dolomite. The dolomite crystals exhibit very fine, inequigranular and anhedral to euhedral crystallization textures.

F9 is present in TFA23 and TFA24 (Table 2; Figure 7L,M). TFA23 comprises dolomitic limestone (with a dolomite content of 25%–50%) or dolomite-bearing limestone (with a dolomite content of 5%–25%) interbedded with dolomite-bearing peloidal limestone and is found in the middle–upper part of the Xiaoerbulake Formation in the XH1 well block and in the top and lower parts of the Xiaoerbulake Formation in the LT1 well block. TFA24 comprises medium–thick-bedded peloidal dolomite-bearing limestone and peloidal dolomitic limestone interbedded with a thin layer of peloidal limy dolostone vertically and is found in the lower part of the Xiaoerbulake Formation in the LT1 well block and in the top of the Xiaoerbulake Formation in the LT3 well block.

Interpretation: The presented characteristics of peloids in F9 indicate that these grains likely resulted from the reworking of weakly lithified carbonate mud located in F8 by storm waves [33,67]. Subsequently, the peloids were recrystallized from micrite-sized calcite crystals during recrystallization. The dominance of peloids in F9, the absence of coarse amalgamated tempestites in F9 and the overlying and underlying strata and the vertical stratal relationship with F8 and F10 demonstrate that unlike F6, F9 was deposited in a relatively low-energy shoal environment of the mid-ramp [1,33,68].

4.1.10. Argillaceous Limestone (F10)

Description: F10 is dark gray in color and thin–thick-bedded (2–8 m thick). Under the microscope, it mainly comprises calcite, terrigenous mud and a small amount of dolomite (Figure 8H). Calcite was completely micritized and locally dolomitized. Calcite and dolomite are characterized by very fine crystalline textures, with crystal size < 0.025 mm. Terrigenous mud was infected with organic matter and bitumen.

F10 usually occurs in thin interbedded layers between mud-bearing, very finely crystalline dolomitic limestone, very finely crystalline dolomitic limestone and very finely crystalline limestone. It is present in TFA25 and TFA26 (Table 2; Figure 7N,O). TFA25 is found in the middle–lower part of the Xiaoerbulake Formation in the QT1–2 and LT3 well blocks, while TFA26 is distributed in the lower part of the Xiaoerbulake Formation in the XH1 well blocks. Vertically, TFA25 and TFA26 lie over black shale or black siliceous shale of the Wusonggeer Formation in the LT3, LT1, QT1–2 and XH1 well blocks in the Tabei Uplift.

Interpretation: The presence of black terrigenous mud and the vertical stratal relationship with black shale or black siliceous shale of a deep outer ramp in the Wusonggeer Formation indicate that F10 was deposited in a deeper water environment, mostly a shallow outer ramp [1,33,69]. The noncarbonate concentrations of terrigenous mud increase as the deposition rate decreases near the far terrigenous end of a mixed siliciclastic–carbonate ramp [69], suggesting a deeper water environment.

4.2. The Distribution of Facies Associations

To better clarify the lateral distribution and spatial evolution of the facies associations of the Xiaoerbulake Formation, we selected two well-based correlation sections oriented NW–SN and EW for systematic analysis (Figures 9 and 10). The interwell facies characteristics of well-based correlation sections mainly follow the work on the sedimentary facies and lithofacies paleogeography of the Xiaoerbulake Formation reported by other researchers [12,14,19,25,28,70]. From the KT1 well block to the ZH1 well block in the Tazhong Uplift, the depositional environment is an inner ramp. The sedimentary thickness gradually decreases, reflecting the affection of the Southwestern Tarim Uplift. In the direction of the Southwestern Tarim Uplift (from the KT1 and ZH1 well blocks to the CT1 well block), the lower part of the Xiaoerbulake Formation is represented by the high-energy shoal facies association. The sedimentary thickness gradually thins and the depositional environment grades into a tidal flat facies association in the H4 well block. The middle part of the Xiaoerbulake Formation is dominated by the lagoon facies association, which is mutated into the sabkha facies association in the H4 well block and is not present in the CT1 well block. The upper part of the Xiaoerbulake Formation is mainly characterized by the high-energy shoal facies association or tidal flat facies association to high-energy shoal facies association. In the region near the Southwestern Tarim Uplift, oolitic shoals and sand-size grain shoals are commonly observed in the high-energy shoal facies association (Figure 9). From the K2 well block in the Bachu Uplift to the LT3 well block in the Tabei Uplift, the sedimentary thickness of the Xiaoerbulake Formation tends to increase, evolving from 195.7 m in Well K2 to 363 m in Well LT1. The depositional environment changes from the inner ramp to the middle ramp and subsequently to the outer ramp. The facies associations evolve from the vertical stacking combination of FA3, FA4 and FA1 and the vertical stacking combination of FA3, FA1, FA2 and FA3 in the K2 and F1 well blocks in the Bachu Uplift to the vertical stacking combination of FA7, FA6 and FA5 in the XH1 and LT1 well blocks and subsequently to FA7 in the LT3 well block (Figure 10).

4.3. Depositional Model

From the facies associations of the Early Cambrian Xiaoerbulake Formation in the Tarim Basin, a depositional model of a carbonate ramp from paleohighs to depressions has been noted (Figure 11). The carbonate ramp in the Xiaoerbulake Formation comprises an inner ramp, a middle ramp and an outer ramp environment. Of these, the inner ramp includes a tidal flat facies association, a lagoon facies association, a sabkha facies association and a high-energy shoal facies association; the middle ramp includes an inter-shoal sea facies association and a low-energy shoal facies association and the outer ramp mainly comprises an open-shelf facies association (Table 2). Jointly controlled by the paleogeographic framework with three paleohighs (Wuqia Paleohigh, Southern Tarim Paleohigh and Keping–Wensu Paleohigh) and two depressions (Southwestern Tarim Depression and North Depression) during the sedimentary period of the Xiaoerbulake Formation [12,14,25,28], the carbonate ramp deposits demonstrate the difference between the depositional environments, facies associations and facies types in different regions of the Tarim Basin. The depositional environment was an inner ramp, a middle ramp and an outer ramp from the Wuqia Paleohigh in the western Tarim Basin to the eastern deep-water basin (Figure 11). The inner ramp occurs above the fair weather wave base and is typically found in the Bachu Uplift, Tazhong Uplift and KT1 well block. Affected by terrigenous inputs, a set of mixed siliciclastic–carbonate rocks of a tidal flat facies association was mainly deposited adjoining the paleohighs. It graded into crystalline dolostone and bindstone of a tidal flat facies association or sabkha facies association, lagoon facies association and high-energy shoal facies association toward the depressions. The middle ramp is located between the fair weather wave base and the storm wave base and is mainly distributed in the middle of the Tabei Uplift, Bachu Uplift and Tazhong Uplift. It is sequentially composed of microcrystalline limestone of inter-shoal facies association and peloidal grainstone of low-energy shoal facies association toward the depressions (Southwestern Tarim Depression and North Depression). The outer ramp occurs in the depositional environment below the normal storm wave base. The seawater in the depositional area is calmer than that of the middle ramp. The distribution position is roughly similar to that of the middle ramp on the plane and different in space. It mainly comprises the open-shelf facies association.

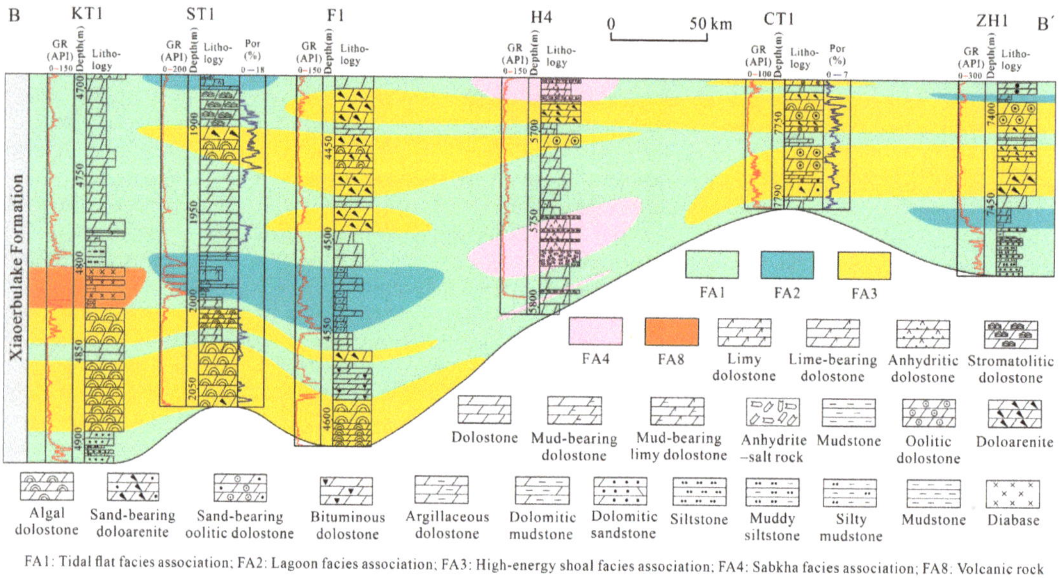

Figure 9. Well-based facies association distribution in the NW–SE direction (KP1 well block–Bachu Uplift–Tazhong Uplift) of the Early Cambrian Xiaoerbulake Formation in the Tarim Basin (the location of cross-well profile B–B′ is shown in Figure 1). Por: porosity, the porosity data from well logging interpretation. B: the starting location of the cross-well profile (B–B′); B′: the end location of the cross-well profile (B–B′).

4.4. Reservoir Pore Systems

According to the microscopic characteristics of thin sections, five main pore types are identified: interparticle pores, intraparticle pores, intercrystalline pores, vuggy pores and moldic pores (Figure 12).

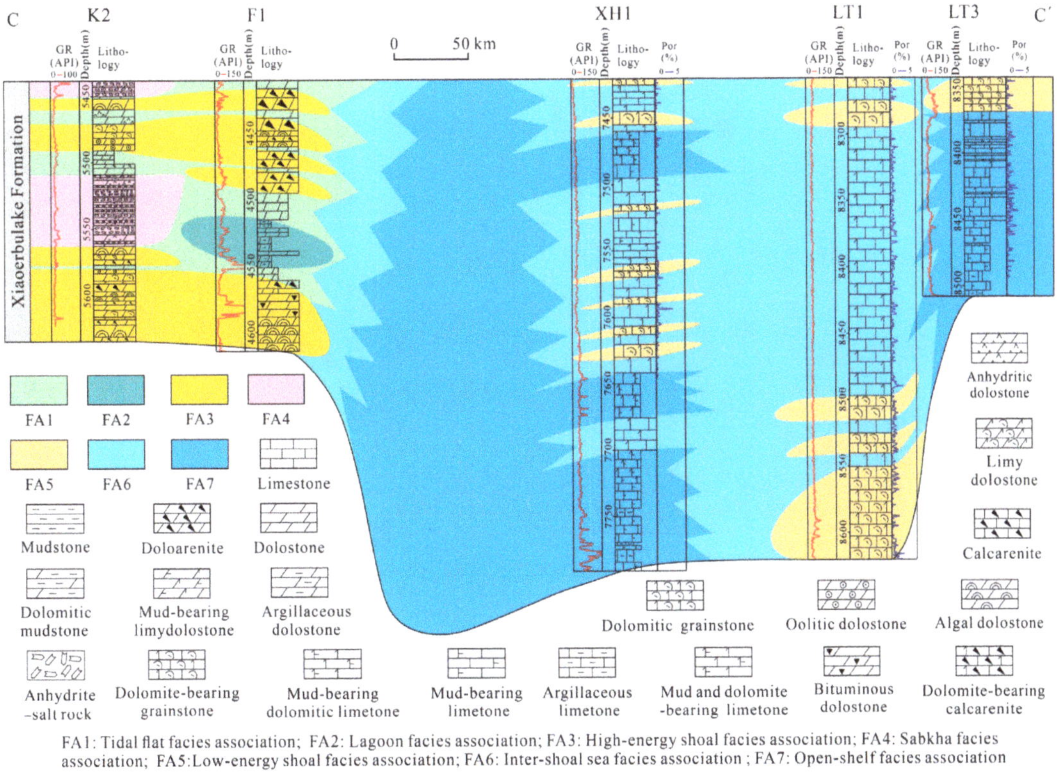

Figure 10. Well-based facies association distribution in EW direction of the Early Cambrian Xiaoerbulake Formation in the Tarim Basin (the location of cross-well profile C–C′ is shown in Figure 1). Por: porosity; the porosity data from well logging interpretation. C: the starting location of the cross-well profile (C–C′); C′: the end location of the cross-well profile (C–C′).

4.4.1. Interparticle Pores

The term "interparticle pore" presented herein is used to describe the pores formed between grains (intergrain), which agrees with the definition provided by [71]. Interparticle pores primarily occur in dolograinstone of high-energy shoal facies association, including sand-bearing dolarenite, dolarenite and oolitic dolostone. The pores occur between ooids and sand-sized grains, mostly exhibiting irregular polygonal and embayed shapes. These pores are filled with bitumen, anhydrite cements and coarsely crystalline dolomite (Figure 12A–C). Dolomite cements exhibit clear crystal faces and euhedral and subhedral crystal shapes. This pore type is characterized by large pore sizes ranging from 0.03 mm to 1 mm, while some solution-enlarged interparticle pores are up to 2 mm in diameter. Interparticle pores are commonly related to the selective dissolution of cements and matrix, strongly controlled by fluid migration pathways and the patchy cementation of both primary and secondary interparticle pores [72].

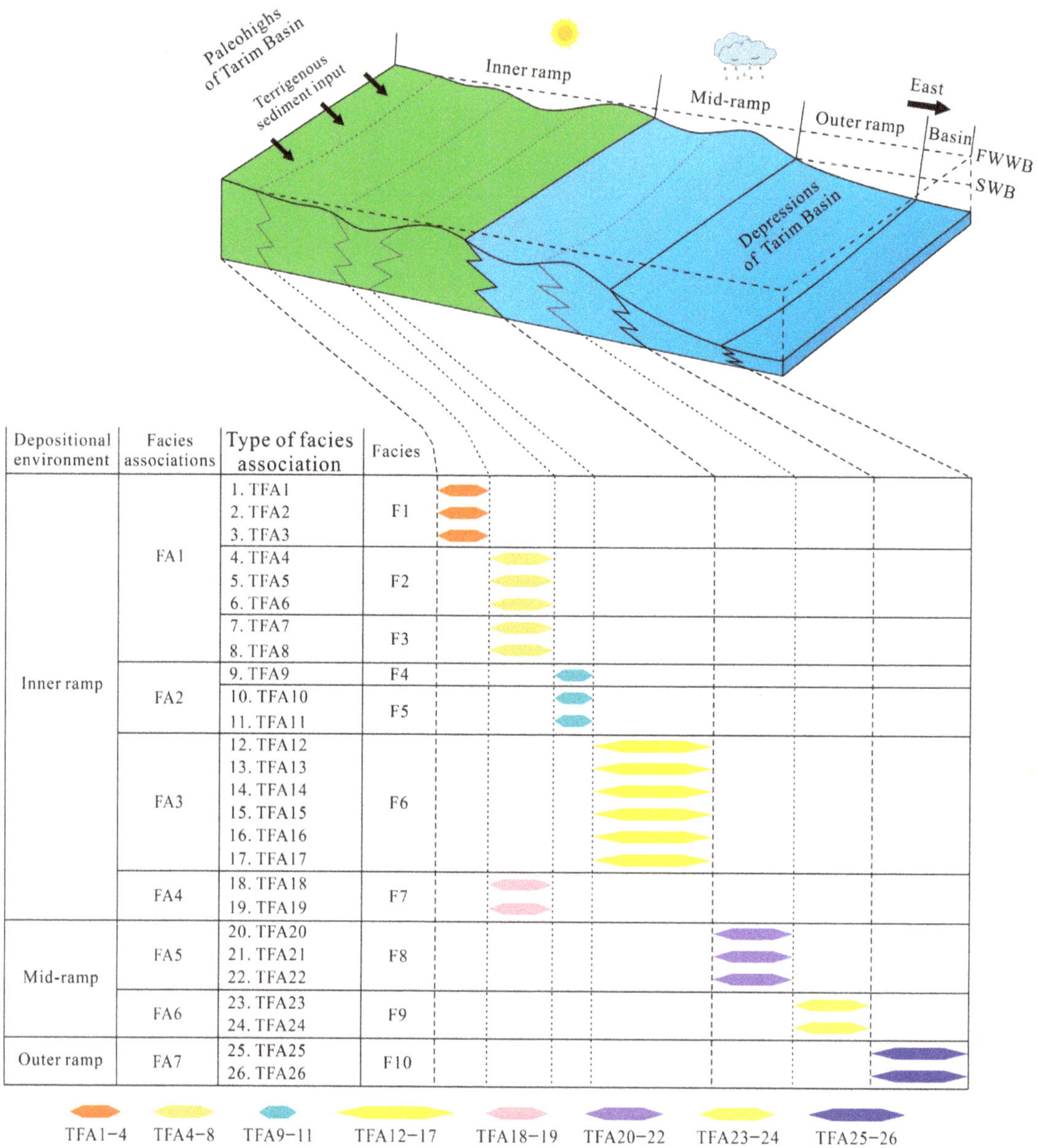

Figure 11. Schematic depositional model of the Early Cambrian Xiaoerbulake Formation in the Tarim Basin showing distribution patterns of the facies and facies associations developed in different parts of an Early Cambrian carbonate ramp.

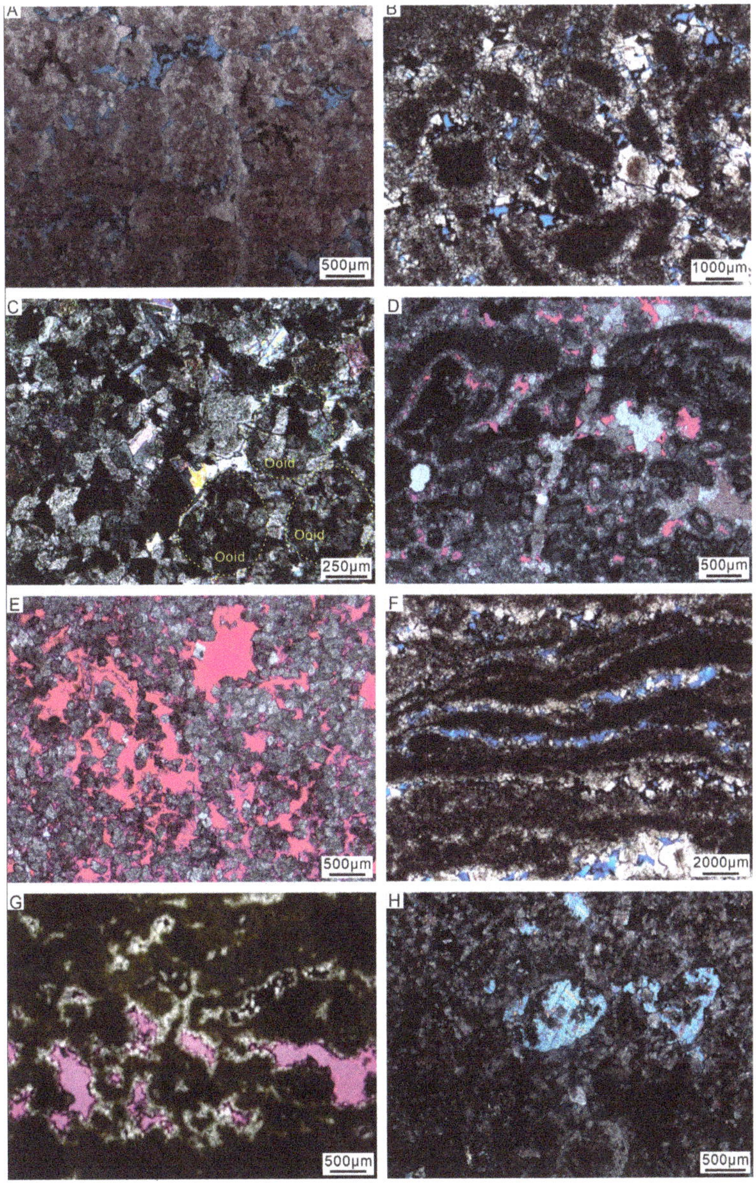

Figure 12. Types of reservoirs and pores of the Early Cambrian Xiaoerbulake Formation in the Tarim Basin. (**A**) Interparticle pores and intraparticle pores of oolitic dolostone, Well CT1, depth 7767.93 m. (**B**) Interparticle pores and intraparticle pores of doloarenite, Well ST1, depth 1886 m. (**C**) Interparticle pores and intraparticle pores of oolitic dolostone filled with anhydrite, Well CT1, depth 7764.99 m. (**D**) Interparticle pores and intraparticle pores of algal doloarenite, Well ST1, depth 1883.07 m. (**E**) Intercrystalline pores and vuggy pores of finely crystalline dolostone, Well ST1, depth 1915.75 m. (**F**) Vuggy pores of stromatolite, Well ST1, depth 1885.6 m. (**G**) Vuggy pores of algal dolostone, Well F1, depth 4599 m. (**H**) Moldic pores of oolitic dolostone completely filled with anhydrite, Well CT1, depth 7764.24 m.

4.4.2. Intraparticle Pores

Intraparticle pores are pore spaces occurring within sand-sized carbonate grains and ooids of the high-energy shoal facies association (Figure 12A,B,D). They form from the selective dissolution of interparticle cements, with sizes ranging from 0.01 to 0.35 mm. These pores are locally filled with anhydrite, suggesting that they likely formed during early dolomitization [73].

4.4.3. Intercrystalline Pores

Various levels of intercrystalline pores are observed under the microscope in the finely to medium crystalline dolomite owing to various degrees of dolomitization [74]. In the studied dataset, intercrystalline pores are developed between euhedral dolomite crystals (Figure 12E). Additionally, in some cases, intercrystalline pores are enlarged by dissolution along the boundaries, creating intercrystalline dissolved pores with irregularly banded and embayed shapes. The sizes of intercrystalline pores range from 0.01 mm to 0.3 mm. These intercrystalline pores occur in finely crystalline dolostone of the high-energy shoal facies association with residual grain texture (Figure 12E).

4.4.4. Vuggy Pores

Vuggy pores are secondary solution pores formed by the dissolution of cement, matrix and grains [71]. They are affected by near-surface meteoric waters and are related to deep burial fluids, mainly controlled by fractures and faults [72,75]. Herein, vugs occur as algal framework dissolution pores and irregular solution-enlarged intercrystalline pores (Figures 3A, 6B and 12E) and interparticle pores (Figure 12A,C,E,F) of the high-energy shoal facies association, with a pore size of 0.2–25 mm.

4.4.5. Moldic Pores

Moldic pores are secondary pores formed by the selective, complete or partial dissolution and recrystallization of grains or crystals [71,72]. Herein, they are commonly associated with partially or completely dissolved and recrystallized ooids and sand-sized grains (Figure 12D, E and H) of the high-energy shoal facies association, with pore sizes of 0.3–0.7 mm. In anhydrite-bearing oolitic dolostone, moldic pores within ooids are generally filled with anhydrite (Figure 12H).

5. Discussion

5.1. Controlling Factors in the Development of Facies and Facies Associations

At the depositional stage of the Xiaoerbulake Formation, in addition to the macro morphologic characteristics of "higher in the south and lower in the north", the basin also exhibited the macro morphologic characteristics of "higher in the west and lower in the east" [14,25,28,76]. Controlled by the continuous subsidence of the Luonan–Yubei Rift, the central basement paleohigh evolved into the Wuqia and Tanan (Southern Tarim) Paleohighs [77]. The original Tabei (northern Tarim) basement paleohigh maintained the pattern of near-EW trend and higher west and lower east during the early period of the Early Cambrian. The Keping–Wensu region was a low paleohigh, while the Lunnan region was an underwater low paleohigh [78]. The Wuqia, Southern Tarim and Keping–Wensu Paleohighs together with the Southwest Tarim Depression and North Depression constituted a new paleogeographic pattern of "three paleohighs and two depressions" [79]. This structural pattern controlled the sedimentary thickness, depositional environment and facies associations of the Early Cambrian Xiaoerbulake Formation. The Xiaoerbulake Formation overlapped with the marginal region of the paleohighs and developed a set of carbonate ramp deposits [25]. From the periphery of the paleohighs to the depressions, the sedimentary thickness gradually increased from 38 m to372 m (Table 1). The depositional environments and facies associations successively developed from FA1 (F1, F2 or F3), FA2 or FA3 of the inner ramp to FA5 and FA6 of the middle ramp and FA7 of the outer ramp (Table 2).

Sea level changes and the paleoclimate also controlled the development of the Xiaoerbulake Formation. The sediments of the Early Cambrian Xiaoerbulake Formation were deposited during a transgression and a regression, the climate was arid and hot and the seawater salinity was relatively higher [21,80]. The relative sea level experienced a rise in the early sedimentary periods of the Xiaoerbulake Formation and subsequently dropped in the middle and late sedimentary period of the Xiaoerbulake Formation [81]. In the Bachu and Tazhong Uplifts, the facies associations evolved from a high-energy shoal facies association or a tidal flat facies association to a lagoonal facies association and subsequently to a high-energy shoal facies association or a tidal flat facies association, represented by Wells ST1, F1 and ZH1. The Xiaoerbulake Formation near the Southwest Depression and North Depression evolved from an open-shelf facies association of the outer ramp to an inter-shoal facies association or a low-energy shoal facies association of the middle ramp from the bottom to the top, represented by Wells LT1, QT1–2 and LT3 (Figures 1A, 9 and 10). During the middle and late stages of the deposition of the Xiaoerbulake Formation, affected by the arid paleoclimate, the salinity of the seawater increased and the anhydrite–salt rock gradually crystallized. The sabkha facies association occurred during the middle stage of the Xiaoerbulake Formation in some well blocks, represented by Wells K2 and H4. Birdseyes of different shapes can be observed in stromatolite (Figure 4A), represented by Wells ST1, CT1 and H4.

5.2. *The Relationship between Facies Associations and Reservoir Quality*

Facies associations and facies control the primary pore structure of the reservoirs and affect the preservation and transformation of pores [66,82]. High-energy shoal facies associations are usually located in the upwarped areas in a ramp. With a slow decline in the sea level, the sedimentary formations of high-energy shoal facies associations were gradually exposed and prone to freshwater leaching, dissolution and dolomitization, forming a dolomitized reservoir with residual particle texture and dissolution pores. The satisfactory compaction resistance of dolomite during this period preserved the original pores. During the burial stage, dissolution played an important role in reconstructing the early reservoirs, and the reconstruction intensity mainly depended on the physical properties of the early reservoirs. Currently, the reservoir pore space is mostly made up by the solution-enlarged pores [14].

In the studied dataset, the inner ramp high-energy shoal facies association exhibits good reservoir quality. Many reservoir pores, including interparticle, intraparticle, intercrystalline, vuggy and moldic pores, are observed in thin sections in the upper part of the Xiaoerbulake Formation. The porosity in Well ST1 (1879–1920 m) ranges from 1.4% to 7.5%, with an average porosity value of 4% [14], and the porosity in the 7726–7792 m of Well CT1 ranges from 2.9% to 4.8%, with an average porosity value of 3.65%. The mid-ramp and outer ramp depositional environments show poor reservoir quality. The porosity from well logging interpretation is lower than 2% (Figures 9 and 10).

5.3. *Diagenetic Controls of Reservoir Quality*

Because dolomitization and anhydrite cementation commonly exert a considerable impact on carbonate reservoir quality, they are major diagenetic processes investigated in carbonate hydrocarbon reservoir rocks [53,83,84].

5.3.1. Effects of Dolomitization

Dolomitization is one of the most important diagenetic processes in the Early Cambrian Xiaoerbulake Formation [28,70,73,83,85]. Multiple geochemical parameters show that the Early Cambrian Xiaoerbulake dolomite is mainly the product of sabkha evaporation dolomitization in a penecontemporaneous environment and seepage–reflux dolomitization in a shallow burial environment. Only a small amount of finely–medium crystalline dolomite was formed in an intermediate and deep burial environment [21,70]. Rapid dolomitization in a penecontemporaneous and shallow burial environment formed fine dolomite crystals with a well-preserved fabric. Consequently, the dolostone in the Early Cambrian Xiaoerbulake Formation in the Tarim Basin mostly occurs with primary sedimentary structures, regardless of outcrops or underground formations. The primary morphology of carbonate grains could be vaguely discernible even after complete dolomitization [12,14]. Owing to good resistance to compaction, these primary rock structures act as a solid rock framework to protect the porosity (residual primary pores and secondary pores formed by freshwater dissolution), transforming dolograinstone into porous dolostones, exhibiting satisfactory reservoir quality after dolomitization [83,86].

5.3.2. Effects of Anhydrite Cementation

Anhydrite cementation is the second most important diagenetic process in the Early Cambrian Xiaoerbulake Formation, following dolomitization. These anhydrites likely originated from the dehydration of gypsum in a burial environment [45,87]. Among the various anhydrite textures, the pore-filling and poikilotopic types play a key role in reservoir quality [64]. They fill bulk pores by spreading within dolograinstone and dolopackstone in a shallow burial environment and considerably decrease reservoir porosity and permeability [64]. Reactive transport simulations in a reflux system demonstrate that anhydrite cements that were related to replacement dolomitization could sharply decrease core plug porosity and considerably reduce porosity by up to 25% [88]. However, breccia and associated fracture porosity are common in anhydrite-bearing sections [73,83]. Breccia-associated fracture porosity caused by a short period of meteoric water-induced dissolution is the predominant type of effective porosity in the anhydrite-bearing dolomudstone reservoir [73]. Additionally, the late anhydrite dissolution is responsible for improving reservoir quality by dissolving most of the early anhydrite cementation and creating new pore spaces [84]. Figure 12G,H show that interparticle pores and moldic pores in oolitic dolostone are fully infilled (cemented) with anhydrite. Herein, anhydrite cementation may undermine reservoir quality.

6. Conclusions

This research involves an integrated analysis of cores, well cuttings, thin section castings, thin sections of core and cuttings and well logging data to study the facies, depositional environment and reservoir quality of the Early Cambrian Xiaoerbulake Formation in the Tarim Basin of Northwest China. The main conclusions are summarized below.

(1) Ten facies types were identified, including mixed siliciclastic–carbonate rock, crystalline dolostone, dolobindstone, dolomudstone, argillaceous dolostone, dolograinstone, anhydrite–salt rock, dolomitized micrite, peloidal grainstone and argillaceous limestone. These facies are distinguished by their thickness, color, texture, structure, grains, fossil type and argillaceous and sand concentration.

(2) Seven facies associations were determined based on ten facies and twenty-six representative types of facies associations. The facies association types indicate that the Early Cambrian Xiaoerbulake Formation in the Tarim Basin represents deposits formed on a ramp on which three depositional environments were developed, including an inner ramp, middle ramp and outer ramp. The inner ramp comprises the tidal flat facies association, lagoon facies association, sabkha facies association and high-energy shoal facies association; the middle ramp comprises the inter-shoal sea facies association and low-energy shoal facies association and the outer ramp mainly comprises the open-shelf facies association.

(3) Facies types, together with dolomitization and anhydrite cementation, are the major factors controlling reservoir quality in the Early Cambrian Xiaoerbulake Formation in the Tarim Basin. The inner ramp high-energy shoal facies association exhibits good reservoir quality, while the mid-ramp and outer ramp depositional environments demonstrate poor reservoir quality. The early dolomitization can partly protect primary porosity and make dolograinstone a high-quality reservoir. Herein, anhydrite

cementation may negatively impact reservoir quality owing to its precipitation in the pore system.

Author Contributions: Conceptualization, Y.Z., J.Z. (Jianfeng Zhang) and J.Z. (Jiankun Zhang); methodology, Y.Z. and J.Z. (Jiankun Zhang); investigation, Y.Z., J.Z. (Jianfeng Zheng) and J.Z. (Jiankun Zhang); data curation, Y.Z., X.L., J.L. and F.H.; writing—original draft preparation, Y.Z., X.L., G.Y. (Guang Yu), J.L., F.H. and G.Y. (Guo Yang); writing—review and editing, J.Z. (Jiankun Zhang); visualization, G.Y. (Gang Yu) and F.H.; supervision, Y.Z. and J.Z. (Jiankun Zhang); project administration, J.Z.(Jianfeng Zheng). All authors have read and agreed to the published version of the manuscript.

Funding: The research was co-funded by the CNPC (China National Petroleum Corporation) Scientific Research and Technology Development Project (Grant Nos. 2021DJ0501, 2021DJ1501), Jiangxi Provincial Natural Science Foundation (No. 20202BAB204035) and Jiangxi Engineering Technology Research Center of Nuclear Geoscience Data Science and System (Grant Nos. JETRCNGDSS202103, JETRCNGDSS202002).

Data Availability Statement: Not applicable.

Acknowledgments: We deeply appreciate two anonymous reviewers and the editor for remarkable constructive comments that significantly improved the manuscript.

Conflicts of Interest: The authors declare no conflict of interest.

References

1. Burchette, T.P.; Wright, V.P. Carbonate ramp depositional systems. *Sediment. Geol.* **1992**, *79*, 3–57. [CrossRef]
2. Ahr, W.M. The carbonate remp: An alternative to the shelf model. *Trans. Gulf Coast Assoc. Geol. Soc.* **1973**, *23*, 221–225.
3. Wright, V.P. Facies sequences on a carbonate ramp: The Carboniferous Limestone of South Wales. *Sedimentology* **1986**, *33*, 221–241. [CrossRef]
4. Read, J.F. Carbonate Platform Facies Models. *AAPG Bull.* **1985**, *69*, 1–21.
5. Bábek, O.; Kalvoda, J.; Cossey, P.; Šimíček, D.; Devuyst, F.X.; Hargreaves, S. Facies and petrophysical signature of the Tournaisian/Viséan (Lower Carboniferous) sea-level cycle in carbonate ramp to basinal settings of the Wales-Brabant massif, British Isles. *Sediment. Geol.* **2013**, *284–285*, 197–213. [CrossRef]
6. Kakemem, U.; Jafarian, A.; Husinec, A.; Adabi, M.H.; Mahmoudi, A. Facies, sequence framework, and reservoir quality along a Triassic carbonate ramp: Kangan Formation, South Pars Field, Persian Gulf Superbasin. *J. Pet. Sci. Eng.* **2021**, *198*, 108166. [CrossRef]
7. Zou, C.; Du, J.; Xu, C.; Wang, Z.; Zhang, B.; Wei, G.; Wang, T.; Yao, G.; Deng, S.; Liu, J.; et al. Formation, distribution, resource potential, and discovery of Sinian–Cambrian giant gas field, Sichuan Basin, SW China. *Petrol. Explor. Dev.* **2014**, *41*, 306–325. [CrossRef]
8. Zhu, G.Y.; Huang, H.P.; Wang, H.T. Geochemical Significance of Discovery in Cambrian Reservoirs at Well ZS1 of the Tarim Basin, Northwest China. *Energy Fuels* **2015**, *29*, 1332–1344. [CrossRef]
9. Trippetta, F.; Durante, D.; Lipparini, L.; Romi, A.; Brandano, M. Carbonate-ramp reservoirs modelling best solutions: Insights from a dense shallow well database in Central Italy. *Mar. Pet. Geol.* **2021**, *126*, 104931. [CrossRef]
10. Tomassetti, L.; Petracchini, L.; Brandano, M.; Trippetta, F.; Tomassi, A. Modeling lateral facies heterogeneity of an upper Oligocene carbonate ramp (Salento, southern Italy). *Mar. Pet. Geol.* **2018**, *96*, 254–270. [CrossRef]
11. Du, J.H.; Pan, W.Q. Accumulation conditions and play targets of oil and gas in the Cambrian subsalt dolostone, Tarim Basin, NW China. *Petrol. Explor. Dev.* **2016**, *45*, 1111–1126. (In Chinese with English Abstract)
12. Zhu, Y.J.; Shen, A.J.; Liu, L.L.; Chen, Y.Q.; Yu, G. Tectonic-sedimentary filling history through the Later Sinian to the Mid-Cambrian in Tarim Basin and its explorational potential. *Acta Sediment. Sin.* **2020**, *38*, 398–410. (In Chinese with English Abstract)
13. Ma, Y.S.; Li, M.W.; Cai, X.Y.; Xu, X.H.; Hu, D.F.; Qu, S.L.; Li, G.S.; He, D.F.; Xiao, X.M.; Zeng, Y.J.; et al. Mechanisms and exploitation of deep marine petroleum accumulations in China: Advances, technological bottlenecks and basic scientific problems. *Oil Gas Geol.* **2020**, *41*, 655–672+683. (In Chinese with English Abstract)
14. Zhu, Y.J.; Ni, X.F.; Liu, L.L.; Qiao, Z.F.; Chen, Y.Q.; Zheng, J.F. Depositional differentiation and reservoir potential and distribution of ramp systems during post-rift period: An example from the Lower Cambrian Xiaoerbulake Formation in the Tarim Basin, NW China. *Acta Sediment. Sin.* **2019**, *37*, 1044–1057. (In Chinese with English Abstract)
15. Zheng, J.F.; Zhu, Y.J.; Huang, L.L.; Yang, G.; Hu, F.J. Geochemical Characteristics and Their Geological Significance of Lower Cambrian Xiaoerblak Formation in Northwestern Tarim Basin, China. *Minerals* **2022**, *12*, 781. [CrossRef]
16. He, Z.L.; Ma, Y.S.; Zhu, D.Y.; Duan, T.Z.; Geng, J.H.; Zhang, J.T.; Ding, Q.; Qian, Y.X.; Wo, Y.J.; Gao, Z.Q. Theoretical and technological progress and research direction of deep and ultra-deep carbonate reservoirs. *Oil Gas Geol.* **2021**, *42*, 533–546. (In Chinese with English Abstract)

17. Brandano, M.; Tomassetti, L.; Trippetta, F.; Ruggieri, R. Facies Heterogeneities and 3D Porosity Modelling in an Oligocene (Upper Chattian) Carbonate Ramp, Salento Peninsula, Southern Italy. *J. Pet. Geol* **2020**, *43*, 191–208. [CrossRef]
18. Bai, Y.; Luo, P.; Zhou, C.M.; Zhai, Z.G.; Wang, S.; Yang, Z.Y.; Wang, S. Sequence division and platform sedimentary evolution model of the Lower Cambrian Xiaoerbulak Formation in the NW Tarim Basin. *Oil Gas Geol.* **2017**, *38*, 152–164. (In Chinese with English Abstract)
19. Hu, M.Y.; Sun, C.Y.; Gao, D. Characteristics of tectonic-lithofacies paleogeography in the Lower Cambrian Xiaoerbulake Formation, Tarim Basin. *Oil Gas Geol.* **2019**, *40*, 12–23. (In Chinese with English Abstract)
20. Zheng, J.F.; Yuan, W.F.; Huang, L.L.; Pan, W.Q.; Qiao, Z.F.; Yang, G. Sedimentary facies model and its exploration significance of the Lower Cambrian Xiaoerblak Formation in Xiaoerblak area, Tarim Basin. *J. Palaeogeogr.* **2019**, *21*, 589–602. (In Chinese with English Abstract)
21. Zheng, J.F.; Huang, L.L.; Yuan, W.F.; Zhu, Y.J.; Qiao, Z.F. Geochemical features and its significance of sedimentary and diageneticenvironment in the Lower Cambrian Xiaoerblak Formation of Keping area, Tarim Basin. *Nat Gas Geosci.* **2020**, *31*, 698–709. (In Chinese with English Abstract)
22. Pan, W.Q.; Chen, Y.Q.; Xiong, Y.X.; Li, B.H.; Xiong, R. Sedimentary facies research and implications to advantaged exploration regions on Lower Cambrian source rocks, Tarim basin. *Oil Gas Geol.* **2015**, *26*, 1224–1232. (In Chinese with English Abstract)
23. Zhu, D.; Meng, Q.; Jin, Z.; Liu, Q.; Hu, W. Formation mechanism of deep Cambrian dolomite reservoirs in the Tarim basin, northwestern China. *Mar. Pet. Geol.* **2015**, *59*, 232–244. [CrossRef]
24. He, D.F.; Jia, C.Z.; Li, D.S.; Zhang, C.J.; Meng, Q.R.; Shi, X. Formation and evolution of polycyclic superimposed Tarim Basin. *Oil Gas Geol.* **2005**, *26*, 64–77. (In Chinese with English Abstract)
25. Wei, G.; Zhu, Y.; Zheng, J.; Yu, G.; Ni, X.; Yan, L.; Tian, L.; Huang, L. Tectonic-lithofacies paleogeography, large-scale source-reservoir distribution and exploration zones of Cambrian subsalt formation, Tarim Basin, NW China. *Petrol. Explor. Dev.* **2021**, *48*, 1289–1303. [CrossRef]
26. Wu, L.; Guan, S.W.; Feng, X.Q.; Ren, R.; Zhang, C.Y. Discussion on stratigraphic division of the Nanhuan and Sinian of the Tarim Basin and its surrounding region. *Acta Petrol. Sin.* **2020**, *36*, 3427–3441. (In Chinese with English Abstract)
27. Shi, K.B.; Liu, B.; Tian, J.C.; Pan, W.Q. Sedimentary characteristics and lithofacies paleogeography of Sinian in Tarim Basin. *Acta Pet. Sin.* **2016**, *37*, 1343–1360. (In Chinese with English Abstract)
28. Zhu, Y.J.; Zheng, J.F.; Cao, P.; Qiao, Z.F.; Yu, G.; Liu, L.L.; Huang, L.L. Tectonic-Lithofacies Paleogeography and Sizeable Reservoir Potential in the Early Cambrian Postrift Depressed Carbonate Ramp, Tarim Basin, NW China. *Lithosphere* **2021**, *2021*, 8804537. [CrossRef]
29. Wang, Z.; Gao, Z.; Fan, T.; Shang, Y.; Qi, L.; Yun, L. Structural characterization and hydrocarbon prediction for the SB5M strike-slip fault zone in the Shuntuo Low Uplift, Tarim Basin. *Mar. Pet. Geol.* **2020**, *117*, 104418. [CrossRef]
30. Deng, S.; Liu, Y.Q.; Liu, J.; Han, J.; Wang, B.; Zhao, R. Structural styles and evolution models of intracratonic strike-slip faults and the implications for reservoir exploration and appraisal: A case study of the Shunbei area, Tarim Basin. *Geotect Metal.* **2020**, *45*, 1111–1126. (In Chinese with English Abstract)
31. Folk, R.L. Practical Petrographic Classification of Limestones. *AAPG Bull.* **1959**, *43*, 1–38.
32. Dunham, R.J. Classification of carbonate rocks according to their depositional texture. In *Classification of Carbonate Rocks (AAPG Memoir 1)*; Ham, W.E., Ed.; American Association of Petroleum Geologists: Tulsa, OK, USA, 1962; pp. 108–121.
33. Flügel, E. *Microfacies of Carbonate Rocks: Analysis, Interpretation and Application*, 2nd ed.; Springer: New York, NY, USA, 2010; pp. 73–900.
34. Friedman, G.M. Terminology of crystallization textures and fabrics in sedimentary rocks. *J. Sediment. Res.* **1965**, *35*, 643–655.
35. Sibley, D.F.; Gregg, J.M. Classification of dolomite rock textures. *J. Sediment. Res.* **1987**, *57*, 967–975.
36. Wilson, J.L. *Carbonate Facies in Geologic History*; Springer: Berlin, Heidelberg; New York, NY, USA, 1975; pp. 20–347.
37. Lucia, F.J. Rock-Fabric/Petrophysical Classification of Carbonate Pore Space for Reservoir Characterization. *AAPG Bull.* **1995**, *79*, 1275–1300.
38. Wu, F.; Xie, X.; Zhu, Y.; Coletti, G.; Betzler, C.; Cui, Y.; Bai, H.; Chen, B.; Shang, Z. Early development of carbonate platform (Xisha Islands) in the northern South China Sea. *Mar. Geol.* **2021**, *441*, 106629. [CrossRef]
39. Christiano-De-Souza, I.; Ricardi-Branco, F.; Silva, A.; El-Dash, L.; Faria, R. New approach for the study of paleofloras using geographical information systems applied to Glossopteris Flora. *Braz. J. Geol.* **2014**, *44*, 681–689. [CrossRef]
40. Ng, C.; Vega, C.S.; Maranhão, M.d.S.A.S. Mixed carbonate-siliciclastic microfacies from Permian deposits of Western Gondwana: Evidence of gradual marine to continental transition or episodes of marine transgression? *Sediment. Geol.* **2019**, *390*, 62–82. [CrossRef]
41. Rivers, J.M.; Dalrymple, R.W.; Yousif, R.; Al-Shaikh, I.; Butler, J.D.; Warren, C.; Skeat, S.L.; Bari, E. Mixed siliciclastic-carbonate-evaporite sedimentation in an arid eolian landscape: The Khor Al Adaid tide-dominated coastal embayment, Qatar. *Sediment. Geol.* **2020**, *408*, 105730. [CrossRef]
42. Farouk, S.; Al-Kahtany, K.; Tawfik, M. Upper Cretaceous (Coniacian -Campanian) siliciclastic/carbonate ramp depositional sequences and relative sea-level changes in Sinai, Egypt: Implications from facies analysis. *Mar. Pet. Geol.* **2021**, *131*, 105183. [CrossRef]
43. Halfar, J.; Ingle, J.C.; Godinez-Orta, L. Modern non-tropical mixed carbonate-siliciclastic sediments and environments of the southwestern Gulf of California, Mexico. *Sediment. Geol.* **2004**, *165*, 93–115. [CrossRef]

44. Zhang, Y.; Pan, W.; Zhu, B.; Li, W.; Yang, L.W.; Chen, Y.; Yang, T. Recrystallization of dolostones in the Cambrian Xiaoerbrak Formation, Tarim Basin and possible link to reservoir development. *Mar. Pet. Geol.* **2022**, *136*, 105452. [CrossRef]
45. Testa, G.; Lugli, S. Gypsum–anhydrite transformations in Messinian evaporites of central Tuscany (Italy). *Sediment. Geol.* **2000**, *130*, 249–268. [CrossRef]
46. Huang, S.J.; Yang, J.J.; Zhang, W.Z.; Huang, Y.M.; Liu, C.X.; Xiao, L.P. Effects of gypsum (or anhydrite) on dissolution of dolomite under different temperatures and pressures of epigenesis and burial diagenesis. *Acta Sediment. Sin.* **1996**, *14*, 103–109. (In Chinese with English Abstract)
47. Murray, R.C. Origin and diagenesis of gypsum and anhydrite. *J. Sediment. Res.* **1964**, *34*, 512–523.
48. Shinn, E.A. Practical significance of birdseye structures in carbonate rocks. *J. Sediment. Res.* **1968**, *38*, 215–223. [CrossRef]
49. Xu, L.; Chen, T.H.; Gao, Y.; Chen, P.; Xie, Q.Q.; Zhou, Y.F. Genesis of bird's-eye structure in the Triassic Dongmaanshan Formation in Chaohu, Anhui Province. *Earth Sci. Front.* **2020**, *27*, 294–301. (In Chinese with English Abstract)
50. Riding, R. Microbialites, stromatolites, and thrombolites. In *Encyclopedia of Geobiology (Encyclopedia of Earth Sciences Series)*; Reitner, J., Thiel, V., Eds.; Springer: Dordrecht, The Netherlands, 2011; pp. 635–654.
51. Zhang, Y.; Li, J.; Chen, L.; Wei, Y.; Shi, Q.; Wang, D.; Wu, Q.; Song, L.; Tian, M.; Kuang, H.; et al. Manganese carbonate stromatolites of the Ediacaran Doushantuo Formation in Chengkou, northern Yangtze Craton, China. *J. Palaeogeogr.* **2021**, *10*, 356–381. [CrossRef]
52. Pruss, S.B.; Knoll, A.H. Environmental covariation of metazoans and microbialites in the Lower Ordovician Boat Harbour Formation, Newfoundland. *Palaeogeogr. Palaeoclimatol. Palaeoecol.* **2017**, *485*, 917–929. [CrossRef]
53. Amel, H.; Jafarian, A.; Husinec, A.; Koeshidayatullah, A.; Swennen, R. Microfacies, depositional environment and diagenetic evolution controls on the reservoir quality of the Permian Upper Dalan Formation, Kish Gas Field, Zagros Basin. *Mar. Pet. Geol.* **2015**, *67*, 57–71. [CrossRef]
54. Yang, W.; Li, P.; Chen, H.; Liu, Z.; Lan, C.; Xu, Z.; Lu, C.; Zou, H. Origin and significance of carbonate shoal depositional cycles: A case study of the Cambrian Longwangmiao Formation, Sichuan Basin, SW China. *J. Asian Earth Sci.* **2022**, *226*, 105083. [CrossRef]
55. Wang, Z.M.; Xie, H.W.; Chen, Y.Q.; Qi, Y.M.; Zhang, K. Discovery and exploration of Cambrian subsalt dolostone original hydrocarbon reservoir at Zhongshen-1 Well in Tarim Basin. *China Pet. Explor.* **2014**, *19*, 1–13. (In Chinese with English Abstract)
56. Hu, M.; Gao, D.; Wei, G.; Yang, W.; Xie, W. Sequence stratigraphy and facies architecture of a mound-shoal-dominated dolomite reservoir in the late Ediacaran Dengying Formation, central Sichuan Basin, SW China. *Geol. J.* **2019**, *54*, 1653–1671. [CrossRef]
57. Riera, R.; Bourget, J.; Allan, T.; Hakansson, E.; Wilson, M.E.J. Early Miocene carbonate ramp development in a warm ocean, North West Shelf, Australia. *Sedimentology* **2022**, *69*, 219–253. [CrossRef]
58. Pettijohn, F.J.; Potter, P.E.; Siever, R. *Sand and Sandstone*; Springer: Berlin, Germany, 1972; pp. 1–618.
59. Li, F.; Yan, J.X.; Chen, Z.Q.; Ogg, J.G.; Tian, L.; Korngreen, D.; Liu, K.; Ma, Z.L.; Woods, A.D. Global oolite deposits across the Permian-Triassic boundary: A synthesis and implications for palaeoceanography immediately after the end-Permian biocrisis. *Earth Sci. Rev.* **2015**, *149*, 163–180. [CrossRef]
60. Corsetti, F.A.; Kidder, D.L.; Marenco, P.J. Trends in oolite dolomitization across the Neoproterozoic–Cambrian boundary: A case study from Death Valley, California. *Sediment. Geol.* **2006**, *191*, 135–150. [CrossRef]
61. Zempolich, W.G.; Baker, P.A. Experimental and natural mimetic dolomitization of aragonite ooids. *J. Sediment. Res.* **1993**, *63*, 596–606.
62. Osburn, M.; Grotzinger, J.; Bergmann, K. Facies, stratigraphy, and evolution of a middle Ediacaran carbonate ramp: Khufai Formation, Sultanate of Oman. *AAPG Bull.* **2014**, *98*, 1631–1667. [CrossRef]
63. Strasser, A. Ooids in Purbeck limestones (lowermost Cretaceous) of the Swiss and French Jura. *Sedimentology* **1986**, *33*, 711–727. [CrossRef]
64. Jafarian, A.; Fallah-Bagtash, R.; Mattern, F.; Heubeck, C. Reservoir quality along a homoclinal carbonate ramp deposit: The Permian Upper Dalan Formation, South Pars Field, Persian Gulf Basin. *Mar. Pet. Geol.* **2017**, *88*, 587–604. [CrossRef]
65. Duringer, P.; Vecsei, A. Middle Triassic shallow-water limestones from the Upper Muschelkalk of eastern France: The origin and depositional environment of some early Mesozoic fine-grained limestones. *Sediment. Geol.* **1998**, *121*, 57–70. [CrossRef]
66. Liu, Y.; Hu, M.; Zhang, S.; Zhang, J.; Gao, D.; Xiao, C. Characteristics and impacts on favorable reservoirs of carbonate ramp microfacies: A case study of the Middle—Lower Ordovician in Gucheng area, Tarim Basin, NW China. *Petrol. Explor. Dev.* **2022**, *49*, 107–120. [CrossRef]
67. Badenas, B.; Aurell, M.; Gasca, J.M. Facies model of a mixed clastic-carbonate, wave-dominated open-coast tidal flat (Tithonian-Berriasian, north-east Spain). *Sedimentology* **2018**, *65*, 1631–1666. [CrossRef]
68. Ismail, M.J.; Ettensohn, F.R.; Handhal, A.M.; Al-Abadi, A. Facies analysis of the Middle Cretaceous Mishrif Formation in southern Iraq borehole image logs and core thin-sections as a tool. *Mar. Pet. Geol.* **2021**, *133*, 105324. [CrossRef]
69. Luan, X.C.; Brett, C.E.; Zhan, R.B.; Liu, J.B.; Wu, R.C.; Liang, Y. Microfacies analysis of the Lower-Middle Ordovician succession at Xiangshuidong, southwestern Hubei Province, and the drowning and shelf-ramp transition of a carbonate platform in the Yangtze region. *Paleogeogr. Paleoclimatol. Paleoecol.* **2017**, *485*, 68–83. [CrossRef]
70. Shen, A.; Zheng, J.; Chen, Y.; Ni, X.; Huang, L. Characteristics, origin and distribution of dolomite reservoirs in Lower-Middle Cambrian, Tarim Basin, NW China. *Petrol. Explor. Dev.* **2016**, *43*, 375–385. [CrossRef]

71. Choquette, P.W.; Pray, L.C. Geologic Nomenclature and Classification of Porosity in Sedimentary Carbonates. *AAPG Bull.* **1970**, *54*, 207–250.
72. Lønøy, A. Making sense of carbonate pore systems. *AAPG Bull.* **2006**, *90*, 1381–1405. [CrossRef]
73. Jiang, L.; Worden, R.H.; Cai, C.F.; Shen, A.J.; Crowley, S.F. Diagenesis of an evaporite-related carbonate reservoir in deeply buried Cambrian strata, Tarim Basin, northwest China. *AAPG Bull.* **2018**, *102*, 77–102. [CrossRef]
74. Lai, J.; Liu, S.C.; Xin, Y.; Wang, S.; Xiao, C.W.; Song, Q.Q.; Chen, X.; Yang, K.F.; Wang, G.W.; Ding, X.J. Geological-petrophysical insights in the deep Cambrian dolostone reservoirs in Tarim Basin, China. *AAPG Bull.* **2021**, *105*, 2263–2296. [CrossRef]
75. Friedman, G.M. Unconformities and Porosity Development in Carbonate Strata: Ideas from a Hedberg Conference: Discussion. *AAPG Bull.* **1995**, *79*, 1183–1184.
76. Zhu, G.Y.; Chen, Z.Y.; Chen, W.Y.; Yan, H.H.; Zhang, P.H. Revisiting to the Neoproterozoic tectonic evolution of the Tarim Block, NW China. *Precambrian Res.* **2021**, *352*, 106013. [CrossRef]
77. Wu, G.H.; Li, Q.M.; Xiao, Z.R.; Li, H.H.; Zhang, L.P.; Zhang, X.J. The Evolution Characteristics of Palaeo-Uplifts in Tarim Basin and Its Exploration Directions for Oil and Gas. *Geotect Metal.* **2009**, *33*, 124–130. (In Chinese with English Abstract)
78. Chen, Y.Q.; Yan, W.; Han, C.W.; Yang, T.F.; Li, Z. Redefinition on structural paleogeography and lithofacies paleogeography framework from Cambrian to Early Ordovician in the Tarim Basin: A new approach based on seismic stratigraphy evidence. *Nat Gas Geosci.* **2015**, *26*, 1832–1843. (In Chinese with English Abstract)
79. Yan, W.; Wu, G.H.; Zhang, Y.Q.; Yang, G.; Lou, H.; Wang, X.M. Sinian-Cambrian tectonic framework in the Tarim Basin and its influences on the paleogeography of the Early Cambrian. *Geotect Metal.* **2018**, *42*, 455–466. (In Chinese with English Abstract)
80. Ouyang, S.Q.; Lyu, X.X.; Xue, N.; Li, F.; Wang, R. Paleoenvironmental characteristics and source rock development model of the Early-Middle Cambrian: A case of the Keping-Bachu area in the Tarim Basin. *J. China Univ. Min. Technol.* **2022**, *51*, 293–310. (In Chinese with English Abstract)
81. Zhang, C.Y.; Guan, S.W.; Wu, L.; Ren, R.; Xiong, L.Q. Geochemical characteristics and its paleo-environmental significance of the Lower Cambrian carbonate in the northwestern Tarim Basin: A case study of Well Shutan-1. *Bull. Geol. Sci. Technol.* **2021**, *40*, 99–111. (In Chinese with English Abstract)
82. Gao, D.; Hu, M.Y.; Li, A.P.; Yang, W.; Xie, W.R.; Sun, C.Y. High-frequency sequence and microfacies and their impacts on favorable reservoir of Longwangmiao Formation in Central Sichuan Basin. *Earth Sci.* **2021**, *46*, 3520–3534. (In Chinese with English Abstract)
83. Jiang, L.; Cai, C.; Worden, R.H.; Crowley, S.F.; Jia, L.; Zhang, K.; Duncan, I.J. Multiphase dolomitization of deeply buried Cambrian petroleum reservoirs, Tarim Basin, north-west China. *Sedimentology* **2016**, *63*, 2130–2157. [CrossRef]
84. Mohammed-Sajed, O.K.; Glover, P.W.J. Influence of anhydritisation on the reservoir quality of the Butmah Formation in north-western Iraq. *Mar. Pet. Geol.* **2022**, *135*, 105391. [CrossRef]
85. Jiang, L.; Xu, Z.H.; Shi, S.Y.; Liu, W. Multiphase dolomitization of a microbialite-dominated gas reservoir, the middle Triassic Leikoupo Formation, Sichuan Basin, China. *J. Pet. Sci. Eng.* **2019**, *180*, 820–834. [CrossRef]
86. Zhao, W.Z.; Shen, A.J.; Zheng, J.F.; Qiao, Z.F.; Wang, X.F.; Lu, J.M. The porosity origin of dolostone reservoirs in the Tarim, Sichuan and Ordos basins and its implication to reservoir prediction. *Sci. China-Earth Sci.* **2014**, *57*, 2498–2511. [CrossRef]
87. Hardie, L.A. The Gypsum–anhydrite equilibrium at one atmosphere pressure. *Am. Mineral.* **1967**, *52*, 171–200.
88. Jones, G.D.; Xiao, Y. Dolomitization, anhydrite cementation, and porosity evolution in a reflux system: Insights from reactive transport models. *AAPG Bull.* **2005**, *89*, 577–601. [CrossRef]

minerals MDPI

Article

Genesis of Dolomite Reservoir in Ediacaran Chigbrak Formation of Tarim Basin, NW China: Evidence from U–Pb Dating, Isotope and Element Geochemistry

Jianfeng Zheng [1,2,*], Hui Wang [1,2,*], Anjiang Shen [1,2], Xianying Luo [1,2], Zhao Cheng [3] and Kun Dai [4]

1 PetroChina Hangzhou Research Institute of Geology, Hangzhou 310023, China; shenaj_hz@petrochina.com.cn (A.S.); luoxy_hz@petrochina.com.cn (X.L.)
2 Key Laboratory of Carbonate Reservoirs CNPC, Hangzhou 310023, China
3 PetroChina Tarim Oil Field Company, Korla 841000, China; chengzh-tlm@petrochina.com.cn
4 School of Earth Sciences, China University of Petroleum (Beijing), Beijing 102249, China; thous_nite@foxmail.com
* Correspondence: zhengjf_hz@petrochina.com.cn (J.Z.); wangh_hz@petrochina.com.cn (H.W.)

Abstract: The Chigbrak Formation in the Upper Ediacaran is one of the important exploration targets in the Tarim Basin, NW China. However, no significant discoveries have been made in this field, and unclear reservoir genesis is one of the important factors restricting exploration breakthrough. This study examined the outcrops of the Aksu area in northwestern Tarim Basin by using systematic descriptions of petrologic features in the Upper Ediacaran Chigbrak Formation. Samples were selected for tests of stable carbon and oxygen isotopic compositions, strontium isotopic compositions, rare earth elements, LA–ICP–MS element mapping and U–Pb dating. It was found that (1) the Chigbrak Formation is mainly composed of dolomitic microbialite, with average values of $\delta^{13}C$ (PDB), $\delta^{18}O$ (PDB) and $^{87}Sr/^{86}Sr$ of 3.50‰, 2.95‰ and 0.709457, and has similar geochemical characteristics to the coeval seawater. The dolomites have also been characterized by a medium degree of cation ordering (avg. 0.68), a low content of ΣREEs (avg. 9.03 ppm) and a chondrite standardized curve of REEs showing enrichment of LREE and depletion of HREE. The U–Pb ages range from 538 to 618 Ma, corresponding to the age of Ediacaran period. (2) Dolomitization occurred in a marine diagenetic environment during the penecontemporaneous period, with seawater as the dolomitization fluid. (3) Vugs are the dominant pore type of Chigbrak Formation, and they are the products of the dissolution of meteoric water in penecontemporaneous period. (4) The main controlling factors of reservoir were lithofacies, meteoric water dissolution controlled by fourth– or fifth–order sequences and tectonic movement, and early dolomitization. The research results are of great significance to the dolomite reservoir prediction of the Upper Ediacaran Chigbrak Formation of the Tarim Basin.

Keywords: dolomite; reservoir genesis; U–Pb dating; isotope and element geochemistry; Ediacaran; Tarim Basin

Citation: Zheng, J.; Wang, H.; Shen, A.; Luo, X.; Cheng, Z.; Dai, K. Genesis of Dolomite Reservoir in Ediacaran Chigbrak Formation of Tarim Basin, NW China: Evidence from U–Pb Dating, Isotope and Element Geochemistry. *Minerals* **2023**, *13*, 725. https://doi.org/10.3390/min13060725

Academic Editors: Hamzeh Mehrabi, Vahid Tavakoli and Alfons M. Van den Kerkhof

Received: 13 April 2023
Revised: 18 May 2023
Accepted: 23 May 2023
Published: 25 May 2023

1. Introduction

In recent years, dramatic breakthroughs have been achieved in the Precambrian hydrocarbon exploration globally. In Russia, about 50 oil and gas fields have been discovered in the Meso–Neoproterozoic Vendian and the Mesoproterozoic Riphean within the Siberian craton, with a total of proved and probable reserves of about 25.12×10^8 m^3. In southern Oman, nine Precambrian carbonate reservoirs have been found within the Havel–Kloosterman area, with proved oil reserves of about 3.5×10^8 m^3 [1]. In China, a Proterozoic buried hill reservoir has been found in Renqiu of the Jizhong depression, with proved oil and gas reserves of about 6.15×10^8 m^3. The Anyue gas field producing from the Upper Ediacaran Dengying Formation has been found in the Leshan–Longnüsi paleo–uplift of the Sichuan Basin, with proved gas reserves of 5940×10^8 m^3 [2].

In the Cambrian pre–salt ultra–deep (>6500 m) strata in the Tarim Basin, two sets of thick dolomite reservoirs have been identified, including the Upper Ediacaran Chigbrak Formation and the Middle–Lower Cambrian, which form favorable reservoir–caprock assemblages with the Cambrian–Precambrian high quality source rocks and the Cambrian gypsum–salt caprocks. Their oil and gas resources are expected to reach about 10.4×10^9 m^3, showing huge exploration potential [3–5]. Breakthroughs in the Middle–Lower Cambrian were successful in obtaining economically viable oil and gas flows from Wells ZS1 and LT1 in 2012 and in 2019, respectively [6,7]. In contrast, the Upper Ediacaran is still in an exploration stage without any large scale reservoirs found. Only 12 wells have been drilled in the Upper Ediacaran Chigbrak Formation, including LT1, QT1, KT1, XH1, TD2 and DT1. The limited well data limit the geological understandings of the Chigbrak Formation. In recent years, an increasing number of studies have been conducted, and some important findings have been obtained. For example, the passive continental margin environment that was formed in the early Ediacaran continued in the uplifted zone of the Bachu–Central Tarim–Southeastern Tarim area during the late Ediacaran, and the Chigbrak Formation developed within carbonate ramp depositional systems [8–10]. In the outcrop area at the northwestern margin of this basin, the Chigbrak Formation is mainly composed of dolomitic stromatolites, thrombolites, etc., and is dominated by vuggy porosity [11–13]. Sedimentary facies, weathering crust karstification and hydrothermal dissolution have been identified as the main factors controlling the genesis of reservoirs [14–16]. Controversies on the genesis of dolomite reservoirs [17–19] and limited knowledge of the genetic mechanism of reservoir pores in the Chigbrak Formation restrict the prediction of favorable reservoirs.

This paper discusses the genesis of dolomite and the reservoir control factors in the Upper Ediacaran Chigbrak Formation in the Aksu outcrop area at the northwestern margin of the Tarim Basin by using 175 samples and through thin section observations, tests of stable carbon and oxygen isotopic compositions, strontium isotopic composition, rare earth elements, laser ablation inductively–coupled–plasma mass spectrometry (LA–ICP–MS) element mapping and U–Pb dating. The research results are of great significance in identifying the genesis and distribution of reservoirs in the Chigbrak Formation of the Tarim Basin.

2. Geological Setting

The Tarim Basin in southern Xinjiang, China (Figure 1a), is a multi–cycle petroliferous basin superimposed by the Paleozoic craton basin and the Mesozoic and Cenozoic foreland basins [20], covering an area of 56×10^4 km^2. According to the division of present–day structural units, the basin includes seven first–order structural units, i.e., the Tabei Uplift, the Central Uplift, the Southeast Uplift, the Kuqa Depression, the North Depression, the Southeast Depression and the Southwest Depression (Figure 1b). Tectonic evolution of the Tarim landmass in the Precambrian period was closely related to the amalgamation–breakup cycles of the Colombia supercontinent and the Rodinia supercontinent [21,22]. At the end of the Neoproterozoic period, affected by the breakup of the Rodinia supercontinent, the continental extension that dominated the Cryogenian–early Ediacaran period led to the rift system development stage in the Tarim landmass, when continental rift basins were formed in its interior and margin and a set of thousand–meter–thick clastic rocks with basic igneous rock was deposited. In the late Ediacaran period, along with the continuous spreading of the rift, depression appeared due to the subsidence of the peripheral areas of the ocean basin, forming the carbonate platform–dominated depressed basin [23,24]. The vast areas to the north of the Central Uplift were covered by the ramp–type carbonate platform sedimentary system [25]. The thick dolomites of the Chigbrak Formation were formed penecontemporaneously, and their sedimentary differentiation was controlled by the earlier paleogeomorphology and differential subsidence of fault. At the end of the Ediacaran period, intensive uplifting within the Tarim landmass induced by the Kalpin movement led to weathering and subsequent dissolution of the dolomites at the top of the Chigbrak Formation. Thus, a regional unconformity was formed between Ediacaran and Cambrian.

Structurally, the study area is a part of the eastern segment of Kalpin fault–uplift zone in the North Tarim Uplift [26,27], and has a complete Ediacaran outcrop (Figure 1c). The Sugetbrak Formation is in conformable contact with the Chigbrak Formation (Figure 2a), the Lower Cambrian Yurtus Formation has a parallel unconformable contact with the Chigbrak Formation (Figure 2b) and the Scholbrak Formation is in conformable contact with the Yurtus and Usugar Formations [28]. In Aksu outcrop area, the Chigbrak Formation is about 173.5 m thick, and divided into four members from bottom to top according to color, lithology and sedimentary structure (Figure 3). Member Ch–1 at the bottom is 42.6 m thick, and mainly composed of gray and light gray dolomitic stromatolites with horizontal layered (Figure 4a), columnar (Figure 4b) and hummocky structures, intercalated with dolomitic oolite (Figure 4c), grainstone with a bonding structure (Figure 4d) and thin layers of yellowish gray and grayish purple argillaceous dolomite, quartz sandstone and argillite (Figure 4e,f). It shows the depositional characteristics of upper intertidal–supratidal environment. Member Ch–2 is 27.3 m thick, and mainly composed of medium–thin layers of grey dolomitic stromatolites with horizontal and wavy structures alternated with dolomitic oolite, grainstone with a bonding structure and dolomitic thrombolites. It shows the depositional characteristics of relatively high energy intertidal environment. Member Ch–3 is 93.1 m thick, and mainly composed of medium–thick layered dolomitic thrombolites with horizontal bedding (Figure 4g) alternated with dolomitic foamy microbialite (microscopically, it has no granular, clotted structure, but looks like foam) (Figure 4h), and transitions up–section into dolomitic grainstone with a bonding structure at its top. It has the depositional characteristics of a high energy subtidal environment. Member Ch–4 is 10.5 m thick, and mainly develops solution–brecciated dolomite as a result of strong weathering crust karstification (Figure 4i,j). Similar to member Ch–3, its original rocks mainly include dolomitic thrombolites and foamy microbialite. According to the gamma ray (GR), different members have different response characteristics (Figure 3): member Ch–1 has the highest GR values with obvious fluctuations, member Ch–2 and member Ch–3 have low GR values, and member Ch–4 has slightly higher GR values than member Ch–2 and member Ch–3. Pores are developed extensively in all four members, especially acicular–dissolved pores, millimeter–size dissolved pores and centimeter–decimeter–size vugs, which are common. The acicular–dissolved pores (Figure 4k) are developed in dolomitic grainstone and foamy microbialite, the millimeter–size dissolved pores and centimeter–size vugs (Figure 4l) are mainly observed in dolomitic stromatolites and dolomitic thrombolites, and the decimeter–size vugs are distributed in the brecciated dolomite in member Ch–4 (Figure 4m). In the upper of member Ch–3, the vugs are distributed along the layers, and often filled with multi–phase cements. Due to the fibrous dolomite cement and laminated dolomite cement have the characteristics of isopachous rim, the edges of vugs have a "laces structure" (Figure 4n,o).

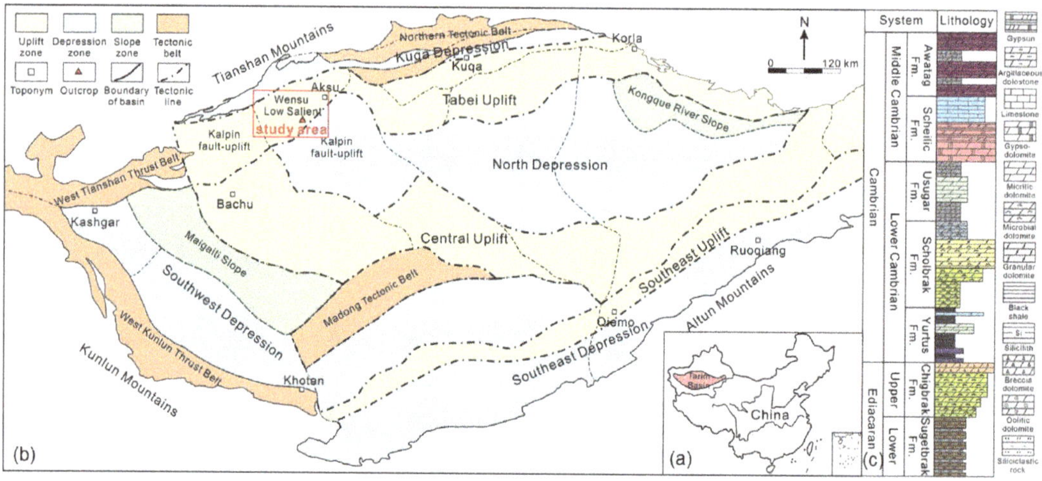

Figure 1. Geological setting map of study area: (**a**) location of the Tarim Basin in northwestern China indicated with pink zone; (**b**) tectonic location of the study area indicated with a red line rectangle; (**c**) stratigraphic sequence from Ediacaran to middle Cambrian.

Figure 2. Photographs of Chigbrak Formation in the outcrop: (**a**) bottom characteristics (the red line is the boundary between Chigbrak Formation and Sugetbrak Formation); (**b**) top characteristics (the red curve is the boundary between Chigbrak Formation and Yurtus Formation).

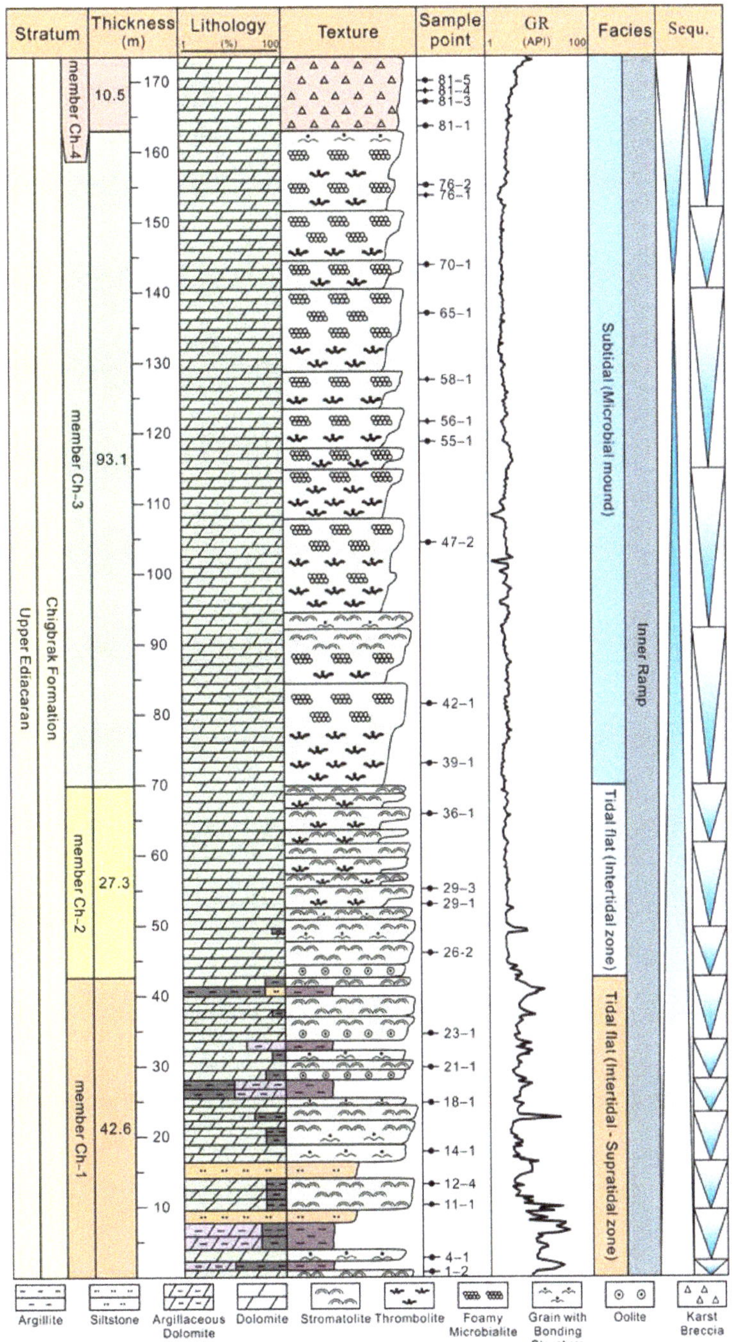

Figure 3. Stratigraphic column of Chigbrak Formation.

Figure 4. Photographs of dolomite in the Chigbrak Formation: (**a**) dolomitic stromatolite (vertical perspective), member Ch–1, outcrop image; (**b**) dolomitic stromatolite, with framework pores, member Ch–1, blue casting thin section PPL image; (**c**) dolomitic oolite, with intergranular pores, member Ch–1, blue casting thin section PPL image; (**d**) dolomitic grainstone with bonding structure, with intergranular pores and intragranular pores, member Ch–1, blue casting thin section PPL image; (**e**) stromatolite dolomite is interbedded with thin layered yellowish–gray, grayish–brown argillaceous dolomite and quartz sandstone, member Ch–1, outcrop image; (**f**) micrite dolomite is interbedded with dolomitic quartz sandstone, blue casting thin section PPL image; (**g**) dolomitic thrombolite, with framework pores, member Ch–3, blue casting thin section PPL image; (**h**) dolomitic foamy microbialite, with dissolved pores, member Ch–3, blue casting thin section PPL image; (**i**) brecciated dolomite, with interbreccia pores and fractures, member Ch–4, outcrop image; (**j**) brecciated dolomite, breccias are better rounded, member Ch–4, outcrop image; (**k**) dolomitic foamy microbialite, with acicular–dissolved pores, member Ch–3, outcrop image; (**l**) dolomitic thrombolite, vugs with layered distribution, member Ch–3, outcrop image; (**m**) brecciated dolomite, with decimeter–size vugs, member Ch–4, outcrop image; (**n**) dolomitic thrombolite, vugs with "lace structure" are developed along the bedding, member Ch–3, blue casting thin section PPL image; (**o**) dolomitic thrombolite, vugs with "lace structure" are developed along the bedding, member Ch–4, sample section image.

3. Samples and Methods

A total of 104 hand specimens were taken from the Upper Ediacaran Chigbrak Formation in the Xigou section. All of them are fresh and have never been weathered or altered. Casting thin sections were prepared for all of the samples, correspondingly. Based on macroscopic characteristics of these hand specimens and the microscopic observation results of casting thin sections, 22 dolomite samples were selected for whole rock detection and analysis of stable carbon and oxygen isotopic compositions ($\delta^{13}C$ and $\delta^{18}O$), strontium isotopic composition ($^{87}Sr/^{86}Sr$), degree of dolomite cation ordering and rare earth elements (REEs) content. In order to minimize the interaction between carbonate cements and host rocks, single structural components were drilled from all samples using a small micro–sampling drill and ground to a powder of 200 mesh with an agate mortar. Then, the sample powder was packaged with transparent drawing paper. In addition, 4 porous samples were chosen for $\delta^{13}C$, $\delta^{18}O$ and $^{7}Sr/^{86}Sr$ analysis, laser ablation U–Pb dating and LA–ICP–MS element mapping (Figure 5). The elements of imaging are Mg, Ca, Mn, Fe, Sr, Ba, Th, U, Y, La, Ce, Eu and Lu. For $\delta^{13}C$ and $\delta^{18}O$ analysis, 60 μm–thick thin sections were prepared, and for $^{7}Sr/^{86}Sr$ analysis, laser ablation U–Pb dating and LA–ICP–MS element mapping, 100 μm–thick thin sections were prepared. Both of them were single–side polished and then purified in a super–clean laboratory.

Laser ablation $\delta^{13}C$, $\delta^{18}O$ and $^{7}Sr/^{86}Sr$ analysis, whole rock $\delta^{13}C$, $\delta^{18}O$ and $^{7}Sr/^{86}Sr$ measurements and the determination of order degree of dolomite were carried out in Key Laboratory of Carbonate Reservoirs, CNPC (China National Petroleum Corporation), Hangzhou, China. A Nd.YAG laser ablation system and a RESOlution S–155 laser ablation system were used for $\delta^{13}C$, $\delta^{18}O$ and $^{7}Sr/^{86}Sr$ analysis, with laser spot diameters of 20 μm and 160 μm, respectively. Measurement of C and O isotopes was conducted using a Delta V Advantage IRMS with error ranges of $\pm 0.06‰$ and $\pm 0.08‰$, respectively, and the carbonate standard samples of GBW4405 and GBW4406 were used. Measurement of the Sr isotope was carried out using a Neptune Plus MC–ICP–MS, when two self–developed standard samples, Hailuo and XK1, were used, and a precision of 0.01% was obtained. Whole rock Sr isotopes were measured with a Isotope Ratio Mass Spectrometry (Triton Plus, Thermo Scientific, Germany), when the carbonate standard reference material of GBW04411 was used, and a precision better than 0.01% was obtained. Degree of dolomite cation ordering was measured using an X′pert Pro X–ray diffractometer, with a relative error < 10%, and the X–ray diffraction peak ratio of (015)/(110) was used to calculate cation ordering [29]. An ICP–MS method was used to analyze the REEs in Tongwei Analytical Technology Co., Ltd., Guizhou, China, using an inductively coupled plasma mass spectrometry (ICP–MS). Two international standard samples, W–2a and BHVO–2, were used, and a precision and an accuracy > 5% were obtained. Laser ablation U–Pb dating was performed in the Radiogenic Isotopes Laboratory, University of Queensland, Australia, using an ASI RESOlution SE laser ablation system with laser spot diameter of 100 μm, together with a Nu Plasma II MC–ICP–MS, the calibration is done by use of two international standard samples, NIST 614 and WC–1. Iolite 3.6 (University of Melbourne, Australia) was used for processing the LA–ICP–MS data acquired, and Isoplot3.0 (University of California Berkeley, CA, USA) was used to calculate the ages and map the Tera–Wasserburg concordia diagram. LA–ICP–MS trace element mapping was also performed in the Radiogenic Isotopes Laboratory, University of Queensland, Australia. An ASI RESOlution SE laser ablation system, where square laser spot with length of side of 50 μm and speed of movement of 0.05 mm/s, was applied together with a Thermo iCap–RQ ICP–MS, the calibration is done by use of International standard sample NIST 614. Element distribution maps were established with the raw data processed by Iolite 3.6.

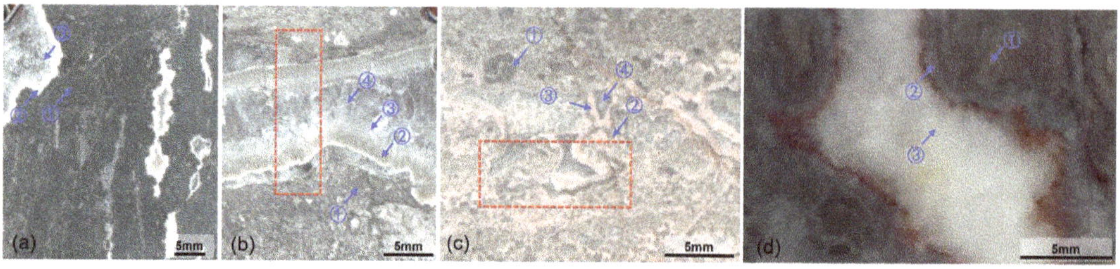

Figure 5. Photographs of test samples in the Chigbrak Formation (the red dotted boxes are the elements mapping area, the numbers with circles are the age, $\delta^{13}C$, $\delta^{18}O$ and $^{87}Sr/^{86}Sr$ test points of different fabrics, corresponds to the numbers with circles in Table 1): (**a**) sample 56–1; (**b**) sample 58–1; (**c**) sample 76–1; (**d**) sample 81–4.

Table 1. $\delta^{13}C$, $\delta^{18}O$, $^{87}Sr/^{86}Sr$ and U–Pb age values of different fabrics in the Chigbrak Formation dolomite.

Sample	Member	Fabrics	$\delta^{13}C$ ‰ (PDB)	$\delta^{18}O$ ‰ (PDB)	$^{87}Sr/^{86}Sr$	U–Pb Age Ma
56–1	Ch-3	① Dolomitic Microbialite (MD)	1.87	−4.26	0.70885	/
		② Fibrous dolomite in the pore (FD)	2.40	−4.81	0.70880	/
		③ Granular dolomite in the pore (GD)	2.28	−7.34	0.70893	/
58–1	Ch-3	①Dolomitic Microbialite (MD)	2.21	−4.73	0.70882	605.4 ± 6.9
		② Fibrous dolomite in the pore (FD)	2.46	−5.08	0.70871	576 ± 17
		③ Bladed dolomite in the pore (BD)	1.59	−6.61	0.70895	553 ± 20
		④ Granular dolomite in the pore (GD)	0.95	−7.86	0.70902	538 ± 25
76–1	Ch-3	① Dolomitic Microbialite (MD)	2.69	−4.29	0.70877	618 ± 37
		② Fibrous dolomite in the pore (FD)	2.12	−5.57	0.70876	558 ± 26
		③ Laminated dolomite in the pore (LD)	2.83	−4.90	0.70894	559 ± 22
		④ Saddle dolomite in the pore (SD)	2.01	−10.39	0.70925	256 ± 12
81–4	Ch-4	① Dolomitic Microbialite (MD)	3.33	−3.95	0.70889	538 ± 28
		② Granular dolomite in the pore (GD)	1.09	−8.14	0.70913	513 ± 20
		③ Saddle dolomite in the pore (SD)	1.95	−9.87	0.70948	172.9 ± 3.9

4. Results

4.1. Stable Carbon and Oxygen Isotopic Compositions

Carbon and oxygen isotopes are related to the fluids causing dolomitization, which is mainly affected by salinity and temperature. Consequently, it serves as an effective indicator of dolomitization fluid properties and diagenetic environment. According to the whole rock $\delta^{13}C$ and $\delta^{18}O$ values (all data are PDB standard) of 22 samples (Table 2) and their scatter spots (Figure 6a), no significant correlation was shown between $\delta^{13}C$ and $\delta^{18}O$ values. This indicates that the samples have experienced weak transformation in the late diagenetic period and remained the basic information of the diagenetic fluids. For the Chigbrak Formation dolomites, $\delta^{13}C$ ranges from 1.31‰ to 6.65‰, with an average of 3.50‰, and $\delta^{18}O$ ranges from −4.81‰ to −1.23‰, with an average of −2.95‰. These values are generally similar to those of the late Proterozoic marine carbonates [30], and also distributed within the value ranges of the late Ediacaran seawater [15]. The values of $\delta^{13}C$ reach the highest in member Ch–1 (4.21‰~6.18‰, avg. 5.50‰) and the lowest in member Ch–4 (1.31‰~2.40‰, avg. 1.93‰), and are almost the same in the other two intervals (1.68‰~3.48‰, avg. 2.47‰). The values of $\delta^{18}O$ reach the lowest in member Ch–4 (–4.65‰~−4.05‰, avg. −4.34‰), are relatively high in member Ch–2 (−2.23‰~−1.39‰, avg. −1.76‰) and vary widely in member Ch–1 (–4.06‰~–1.23‰, avg. –2.56‰) and member Ch–3 (−4.81‰~−2.03‰, avg. −3.50‰).

Table 2. $\delta^{13}C$, $\delta^{18}O$, $^{87}Sr/^{86}Sr$ and order degree values of dolomite (whole rock) in the Chigbrak Formation.

Sample	Lithofacies	Member	$\delta^{13}C$ ‰ (PDB)	$\delta^{18}O$ ‰ (PDB)	$^{87}Sr/^{86}Sr$	Ordering Degree
1–2	Dolomitic Stromatolite (SD)	Ch–1	5.86	−2.67	0.709992	0.68
4–1	Dolomitic Grainstone with clay (GDC)	Ch–1	6.18	−2.71	0.710758	0.62
11–1	Dolomitic Stromatolite (SD)	Ch–1	5.34	−3.05	0.709634	0.61
12–4	Dolomitic Stromatolite (SD)	Ch–1	6.65	−4.06	0.709334	0.67
14–1	Dolomitic Grainstone with bonding structure (GDBS)	Ch–1	5.93	−2.20	0.709359	0.60
18–1	Dolomitic Grainstone with bonding structure (GDBS)	Ch–1	4.49	−2.55	0.709877	0.62
21–1	Dolomitic Stromatolite (SD)	Ch–1	5.35	−1.23	0.710885	0.66
23–1	Dolomitic Oolite (OD)	Ch–1	4.21	−1.99	0.710710	0.64
26–2	Dolomitic Stromatolite (SD)	Ch–2	2.89	−1.44	0.709242	0.63
29–1	Dolomitic Thrombolite (TD)	Ch–2	2.21	−1.39	0.709077	0.68
29–3	Dolomitic Stromatolite (SD)	Ch–2	2.35	−2.23	0.709059	0.70
36–1	Dolomitic Stromatolite (SD)	Ch–2	2.17	−1.98	0.708906	0.70
39–1	Dolomitic Thrombolite (TD)	Ch–3	2.64	−2.03	0.708745	0.67
42–1	Dolomitic Foamy Microbialite (FMD)	Ch–3	3.03	−4.45	0.709093	0.68
47–2	Dolomitic Foamy Microbialite (FMD)	Ch–3	3.48	−2.65	0.709214	0.71
55–1	Dolomitic Thrombolite (TD)	Ch–3	2.15	−2.46	0.709086	0.67
65–1	Dolomitic Foamy Microbialite (FMD)	Ch–3	1.95	−4.81	0.709015	0.69
70–1	Dolomitic Foamy Microbialite (FMD)	Ch–3	1.68	−4.40	0.709178	0.70
76–2	Dolomitic Foamy Microbialite (FMD)	Ch–3	2.65	−3.67	0.709097	0.71
81–1	Brecciated Dolomitic Thrombolite (BTD)	Ch–4	2.09	−4.33	0.709132	0.73
81–3	Brecciated Dolomitic Foamy Microbialite (BFMD)	Ch–4	1.31	−4.65	0.709419	0.73
81–5	Brecciated Dolomitic Thrombolite (BTD)	Ch–4	2.40	−4.05	0.709241	0.77

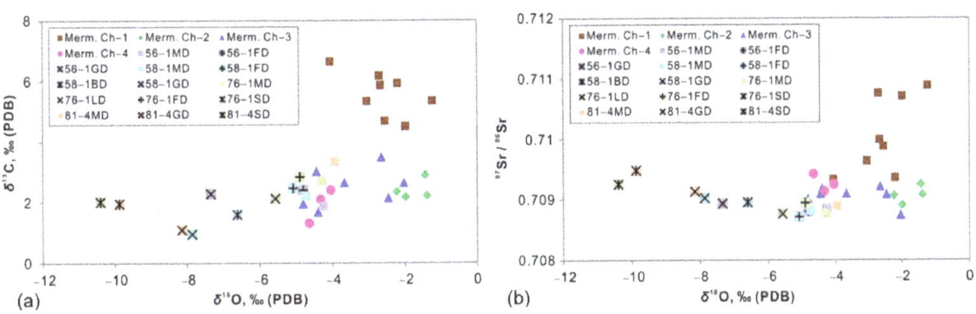

Figure 6. Binary plots of geochemistry in the Chigbrak Formation dolomite: (**a**) $\delta^{13}C$ versus $\delta^{18}O$; (**b**) $^{87}Sr/^{86}Sr$ versus $\delta^{18}O$.

According to the $\delta^{13}C$ and $\delta^{18}O$ values of four samples determined by the laser microsampling method (Table 1) and their scatter plots (Figure 6a), $\delta^{13}C$ values of the host rocks and cements in the pores are generally consistent with those of the whole rock samples from Ch–3, and only slightly smaller in granular dolomite (avg. 1.48‰). The values of $\delta^{18}O$ change greatly in the original rocks and different types of cements, and gradually decrease from the margin to the center of the pores. The averages of $\delta^{18}O$ in the host rocks of dolomitic microbialite, laminated/fibrous dolomite cements, granular/bladed dolomite cements and saddle dolomite cements are −4.31‰, −5.09‰, −7.49‰ and −10.13‰, respectively.

4.2. *Strontium Isotopic Composition*

Strontium isotopic composition in carbonate rocks is affected by diagenetic environment and fluids, and is one of the important indicators of paleoclimate and diagenetic fluids [31]. According to the detected whole rock $^{87}Sr/^{86}Sr$ measurements of 22 samples (Table 2) and the scatter plot between $^{87}Sr/^{86}Sr$ and $\delta^{18}O$ (Figure 6b), the values of $^{87}Sr/^{86}Sr$ range from 0.708745 to 0.710885, with an average of 0.709457. Similar to the values of $\delta^{13}C$, these values reach the highest in member Ch–1 (0.709334~0.710885, avg. 0.710069) and are almost consistent in the remaining intervals (0.708745~0.709419, avg. 0.709107). In general, the values of $^{87}Sr/^{86}Sr$ in the Chigbrak Formation in the study area are slightly higher than those of the coeval seawater, which range between 0.7087 and 0.7090 [32,33]. It is suggested that this is related to dolomite samples generally have higher values of $^{87}Sr/^{86}Sr$ than limestone ones [34]. The relatively high $^{87}Sr/^{86}Sr$ values of member Ch–1 samples may be explained by their sedimentary environment of a tidal flat, where crust–derived Sr was mixed into the sediments.

According to the $^{87}Sr/^{86}Sr$ measurements of the four samples determined by the laser microsampling method (Table 2) and the scatter plot between $^{87}Sr/^{86}Sr$ and $\delta^{18}O$ (Figure 6b), the values of $^{87}Sr/^{86}Sr$ in the host rocks and other types of dolomite in the pores–except for saddle dolomite, whose values of $^{87}Sr/^{86}Sr$ are slightly higher (avg. 0.70936)–are generally distributed within the $^{87}Sr/^{86}Sr$ variation range of the coeval seawater. The values of $^{87}Sr/^{86}Sr$ are approximate in the host rocks of dolomitic microbialite (avg. 0.70883) and laminated/fibrous dolomite cements (avg. 0.70880), slightly higher in granular/bladed dolomite cements (avg. 0.70901) and exhibit an increasing trend from margin to center of the pores.

4.3. *Degree of Dolomite Cation Ordering*

Dolomite is a tripartite mineral whose lattice parameters are affected by composition, temperature and pressure [35]. The ideal crystal structure of dolomite is that Ca^{2+} and Mg^{2+} are arranged alternately along the C–axis and the molar percentage of Ca^{2+} and Mg^{2+} is the same, while in disordered dolomite, Ca^{2+} and Mg^{2+} are completely randomly distributed as the same structure of calcite. Generally, the order degree value greater than 0.8 is defined as high ordering, 0.6~0.8 is subordering, 0.4~0.6 is low ordering. Less than 0.4 is disordered [36,37], which can be used to indicate the crystallization rate, crystallization temperature and evolution of dolomite. According to the measurement results (Table 2), the degree of dolomite cation ordering in the Chigbrak Formation ranges from 0.60 to 0.77, and increases gradually from bottom to top, which is the opposite of the variation trend of $\delta^{18}O$ (PDB) value. Its average value is 0.68, which belongs to the subordering range.

4.4. *Rare Earth Elements*

The relative abundance of REEs in carbonate minerals mainly depends on the content and geochemical properties of REEs in fluid [38]. Thus, it has acted as an effective tool for determining the source of dolomitization fluid. The analysis results (Table 3) reveal that the total REEs (∑REEs) are remarkably higher in member Ch–1 (2.88~22.34 ppm, avg. 13.11 ppm) and member Ch–4 (2.88 ppm~22.34 ppm, avg. 13.11 ppm) than in member Ch–2 and member Ch–3 (0.36 ppm~2.51 ppm, avg. 1.22 ppm). The results were standardized by chondrite, and the standard normalized (SN) element anomalies were calculated by the following equations: $\delta Ce = 2 \times Ce_{SN}/(La_{SN} + Pr_{SN})$ and $\delta Eu = 2 \times Eu_{SN}/(Sm_{SN} + Gd_{SN})$ [39]. Calculating δCe and δEu as larger than 1.2 indicates a positive anomaly, and less than 0.8 indicates a negative anomaly. The standardized REE distribution pattern (Figure 7) shows that the Chigbrak Formation dolomites are characterized by enrichment of LREEs (La–Eu) and depletion of HREEs (Gd–Lu). Member Ch–4 has significant negative anomalies of δCe and δEu, and some samples from member Ch–1 and member Ch–3 have negative anomalies of δEu.

Table 3. Rare earth elements content values of dolomite in the Chigbrak Formation.

Sample	La (ppm)	Ce (ppm)	Pr (ppm)	Nd (ppm)	Sm (ppm)	Eu (ppm)	Gd (ppm)	Tb (ppm)	Dy (ppm)	Ho (ppm)	Er (ppm)	Tm (ppm)	Yb (ppm)	Lu (ppm)	ΣREEs (ppm)
1–2	4.037	8.303	0.985	3.746	0.730	0.188	0.665	0.100	0.562	0.108	0.304	0.044	0.283	0.043	20.10
4–1	3.265	6.807	0.755	2.795	0.513	0.107	0.478	0.076	0.465	0.101	0.305	0.046	0.299	0.045	16.06
11–1	3.746	8.640	1.168	4.559	0.882	0.204	0.851	0.138	0.843	0.178	0.520	0.075	0.467	0.070	22.34
12–4	2.039	4.773	0.525	1.898	0.335	0.086	0.310	0.048	0.279	0.055	0.153	0.021	0.127	0.018	10.67
14–1	3.805	8.594	0.922	3.288	0.629	0.132	0.568	0.094	0.571	0.117	0.344	0.050	0.318	0.045	19.48
18–1	0.842	1.631	0.224	0.829	0.162	0.051	0.176	0.030	0.186	0.040	0.112	0.017	0.108	0.017	4.43
21–1	0.598	1.188	0.147	0.501	0.099	0.029	0.089	0.014	0.086	0.018	0.051	0.008	0.047	0.007	2.88
23–1	1.546	3.657	0.442	1.721	0.338	0.080	0.331	0.052	0.308	0.062	0.176	0.025	0.151	0.022	8.91
26–2	0.457	1.102	0.132	0.446	0.086	0.026	0.080	0.012	0.068	0.013	0.041	0.006	0.038	0.006	2.51
29–1	0.300	0.770	0.089	0.327	0.058	0.017	0.053	0.008	0.046	0.008	0.025	0.003	0.021	0.003	1.73
29–3	0.222	0.645	0.074	0.283	0.051	0.016	0.048	0.007	0.041	0.008	0.023	0.003	0.021	0.003	1.45
36–1	0.070	0.157	0.016	0.056	0.011	0.003	0.013	0.002	0.013	0.003	0.009	0.001	0.008	0.001	0.36
39–1	0.199	0.437	0.041	0.155	0.031	0.008	0.032	0.005	0.033	0.007	0.024	0.004	0.021	0.003	1.00
42–1	0.182	0.320	0.033	0.133	0.027	0.007	0.031	0.005	0.033	0.008	0.024	0.004	0.022	0.003	0.83
47–2	0.123	0.204	0.023	0.096	0.020	0.005	0.027	0.004	0.028	0.007	0.023	0.003	0.021	0.003	0.59
55–1	0.218	0.360	0.045	0.182	0.036	0.011	0.045	0.007	0.045	0.011	0.034	0.005	0.031	0.004	1.03
65–1	0.220	0.388	0.045	0.188	0.036	0.010	0.046	0.007	0.046	0.011	0.036	0.005	0.031	0.005	1.08
70–1	0.457	0.888	0.099	0.358	0.070	0.017	0.071	0.011	0.072	0.017	0.047	0.007	0.042	0.007	2.16
76–2	0.159	0.228	0.030	0.121	0.026	0.006	0.027	0.006	0.033	0.009	0.030	0.005	0.029	0.005	0.71
81–1	7.172	8.965	1.673	7.738	1.956	0.436	2.163	0.323	1.927	0.430	1.254	0.175	1.034	0.150	35.40
81–3	4.507	5.608	1.055	4.342	0.893	0.192	0.914	0.140	0.836	0.177	0.525	0.076	0.458	0.068	19.79
81–5	5.326	6.711	1.238	5.465	1.284	0.283	1.385	0.208	1.242	0.272	0.798	0.112	0.670	0.098	25.09

4.5. U–Pb Dating

U–Pb dating is used to determine the age of the host rocks of ancient carbonate rocks and their cements [40,41]. According to the dating results (Table 1 and Figure 8), for three samples of dolomitic microbialite, the ages of their host rocks range from 538 ± 28 Ma to 618 ± 37 Ma, corresponding to the Ediacaran period (538.8–635 Ma). The ages of laminated/fibrous dolomite cements vary from 558 ± 26 Ma to 576 ± 17 Ma, which is within the age range of the Ediacaran. The ages of granular/bladed dolomite cements range from 513 ± 20 Ma to 553 ± 20 Ma, corresponding to the age range between the second series of Cambrian (509–521 Ma) and the Ediacaran. The ages of saddle dolomite cements are 172.9 ± 3.9 Ma and 256 ± 12 Ma, corresponding to the geological ages of the middle Jurassic Aalenian (170.3–174.1 Ma) and the Upper Permian Lopingian (251.9–259.5 Ma), respectively. It was revealed that the age sequences of these dolomites with different textures in the three samples are consistent with their diagenetic sequences, that is, the age decreases from the host rocks of microbial dolomite, the dolomite at the margin of the pores to the dolomite in the pore center, which also reflected the reliability of all the measurement results.

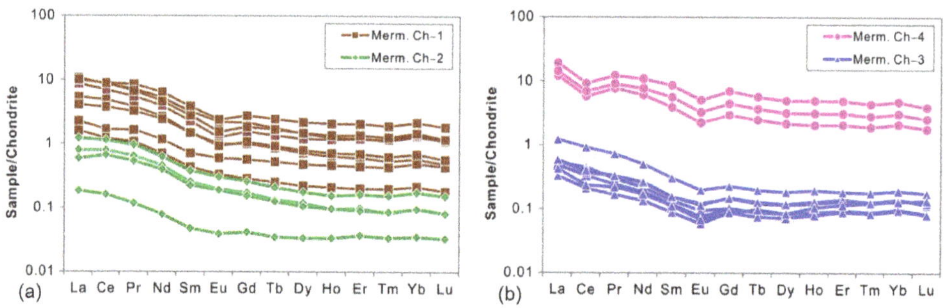

Figure 7. Normalized distribution patterns of rare earth elements in the Chigbrak Formation dolomite: (**a**) member Ch–1 and member Ch–2; (**b**) member Ch–3 and member Ch–4.

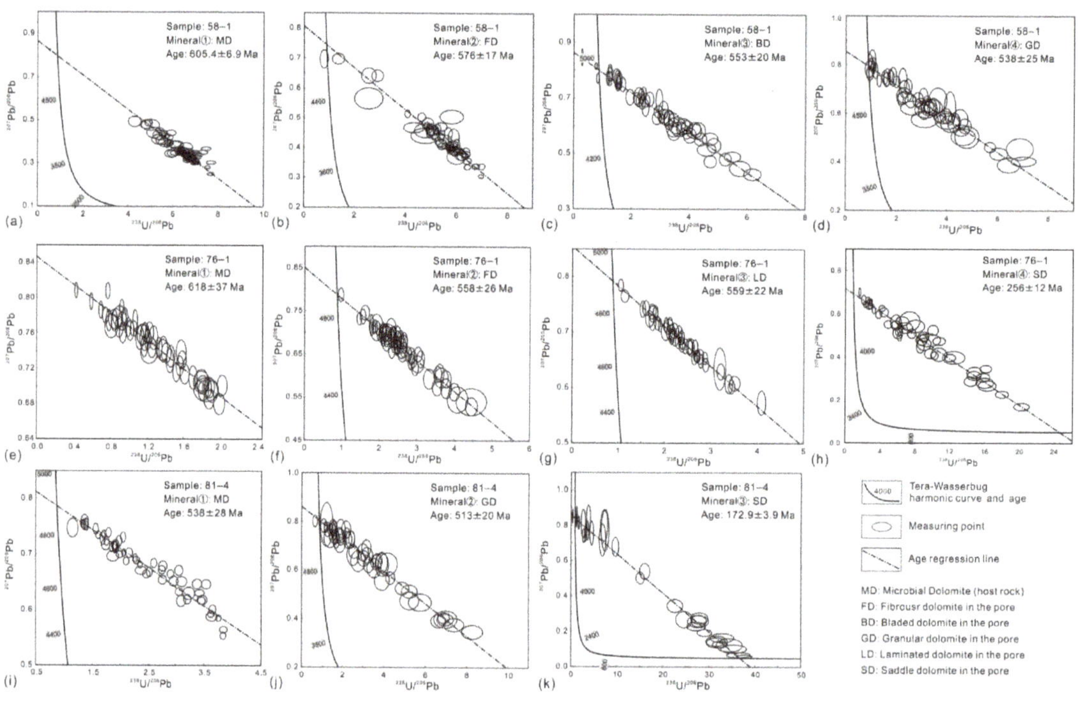

Figure 8. LA–MC–ICP–MS U–Pb concordia diagram of dolomite in the Chigbrak Formation: (**a**) MD of sample 58–1; (**b**) FD of sample 58–1; (**c**) BD of sample 58–1; (**d**) GD of sample 58–1; (**e**) MD of sample 76–1; (**f**) FD of sample 76–1; (**g**) LD of sample 76–1; (**h**) SD of sample 76–1; (**i**) MD of sample 81–4; (**j**) GD of sample 81–4; (**k**) SD of sample 81–4.

4.6. Element Mapping

The composition of multi–phase cements in ancient carbonate rocks always exhibits high heterogeneity. A LA–ICP–MS–based element–mapping technique was applied on the samples to intuitively present the planar variation characteristics of elements within a millimeter–centimeter–scale area, which is crucial to understanding their forming process [42,43].

The plane distributions of Mg, Ca, Mn, Fe, Sr, Ba, Th, U, Y, La, Ce, Eu and Lu were analyzed for sample 58–1 (Figure 9). According to its plane–polarized light (PPL) and cathode luminescence (CL) characteristics and the plane distribution of Mg and Ca, three phases of dolomite cements–including fibrous dolomite (FD), bladed dolomite (BD) and granular dolomite (GD), and one phase of sparry calcite cement were identified in pores. MD has relatively high contents of Fe and Sr and low contents of other elements. FD has a high content of Fe; BD has high contents of Mn, Y, La, Ce, Eu and Lu; and GD has low contents of all kinds of trace elements. Sparry calcite has high content of Sr and low content of Mn, Ba, Th, U and Ce, and the extremely low content of Mn is the cause of its nonluminescence. In summary, the contents of trace elements in FD are close to those in MD.

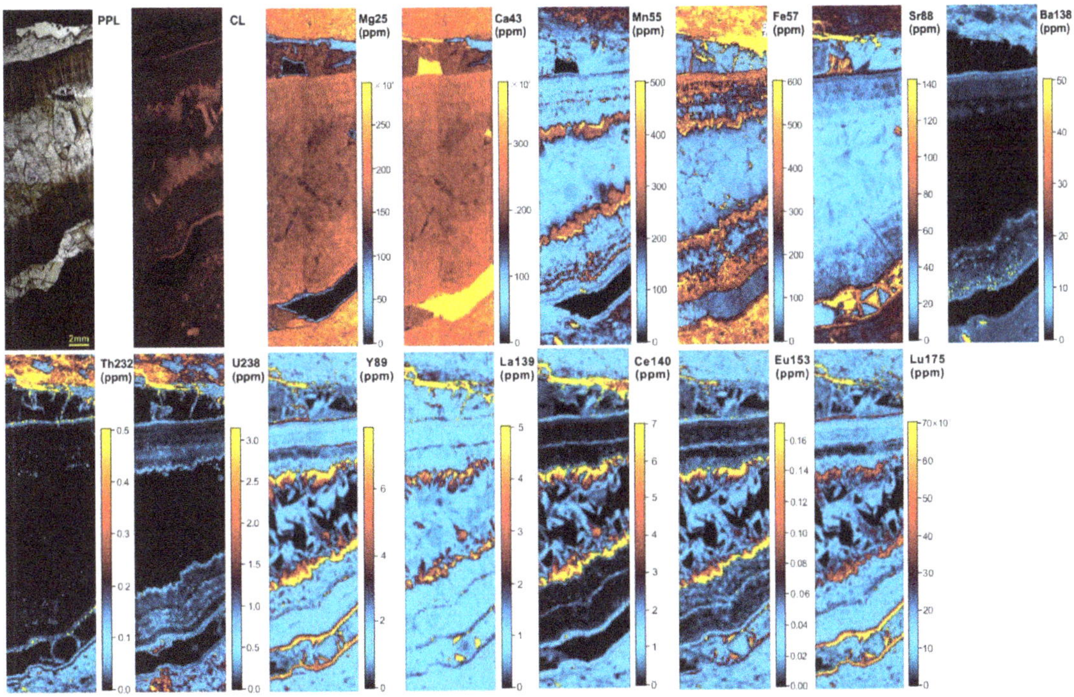

Figure 9. Photomicrographs and element maps of sample 58–1.

The plane distributions of 13 elements were analyzed for sample 76–1 (Figure 10). According to the Mg and Ca plane distribution map, a weird banding effect can be seen which may be the result of imperfect data processing. Although the image resolution is not high, it is still possible to distinguish the Mg and Ca content of the main structure. Combining PPL and CL characteristics and plane distribution of Mg and Ca, three phases of dolomite cements in the pores, including fibrous dolomite (FD), laminated dolomite (LD) and saddle dolomite (SD), and fine dolomite cements in the micro–fractures were identified. The contents of all trace elements are low in MD. Similar to MD, FD also has low contents of all trace elements. LD has relatively high contents of Fe and Mn, and SD has high contents of Y and Eu. Dolomite in the micro–fractures contains abundant Fe, Mn, Y, La, Ce, Eu and Lu, and its cathodoluminescence characteristics have a correlation with the contents of Mn and Fe.

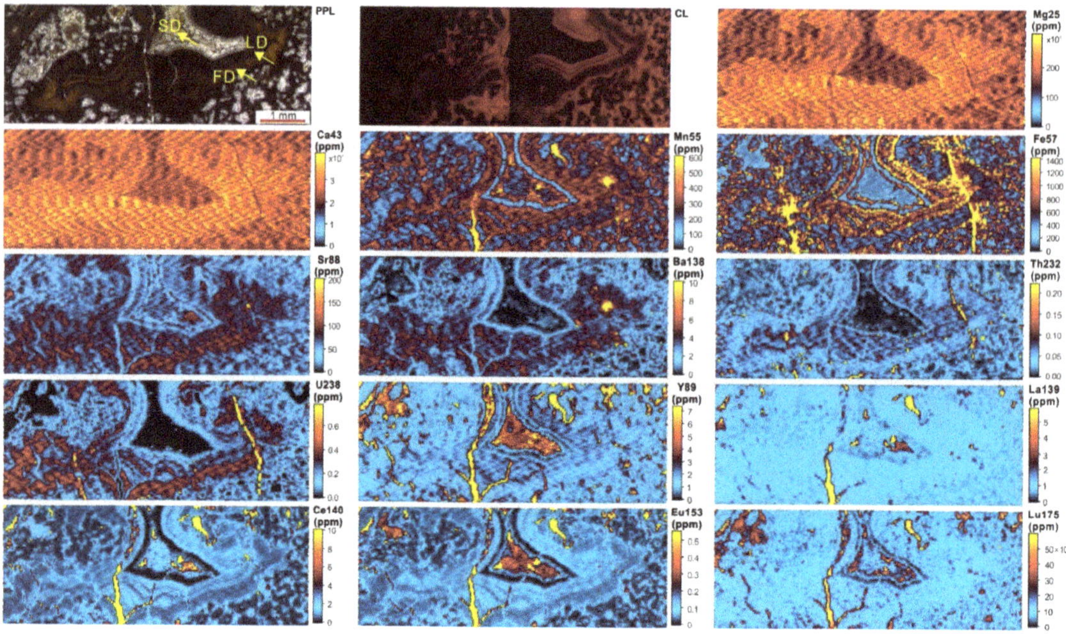

Figure 10. Photomicrographs and element maps of sample 76–1.

5. Discussion

5.1. Genesis of Dolomite

As the Upper Ediacaran Chigbrak Formation is predominantly composed of dolomite; determining the origin of dolomite is critical for understanding the diagenetic evolution of reservoir. The dolomitization of the early diagenetic stage plays an important constructive role in the preservation of primary pores and early dissolved pores. However, dolomitization in the late burial stage had little effect on the early pore preservation, and even played a destructive role [44,45]. Therefore, it is very important to determine its occurrence time. In general, a preliminary estimate of the diagenetic stage when dolomitization occurred based on the petrographic characteristics is feasible. In the early diagenetic stage, dolomitization occurs quickly, the dolomite crystals formed are fine and the original depositional textures of rock was always retained. In the middle–late diagenetic stage, dolomitization occurred relatively slowly and the formed dolomites are coarse, usually larger than fine grain in size [46]. The Chigbrak Formation in the study area is mainly composed of micrite and powder crystal dolomites, and retained the original microorganisms or granular texture well, showing the characteristics of early diagenetic dolomite.

According to the geochemical data obtained by analysis and tests, the Chigbrak Formation has medium degree of dolomite cation ordering (avg. 0.68), indicating that its dolomitization process was characterized by rapid replacement and crystal growth, and occurred in the early burial stage at low temperature [47]. In general, the dolomites have low content of ΣREEs, no anomaly of δCe and negative anomaly or no anomaly of δEu, indicating seawater as the dolomitization fluid. High content of ΣREEs suggests the containing of mud or transformation by hydrothermal fluid. A negative anomaly of δCe is used to identify the mixture of meteoric water into dolomitization fluid, and positive anomaly of δEu is a typical indication of hydrothermal process [48]. The dolomites in the study area are generally characterized by low content of ΣREEs, and negative anomaly or no obvious anomaly of δCe and δEu. This reveals that the dolomitization fluid was seawater, and no hydrothermal fluid was involved in the dolomitization process. The exceptionally high content of ΣREEs (>20 ppm) in some samples from member Ch–1 and member Ch–4 was the result of the mud components in some layers in member Ch–1 and the mixture of atmospheric water into the dolomitization fluid in member Ch–4. This supports the conclusion that member Ch–1 was deposited in tidal flat environment and member Ch–4 experienced exposure in its late depositional stage.

The oxygen isotope composition of dolomite is determined by the salinity and temperature of diagenetic fluid. Therefore, the oxygen isotope value of dolomite is higher in high salinity evaporation environment; On the contrary, the oxygen isotope value of dolomite is lower in high–temperature and high–pressure buried environments [49]. As mentioned above, no obvious correlation was detected between the values of $\delta^{13}C$ (PDB) and $\delta^{18}O$ (PDB) in the samples, which was explained by the weak modification in the late diagenetic stage. These values were distributed within the variation range of $\delta^{13}C$ (PDB) and $\delta^{18}O$ (PDB) values of the late Ediacaran seawater, indicating that the Chigbrak Formation was formed at low temperature with seawater as the dolomitization fluid. Moreover, from bottom to top, the values of $\delta^{18}O$ (PDB) decease first and then increase, corresponding to the depositional environment of a tidal flat with relatively strong evaporation for member Ch–1 and the supergene environment for member Ch–4, where it suffered from exposure and meteoric water dissolution.

Containing clay or the involvement of a hydrothermal process led to an increased ratio of $^{87}Sr/^{86}Sr$ [50]. The values of $^{87}Sr/^{86}Sr$ of the Chigbrak Formation are generally slightly higher than those of the coeval seawater. Two main reasons have been identified for this phenomenon. One is that dolomite samples usually have higher values of $^{87}Sr/^{86}Sr$ than limestone samples [34], and the other is that the excessively high values of 87Sr/86Sr in some samples from member Ch–1 prop up the overall average. The study area is located in Wensu Low Salient [25], and member Ch–1 probably contained some clay because of its depositional environment of tidal flat, which led to obviously higher $^{87}Sr/^{86}Sr$ values due to mixture of crust derived strontium. In addition, the slightly higher $^{87}Sr/^{86}Sr$ values in member Ch–4 are also probably related to the mixture of a small amount of crust derived strontium due to the exposure it suffered from in the late Ediacaran period. The average value of $^{87}Sr/^{86}Sr$ in the remaining intervals is 0.709064, which is close to that of the coeval seawater. This also indicated seawater as the dolomitization fluid for the Chigbrak Formation.

Pores are well developed in the dolomitic microbialite of Chigbrak Formation, and filled with dolomite cements of different phases and different types. Accurate determination of the formation age of these dolomites is achievable by an analytical comparison of the ages between the host rocks of dolomitic microbialite and the first phase dolomite cements at the margin of the pores [40]. It is revealed by the U–Pb dating data that ages of the host rocks of all the dolomitic microbialite samples (538 $\pm$ 28 to 618 $\pm$ 37 Ma) are larger than those of the first phase dolomite cements at the margin of the pores (558 $\pm$ 26 to 576 $\pm$ 17 Ma), and ages of the dolomite cements of the first phase are larger than those of other phases. This suggests a favorable consistency between age sequence and diagenesis sequence, and also reveals that dolomitization has already happened in the depositional stage of the Chigbrak Formation.

On the basis of the petrological, isotope geochemical, element geochemical and geochronological characteristics of carbonate minerals, it has been clarified that the Chigbrak Formation dolomites in the study area were formed in the epidiagenetic environment with seawater as the dolomitization fluid. The findings of the research on paleoenvironment and sedimentary model of the study area [12,19] demonstrate that the sedimentary sequence of inner ramp tidal flat and microbial mound–shoal complex in the setting of carbonate ramp was developed in the depositional stage of the Chigbrak Formation, when the whole study area was a near shore shallow water environment in warm and relatively dry climate. In this geological setting, the relatively strong evaporation led to an increased content of Mg^{2+} in seawater and allowed evaporative dolomitization of the sediments dominated by microorganisms in this near shore shallow water environment.

5.2. Pore Origin and Diagenesis Sequence

Based on the diagenetic process, pores in the carbonate rocks are mainly classified into primary pores formed in the depositional stage, secondary dissolved pores formed in the epidiagenetic stage and secondary dissolved pores formed in the burial stage. An accurate determination of origin of the pores is of great significance in clarifying the main controlling factors and distribution laws of reservoirs, especially that the identification of primary pores and penecontemporaneous epidiagenetic dissolved pores directly influences the prediction of facies controlled reservoirs. In the Chigbrak Formation in the study area, the intergranular pores in dolomitic oolite and grainstone with a bonding structure in member Ch–1 and member Ch–2 were easily identified as primary pores, and the framework pores in some dolomitic stromatolites and thrombolites were also identified as primary pores. Both of them have shared the characteristics of obvious fabric selectivity, homogeneous distribution and small pore size (<1 mm). As described above, the Kalpin Movement at the end of Ediacaran uplifted the study area tectonically, leading to atmospheric water dissolution and the resulted dolomite weathering crust. Therefore, the irregular shaped centimeter to decimeter–size vugs and fractures were clearly identified as the products of atmospheric water karstification near the unconformity surface. It is difficult to clarify the origin of the large pores and vugs predominantly developed in member Ch–3. However, these pores and vugs always have multiple phases of carbonate cements. Thus, based on the diagenesis sequences of cements of different phases within them, their forming mechanism and evolution history could be well understood by analyzing their geochemical characteristics and comparing with the host rocks.

Vugs in member Ch–3 have relatively falt, bedding–parallel occurrence and are developed by multi–cycles. From margin to center of the pores and vugs, one or multiple cements among fibrous dolomite, laminated dolomite, bladed dolomite, granular dolomite, saddle dolomite and sparry calcite occurred successively. Fibrous isopachous rim dolomites are recognized as the first phase cements. It is generally assumed that the carbonate cements with fibrous textures are the products of cementation in seawater in the early diagenetic stage and are originated from magnesium calcite and aragonite with fibrous texture [51]. According to the U–Pb dating data, fibrous isopachous rim dolomites in samples 56–1, 58–1 and 76–1 are slightly younger than the host rocks, and their geological ages vary within the age range of the Ediacaran (>541 Ma). The geochemical data indicate that the fibrous isopachous rim dolomites have similar values of $\delta^{13}C$ (PDB), $\delta^{18}O$ (PDB) and $^{87}Sr/^{86}Sr$ to the host rocks, and also show consistent distribution patterns of trace and rare earth elements as shown in Figures 8 and 9. Based on the description above that the dolomitization fluid for the host rocks was seawater and the dolomitization occurred in the early burial stage, it has been confirmed that these pores and vugs were formed before cementation of the fibrous dolomites, and were the products of penecontemporaneous meteoric water dissolution.

Laminated isopachous rim dolomites have similar values of $\delta^{13}C$ (PDB), $\delta^{18}O$ (PDB) and $^{87}Sr/^{86}Sr$ and consistent U–Pb ages to fibrous dolomites but different contents of Mn and Fe, which resulted in different luminescence intensity under cathodic excitation. Specifically, laminated dolomites present intensive luminescence and fibrous dolomites seldom or never show luminescence, which comprehensively suggest that the cementation fluid was the mixture of seawater and atmospheric water. Compared with fibrous dolomites, bladed/granular dolomites have similar values of $\delta^{13}C$ (PDB) and slightly higher $^{87}Sr/^{86}Sr$, but more negative values of $\delta^{18}O$ (PDB), which indicate their different diagenetic environments and warmer diagenetic fluid for fibrous dolomites. Bladed dolomites have higher contents of Mn and REEs, including Y, La, Ce, Eu and Lu, than granular dolomites, leading to higher luminescence intensity under cathodic excitation, and show the characteristics of a burial diagenetic environment. Judging from their U–Pb ages, bladed dolomites are determined as the products of the early burial diagenetic stage, and granular dolomites as the products of the middle burial diagenetic stage with formation water as the cementation fluid. Saddle dolomites have more negative values of $\delta^{18}O$ (PDB) than the host rocks and other types of cements (avg. −10.13‰), indicating a higher temperature of diagenetic fluid. Their high values of $^{87}Sr/^{86}Sr$ (avg. 0.70936), high contents of Y and Eu and relatively high content of Mn comprehensively reflect that their diagenetic fluid was hydrothermal fluid. In addition, their U–Pb ages indicate they were formed in the late Permian–middle Jurassic period, corresponding to the late Hercynian movement and the tectonic active stage of the early Yanshan movement. This suggests that saddle dolomites are the products of the late burial diagenetic stage with a contribution of at least two hydrothermal processes.

Based on the above analyses, combined with the geological setting of the study area, the diagenesis sequence was established for each important stage that the Chigbrak Formation experienced after deposition (Figure 11). The early diagenetic stage mainly involved micritization under the mixture of seawater and atmospheric water, meteoric water dissolution I, dolomitization, cementation I of fibrous isopachous rim calcite, cementation II of laminated calcite, mechanical compression in an early burial environment, cementation III of bladed calcite and atmospheric water dissolution II (Kalpin Movement period). The middle diagenetic stage mainly included cementation of granular dolomite and recrystallization of dolomite (according to the ordering degree of partial dolomite being relatively high, and coupled with the compositional zoning illustrated by the CL, it is possible that these dolomites have been recrystallized [52]; furthermore, the U–Pb age dates show a range of 80 million years which may potentially reflect dolomite recrystallization). The late diagenetic stage mainly included structural fracturing in a late burial environment, cementation II of saddle dolomite, cementation IV of massive calcite, structural fracturing II and cementation III of saddle dolomite. In all the major diagenesis processes, two phases of meteoric water dissolution and dolomitization played the most important constructive role. The former formed vugs of layered distribution. Although it did not generate pores, the latter produced dolomites with higher compressive strength than limestone, which kept the primary pores and vugs formed in the early diagenetic stage from being destroyed by compaction throughout the long geological history. This understanding provided a reliable basis for clarifying the genesis of reservoirs. Specifically, sedimentary microfacies, penecontemporaneous meteoric water dissolution controlled by fourth– or fifth–order sequences, meteoric water dissolution controlled by tectonic movement and early dolomitization jointly control the development of dolomite reservoirs in the Upper Ediacaran Chigbrak Formation of the Tarim Basin.

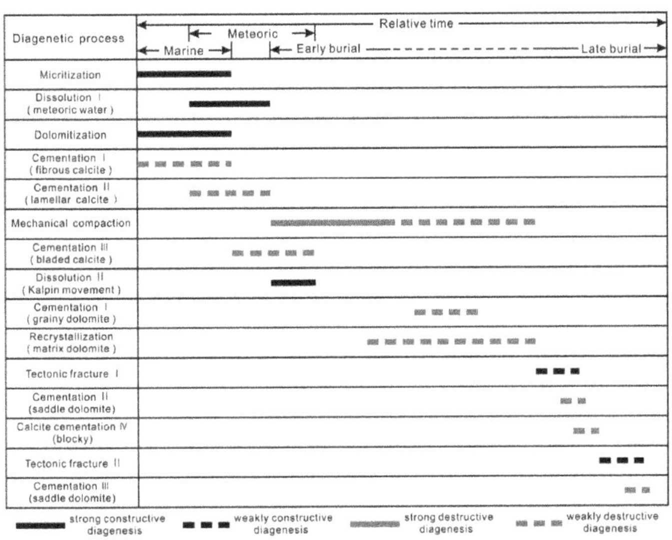

Figure 11. Diagenetic sequence of Chigbrak Formation.

6. Conclusions

(1) The Upper Ediacaran Chigbrak Formation in the study area is mainly composed of dolomitic microbialite, the average values of $\delta^{13}C$ (PDB), $\delta^{18}O$ (PDB) and $^{87}Sr/^{86}Sr$ are 3.50‰ (1.31‰~6.65‰), −2.95‰ (−4.81‰~−1.23‰) and 0.709457 (0.708745~0.710885), which are similar to the characteristics coeval seawater. The dolomites are characterized by a medium degree of cation ordering (0.60~0.77, avg. 0.68) and low content of ΣREEs (0.36 ppm~35.40 ppm, avg. 9.03 ppm), and a chondrite standardized curve of REEs in a right wing type shows enrichment of LREE (La–Eu) and depletion of HREE (Gd–Lu). The ages determined by laser ablation U–Pb dating range from 538 to 618 Ma, corresponding to the age of the Ediacaran period.

(2) Based on the isotope geochemical characteristics, element geochemical characteristics and U–Pb ages, combined with the petrologic characteristics (the dolomites have well preserved depositional textures, and their crystals are finer), it was identified that the dolomite of the Upper Ediacaran Chigbrak Formation was formed penecontemporaneously with marine seawater being the dolomitizing fluid.

(3) Vugs are the dominant pore type of the Upper Ediacaran Chigbrak Formation. Based on the occurrence sequence of six types of cements (fibrous dolomite, laminated dolomite, bladed dolomite, granular dolomite, saddle dolomite and sparry calcite) in the vugs and their values of $\delta^{18}O$ (PDB), $^{87}Sr/^{86}Sr$, elements content and U–Pb ages, it has been concluded that the vugs are the products of penecontemporaneous dissolution of meteoric water.

(4) Lithofacies, sedimentary microfacies, penecontemporaneous meteoric water dissolution controlled by fourth– or fifth–order sequences, meteoric water dissolution controlled by tectonic movement and early dolomitization are confirmed as the major controlling factors for the development of dolomite reservoirs in the Upper Ediacaran Chigbrak Formation.

Author Contributions: Conceptualization, J.Z. and A.S.; methodology, J.Z. and H.W.; software, X.L. and K.D.; validation, Z.C.; formal analysis, J.Z.; investigation, J.Z. and H.W.; resources, A.S. and Z.C.; data curation, H.W. and X.L.; writing–original draft preparation, J.Z. and H.W.; writing–review and editing, J.Z.; visualization, X.L. and K.D.; supervision, J.Z.; project administration, J.Z.; funding acquisition, J.Z. and H.W. All authors have read and agreed to the published version of the manuscript.

Funding: This research was jointly funded by the Scientific Research & Technology Development Project of PetroChina Company Limited (Grant No. 2021DJ0501) and Basic science and strategic reserve technology fund project of CNPC (Grant No. 2022D–HZ02).

Data Availability Statement: Data are available upon reasonable request. The data that support the findings of this study are available on request from the corresponding author. The data are not publicly available due to privacy or ethical restrictions.

Acknowledgments: We would thank Yongquan Chen and Jingao Zhou for their valuable suggestions, Anping Hu and Feng Liang for their guidance in the experiments. We would also like to thank Yongjin Zhu, Guang Yu, Rong Xiong, Jianhua Dong and Yu Liu for the help in field work. We acknowledge this generous funding from PetroChina Company Limited.

Conflicts of Interest: The authors declare no conflict of interest.

References

1. Wang, T.G.; Han, K.Y. On Meso–Neoproterozoic primary petroleum resources. *Acta Pet. Sin.* **2011**, *32*, 1–7. [CrossRef]
2. Tian, X.W.; Peng, H.L.; Wang, Y.L.; Yang, D.L.; Sun, Y.T.; Zhang, X.H.; Wen, L.; Luo, B.; Hong, H.T.; Wang, W.Z.; et al. Analysis of reservoir difference and controlling factors between the platform margin and the inner area of the fourth member of Sinian Dengying Formation in Anyue Gas Field, central Sichuan. *Nat. Gas Geosci.* **2020**, *31*, 1225–1238.
3. Wang, Z.M.; Xie, H.W.; Chen, Y.Q.; Qi, Y.M.; Zhang, K. Discovery and exploration of Cambrian subsalt dolomite original hydrocarbon reservoir at Zhongshen–1 well in Tarim Basin. *China Pet. Explor.* **2014**, *19*, 1–13. [CrossRef]
4. Zhu, G.Y.; Chen, F.R.; Wang, M.; Zhang, Z.Y.; Ren, R.; Wu, L. Discovery of the Lower Cambrian high–quality source rocks and deep oil and gas exploration potential in the Tarim Basin, China. *Bull. Am. Assoc. Pet. Geol.* **2018**, *102*, 2123–2151. [CrossRef]

5. Zhu, G.Y.; Li, T.T.; Zhang, Z.Y.; Zhao, K.; Zhang, K.J.; Chen, W.Y.; Yan, H.H.; Wang, P.J. Distribution and geodynamic setting of the Late Neoproterozoic–EarlyCambrian hydrocarbon source rocks in the South China and Tarim Blocks. *Asian J. Earth Sci.* **2020**, *201*, 104504. [CrossRef]
6. Yang, H.J.; Chen, Y.Q.; Tian, J.; Du, J.H.; Zhu, Y.F.; Li, H.H.; Pan, W.Q.; Yang, P.F.; Li, Y.; AN, H.T. Great discovery and its significance of ultra–deep oil and gas exploration in well Luntan–1 of the Tarim Basin. *China Pet. Explor.* **2020**, *25*, 62–72. [CrossRef]
7. Zhang, T.F.; Huang, L.L.; Ni, X.F.; Xiong, R.; Yang, G.; Meng, G.R.; Zheng, J.F.; Chen, W. Lithological combination, genesis and exploration significance of the Lower Cambrian Wusonggeer Formation of Kalpin area in Tarim Basin: Insight through the deepest Asian onshore well–Well Luntan 1. *Oil Gas Geol.* **2014**, *41*, 928–940. [CrossRef]
8. Zhou, X.B.; Li, J.H.; Fu, C.J.; Li, W.S.; Wang, H.H. A discussions on the Cryogenian–Cambrian tectonic–sedimentary event and tectonic setting of northern Tarim Basin. *Geol. China* **2012**, *39*, 900–911. [CrossRef]
9. Shi, K.B.; Liu, B.; Tian, J.C.; Pan, W.Q. Sedimentary characteristics and lithofacies paleogeography of Sinian in Tarim basin. *Acta Pet. Sin.* **2016**, *37*, 1343–1360. [CrossRef]
10. Wei, G.Q.; Zhu, Y.J.; Zheng, J.F.; Yu, G.; Ni, X.F.; Yan, L.; Tian, L.; Huang, L.L. Tectonic–lithofacies paleogeography, large–scale source–reservoir distribution and exploration zones of Cambrian subsalt formation, Tarim Basin, NW China. *Pet. Explor. Dev.* **2021**, *48*, 1114–1126. [CrossRef]
11. Yan, W.; Yang, G.; Yi, Y.; Zuo, X.J.; Wang, X.M.; Lou, H.; Rao, H.W. Characteristics and genesis of Upper Sinian dolomite reservoirs in Keping area, Tarim Basin. *Acta Pet. Sin.* **2019**, *40*, 295–307+321. [CrossRef]
12. Zheng, J.F.; Shen, A.J.; Yang, H.X.; Zhu, Y.J.; Liang, F. Geochemistary and geochronology characteristics and their geological significance of microbial dolomite in Upper Sinian, NW Tarim Basin. *Acta Pet. Sin.* **2021**, *37*, 2189–2202. [CrossRef]
13. Liu, Y.; Zheng, J.F.; Zeng, J.H.; Zhu, Y.J. Micro–characterization of microbial dolomite reservoir of Upper Sinian Qigeblak Formation in Keping area, Tarim Basin. *Nat. Gas Geosci.* **2022**, *33*, 49–62. [CrossRef]
14. Li, P.W.; Luo, P.; Song, J.M.; Jin, Y.F.; Wang, G.Q. Characteristics of Upper Sinian dolomite reservoir in the northwestern margin of Tarim Basin. *Mar. Pet. Geol.* **2015**, *20*, 1–12. [CrossRef]
15. Shi, S.Y.; Liu, W.; Huang, Q.Y.; Wang, T.S.; Zhou, H.; Wang, K.; Ma, K. Dolostone reservoir characteristic and its formation mechanism in Qigebulake formation, Sinian, north Tarim Basin. *Nat. Gas Geosci.* **2017**, *28*, 1226–1234. [CrossRef]
16. Yang, H.X.; Hu, A.P.; Zheng, J.F.; Liang, F.; Luo, X.Y.; Feng, Y.X.; Sheng, A.J. Application of mapping and dating techniques in the study of ancient carbonate reservoirs: A case study of Sinian Qigebrak Formation in northwestern Tarim Basin, NW China. *Pet. Explor. Dev.* **2020**, *47*, 935–946. [CrossRef]
17. Qian, Y.X.; He, Z.L.; Li, H.L.; Chen, Y.; Jin, T.; Sha, X.G.; Li, H.Q. Discovery and discussion on origin of botryoidal dolostone in the Upper Sinian in North Tarim Basin. *J. Palaeogeogr.* **2017**, *19*, 197–210. [CrossRef]
18. Ma, X.H.; Wang, N.M.; Tian, X.B. Genesis mechanism and reservoir model of Sinian dolomite in eastern Tarim area. *Fault–Block Oil Gas Field.* **2019**, *26*, 566–570.
19. Zheng, J.F.; Liu, Y.; Zhu, Y.J.; Liang, F. Geochemical features and its geological significances of the Upper Sinian Qigeblak Formation in Wush area, Tarim Basin. *J. Palaeogeogr.* **2021**, *23*, 983–998. [CrossRef]
20. Jia, C.Z. Structural characteristics and oil/gas accumulative regularity in Tarim Basin. *Xinjiang Pet. Geol.* **1999**, *20*, 177–183. [CrossRef]
21. He, D.F.; Jia, C.Z.; Li, D.S.; Zhang, C.J.; Meng, Q.R.; Shi, X. Formation and evolution of polycyclic superimposed Tarim Basin. *Oil Gas Geol.* **2005**, *26*, 64–77. [CrossRef]
22. Zhai, M.G. The main old lands in China and assembly of Chinese unified continent. *Sci. China Earth Sci.* **2013**, *43*, 1583–1606. [CrossRef]
23. Yang, Y.K.; Shi, K.B.; Liu, B.; Qin, S.; Wang, J.Q.; Zhang, X.F. Tectono–sedimentary evolution of the Sinian in the Northwest Tarim Basin. *Chin. J. Geol.* **2014**, *49*, 19–29. [CrossRef]
24. Wu, L.; Guan, S.W.; Yang, H.J.; Ren, R.; Zhu, G.Y.; Jin, J.Q.; Zhang, C.Y. The paleogeographic framework and hydrocarbon exploration potential of Neoproterozoic rift basin in northern Tarim basin. *Acta Pet. Sin.* **2017**, *38*, 129–138. [CrossRef]
25. Zhu, Y.J.; Zheng, J.F.; Cao, P.; Qiao, Z.F.; Yu, G.; Liu, L.L.; Huang, L.L. Tectonic–Lithofacies paleogeography and sizeable reservoir potential in the Early Cambrian postrift depressed carbonate ramp, Tarim Basin, NW China. *Lithosphere* **2021**, *2021*, 8804537. [CrossRef]
26. Guan, S.W.; Zhang, C.Y.; Ren, R.; Zhang, S.C.C.; Wu, L.; Wang, L.; Ma, P.L.; Han, C.W. Early Cambrian syndepositional structure of the northern Tarim Basin and a discussion of Cambrian subsalt and deep exploration. *Pet. Explor. Dev.* **2019**, *46*, 1141–1152. [CrossRef]
27. Chen, Y.Q.; Yan, W.; Han, C.W.; Yan, L.; Ran, Q.G.; Kang, Q.; He, H.; Ma, Y. Structural and sedimentary basin transformation at the Cambrian/Neoproterozoic interval in Tarim Basin: Implication to subsalt dolostone exploration. *Nat. Gas Geosci.* **2019**, *30*, 39–50. [CrossRef]
28. Zheng, J.F.; Pan, W.Q.; Shen, A.J.; Yuan, W.F.; Huang, L.L.; Ni, X.F.; Zhu, Y.J. Reservoir geological modeling and significance of cambrian Xiaoerblak formation in Keping outcrop area, Tarim basin, NW China. *Pet. Explor. Dev.* **2020**, *47*, 499–511. [CrossRef]
29. Goldsmith, J.R.; Graf, D.L. Structural and compositional variations in some natural dolomites. *J. Geol.* **1958**, *66*, 678–693. [CrossRef]
30. Zempolich, W.G.; Wilkinson, B.H.; Lohmann, K.C. Diagenesis of late Proterozoic carbonates: The beck spring dolomite of eastern California. *J. Petrol.* **1988**, *58*, 656–672. [CrossRef]

31. Palmer, M.R.; Edmond, J.M. The strontium isotope budget of the modern ocean. *Earth Planet Sci. Lett.* **1989**, *92*, 11–26. [CrossRef]
32. Burns, S.J.; Haudenschild, U.; Matter, A. The strontium isotopic composition of carbonates from the late Precambrian (~560–540 Ma) Huqf Group of Oman. *Chem. Geol.* **1994**, *111*, 269–282. [CrossRef]
33. Halverson, G.P.; Dud, S.F.; Maloof, A.C.; Bowring, S.A. Evolution of the $^{87}Sr/^{86}Sr$ composition of Neoproterozoic seawater. *Palaeogeogr. Palaeoclimatol. Palaeoecol.* **2007**, *256*, 103–129. [CrossRef]
34. Yang, J.D.; Sun, W.G.; Wang, Z.Z.; Xue, Y.S.; Tao, X.C. Variations in Sr and C isotopes and Ce anomalies in successions from China: Evidence for the oxygenation of Neoproterozoic seawater? *Precambrian Res.* **1999**, *93*, 215–233. [CrossRef]
35. Merlini, M.; Sapelli, F. High–temperature and high–pressure behavior of carbonates in the ternary diagram $CaCO_3$–$MgCO_3$–$FeCO_3$. *Am. Mineral.* **2006**, *101*, 1423–1430. [CrossRef]
36. Deng, M.; Qian, G.R.; Tang, M.S. Ordered index and dedolomitization of dolomite crystals. *J. Nanjing Univ. Chem. Technol.* **2001**, *23*, 1–5. [CrossRef]
37. Zheng, J.F.; Shen, A.J.; Qiao, Z.F.; Wu, X.N.; Zhang, T.F. Characteristics and pore genesis of dolomite in the Penglaiba Formation in Keping–Bachu outcrop area. *Acta Pet. Sin.* **2014**, *35*, 664–672. [CrossRef]
38. Lottermoser, B.G. Rare earth elements and hydrothermal ore formation processes. *Ore Geol. Rev.* **1992**, *7*, 25–41. [CrossRef]
39. Bau, M.; Dulski, P. Distribution of yttrium and rare–earth elements in the Penge and Kuruman iron–formations, Transvaal Supergroup, South Africa. *Precambrian Res.* **1996**, *79*, 37–55. [CrossRef]
40. Shen, A.J.; Hu, A.P.; Chen, T.; Liang, F.; Pan, W.Q.; Feng, Y.X.; Zhao, J.X. Laser ablation in situ U–Pb dating and its application to diagenesis–porosity evolution of carbonate reservoirs. *Pet. Explor. Dev.* **2019**, *46*, 1062–1074. [CrossRef]
41. Zheng, J.F.; Zhu, Y.J.; Huang, L.L.; Yang, G.; Hu, F.J. Geochemical Characteristics and Their Geological Significance of Lower Cambrian Xiaoerblak Formation in Northwestern Tarim Basin, China. *Minerals* **2022**, *12*, 781. [CrossRef]
42. Ulrich, T.; Kamber, B.S.; Jugo, P.J.; Tinkham, D.K. Imaging elementdistribution patterns in minerals by laser ablation–inductively coupled plasma–mass spectrometry (LA–ICP–MS). *Can. Mineral.* **2009**, *47*, 1001–1012. [CrossRef]
43. Petrus, J.A.; Chew, D.M.; Leybourne, M.I.; Kamber, B.S. A new approach to laser–ablation inductively–coupled–plasma mass–spectrometry (LA–ICP–MS) using the flexible map interrogation tool 'Monocle'. *Chem. Geol.* **2017**, *463*, 76–93. [CrossRef]
44. Zhao, W.Z.; Shen, A.J.; Zheng, J.F.; Qiao, Z.F.; Wang, X.F.; Lu, J.M. The porosity origin of dolostone reservoirs in the Tarim, Sichuan and Ordos basins and its implication to reservoir prediction. *Sci. China Earth Sci.* **2014**, *57*, 2498–2511. [CrossRef]
45. Shen, A.J.; Zheng, J.F.; Chen, Y.Q.; Ni, X.F.; Huang, L.L. Characteristics, origin and distribution of dolomite reservoirs in Lower–Middle Cambrian, Tarim Basin, NW China. *Pet. Explor. Dev.* **2016**, *43*, 340–349. [CrossRef]
46. Zhao, W.Z.; Shen, A.J.; Qiao, Z.F.; Pan, L.Y.; Hu, A.P.; Zhang, J. Genetic types and distinguished characteristics of dolomite and the origin of dolomite reservoirs. *Pet. Explor.* **2018**, *45*, 923–935. [CrossRef]
47. Zheng, J.F.; Shen, A.J.; Liu, Y.F.; Chen, Y.Q. Using Multi–parameter Comprehensive Method to Identify the Genesis of Lower Paleozoic Dolomite in Tarim Basin. *Acta Pet. Sin.* **2012**, *32*, 731–737. [CrossRef]
48. Hu, W.X.; Chen, Q.; Wang, X.L.; Cao, J. REE models for the discrimination of fluids in the formation and evolution of dolomite reservoirs. *Oil Gas Geol.* **2010**, *31*, 810–818. [CrossRef]
49. Keith, M.H.; Weber, J.N. Carbon and oxygen isotopic composition of selected limestones and fossils. *Geochim. Cosmochim. Acta* **1964**, *28*, 1787–1816. [CrossRef]
50. Banner, J.L. Application of the trace element and isotope geochemistry of strontium to studies of carbonate diagenesis. *Sedimentology* **1995**, *42*, 805–824. [CrossRef]
51. Dickson, J. Crystal growth diagrams as an aid to interpreting the fabrics of calcite aggregates. *J. Petrol.* **1993**, *63*, 1–17. [CrossRef]
52. Manche, C.J.; Kaczmarek, S.E. A global study of dolomite stoichiometry and cation ordering through the Phanerozoic. *J. Sediment. Res.* **2021**, *9*, 520–546. [CrossRef]

minerals MDPI

Article

Results of the Study of Epigenetic Changes of Famennian–Tournaisian Carbonate Rocks of the Northern Marginal Shear Zone of the Caspian Syneclise (Kazakhstan)

Valeriy Korobkin [1,*], Akhan Chaklikov [1,*], Zhamal Tulemissova [1], Iskander Samatov [2] and Yelena Dobrovolskaya [1]

[1] School of Geology, Kazakh-British Technical University, Almaty 050000, Kazakhstan
[2] Institute of Geological Sciences of K.I. Satpayev, Almaty 050010, Kazakhstan
* Correspondence: korobkin_vv@kbtu.kz (V.K.); a.chaklikov@kbtu.kz (A.C.);
Tel.: +7-777-136-88-11 (V.K.); +7-701-999-51-10 (A.C.)

Abstract: Various genetic and morphological types of voids in carbonate reservoirs make it difficult to diagnose them, which can be seen in the determination of reservoir properties in the northern marginal shear zone of the Caspian Syneclise. A macro- and microscopic study of rocks was carried out by staining carbonates in thin sections with alizarin (determination of the mineral composition, structure, texture, void and fracture spaces, rock genesis). Instrumental methods (X-ray, DTA—differential thermal analysis, TGA—thermo-gravimetric analysis, and probe microanalysis) established the composition of rocks, the nature of their diagenetic transformations, and the formation of void space. The elemental and oxide composition of a number of samples was carried out using the X-ray probe microanalysis method, and mineral formations with intermediate thermochemical properties were found. The results of X-ray, DTA, and TGA measurements and the data of probe microanalysis made it possible to reveal thermally inert formations of oxides of calcium, magnesium, silicon, iron, and other compounds in the composition of carbonates. A relatively low-cost express method was used to determine the material composition and the nature of epigenetic changes and to obtain data on the void space as a result of the development of tectonic fracturing and diagenetic processes of leaching and secondary mineral formation in bedded carbonate reservoirs.

Keywords: north marginal shear zone; Caspian Syneclise; carbonate reservoir; diagenetic transformations; shear and tension joints; microscopic analysis; X-ray diffractometry; thermal analysis; probe microanalysis

Citation: Korobkin, V.; Chaklikov, A.; Tulemissova, Z.; Samatov, I.; Dobrovolskaya, Y. Results of the Study of Epigenetic Changes of Famennian–Tournaisian Carbonate Rocks of the Northern Marginal Shear Zone of the Caspian Syneclise (Kazakhstan). *Minerals* **2023**, *13*, 249. https://doi.org/10.3390/min13020249

Academic Editors: Hamzeh Mehrabi, Vahid Tavakoli and Jia-Xi Zhou

Received: 28 December 2022
Revised: 8 February 2023
Accepted: 8 February 2023
Published: 9 February 2023

1. Introduction

Carbonate reservoirs play an important role in global hydrocarbon production. The purpose of this study was to select the best methods for studying epigenetic changes in the bedded carbonate reservoirs of the Famennian–Tournaisian reservoirs of the Northern marginal zone of the Caspian Syneclise. Therefore, the primary task for researching these reservoirs is studying the void space and preserving it as a part of the process of further geological evolution. The diversity of genetic types of voids in carbonate rocks determines if a complex reservoir will be created within them. At the early stage of accumulation formation (syngenesis and early diagenesis), a combination of basic parameters such as carbonate deposition depth, oxygen regime, hydrodynamic activity, the composition of seawater and pore water, and the influence of drift sources is important. Subsequent transformations associated with epigenesis are caused by tectonic rearrangements that activate the movements of fluids, including hydrocarbons. As a result of these transformations, the primary structure of the rock can be altered completely [1–6].

The Caspian Syneclise is located in the southeastern part of the East European Craton. Most of its area (about two-thirds) belongs to the Republic of Kazakhstan. The Caspian Syneclise is a part of the main oil and gas province of the country. The sedimentary cover of the syneclise has a vertical thickness ranging from 6 to 24 km and is subdivided into three lithological-stratigraphic units: (1) pre-salt (Mesoproterozoic–Permian and Artinian stages); (2) saline (Permian and Kungurian stages); and (3) post-salt (Guadalupian section of the Permian–Mesozoic–Cenozoic stage). A characteristic feature of the upper part of the pre-salt section is the presence of an extensive carbonate massif. The reservoirs are carbonate rocks and, mainly, organogenic limestones from the Devonian and Carboniferous age, in which the main oil and gas fields are localized. For a long time, the properties of massive carbonate reservoirs of such known and unique deposits, including the Karachaganak, Tengiz, Kashagan, and Zhanazhol fields, as well as other objects, have been studied (Figure 1) [1,2,4–6]. On the other hand, many deposits that are located in the Chinarevskoye and Lobodin-Teplov groups along the northern border zone of the Caspian Syneclise, which have bedded and bedded–massive reservoirs, have not received enough attention. The objective of our research was to the study the material composition and reservoir properties of carbonate rocks of the northern marginal zone of the Caspian Syneclise.

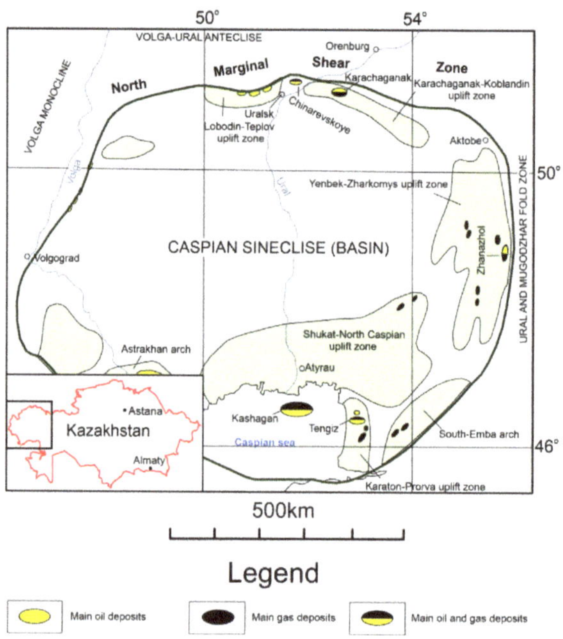

Figure 1. Modified overview diagram of the Caspian Syneclise (Basin) [1,4]. In the northern part of the scheme, there is a northern marginal shear zone separating the Caspian Syneclise from the Volga-Ural Anteclise.

In the outer-edge zone of the northern part of the Caspian Syneclise (oil and gas province), the issues of epigenetic changes of the Famennian–Tournaisian reservoirs can be considered by studying, for example, the Lobodin-Teplov and Chinarevskoye groups of deposits (LTCGD) (Figures 1–3). There are two industrially productive stratigraphic levels of oil and gas content: (1) Givetian–Frasnian gas condensate and (2) Famennian–Tournaisian gas and oil. The trap has a complex combined geological structure. During the Givetian–Tournaisian period, a carbonate formation was formed, which was subsequently converted into an anticline with a complicated sublatitudinal discharge (Figure 2) [5–9].

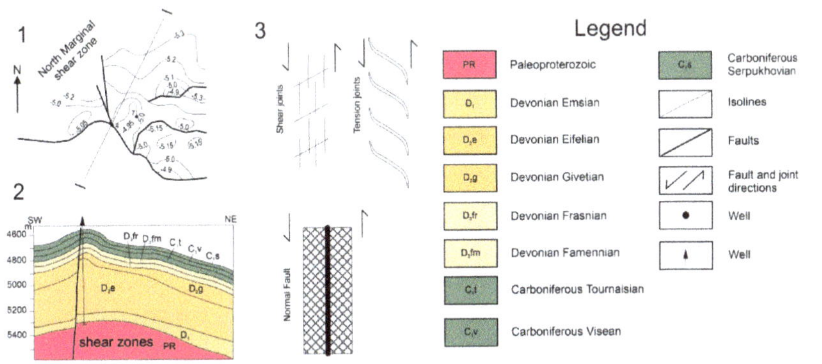

Figure 2. Northern border zone near the Chinarevskoye and Lobodin-Teplov groups: (**1**)—structural diagram; (**2**)—geological section along the I–I line [2]; (**3**)—diagram explaining the formation of joints in the suture zone.

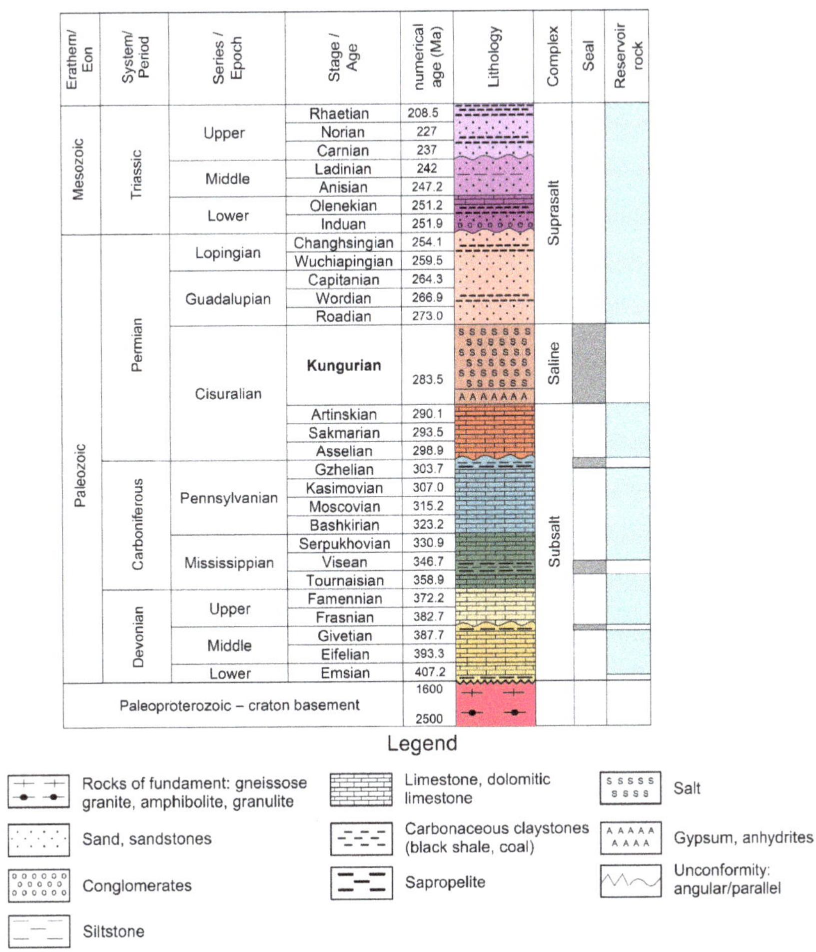

Figure 3. Lithological and stratigraphic diagram of the northern border zone of the Caspian Basin; the modern geochronological edition.

At the LTCGD, the carbonate rocks of the Tournaisian and upper Famennian stages were formed under the conditions of the marine shelf with a variable porosity from 6% to 12%, an average porosity of the gas-saturated part of the reservoir of 8.9%, and an average porosity of the oil-saturated part of up to 8.1%. The Tournaisian oil reservoir with a thin gas cap was discovered and tested in well 10; the GOC was installed at a depth of 4318 m, and the thickness of the gas-bearing stratum is 43 m. The OWC is probably in the range of 4503 m. The established height of the oil reservoir is 77 m, and the effective gas- and oil-saturated thicknesses in the well reach 12.4 m and 61.4 m, respectively [2,10].

In the area of the LTCGD, carbonate sedimentation was carried out on the raised ledge of the foundation, which was formed as a result of the Breton folding phase (according to Hans Stille) that occurred on the border of the Famennian and Tournaisian stages. Fine and coarse clastic sediments accumulated in the depressions surrounding the China–Revskoye uplift. The offshore stage of carbonate offshore sedimentation under the conditions of the passive continental margin of the main basin and subsequent processes created favorable conditions for the development of reservoirs.

In general, the studied reservoir rocks of the Famennian–Tournaisian stage are represented by limestones, dolomites, and calcareous dolomites, with rare inclusions of anhydrite and gypsum. The void space is represented by leaching pores of various shapes and sizes. There are also caverns confined to joints, on the walls of which large dolomite crystals are noted [1,5]. Bioclastic algal limestones with stylolites have widely developed among the rocks. The rocks are fractured (closed joints filled with secondary minerals that are open, subhorizontal, and subvertical), and unevenly porous, as seen in Figure 4. The gross porosity averages 8%–9%.

Figure 4. Carbonate reservoirs: (**1**)—limestone, depth of 4390 m; (**2**)—dolomite, depth of 5083–5090 m; (**3**)—dolomite limestone with signs of hydrocarbons, depth of 4943–4952 m; (**4**)—limestone, depth of 4367 m; (**5**)—dolomite, depth of 4616–4621 m; (**6**)—limestone, depth of 2802 m; (**7**)—limestone, depth of 4334 m; (**8**)—limestone, depth of 4390 m (stylolites); (**9**)—limestone, depth of 4924 m.

2. Research Methods

2.1. Macro- and Microscopic Description of Rocks

Macro- and microscopic studies of rocks were carried out for the determination of mineral composition, structure and texture, void space, and rock genesis. The analysis of the reservoir rocks was carried out on R-312 series and Leica FLOORS polarization microscopes (model DMLSP, manufacturer—Leica Microsystems, Wetzlar, Germany). For the diagnosis of carbonates, the staining of a thin section of the carbonates with alizarin red was performed according to the special research method of I. Mitchell [11], allowing us to distinguish calcite from dolomite.

2.2. Instrumental Research Methods

The material composition of the rocks was studied via thermal (DTA and TGA) and X-ray diffractometric analysis methods. Control of the elemental, oxide, and carbon composition of several samples was carried out using radiometric probe microanalysis.

Thermal analysis was carried out on a thermal balance unit: a derivatograph (model: Q-1500D-grade, Budapest, Hungary) created by F. Paulik, J. Paulik, and L. Erdei. The method was based on the device recording changes in the thermochemical and physical parameters of the substance that may be caused by its heating. The thermochemical state of the sample was described by the following curves: T (temperature), DTA (differential thermoanalytical), TG (thermogravimetric), and DTG (differential thermogravimetric, which is a derivative of the TG function). The analysis was performed in the open air, within a temperature range from 20 °C to 1000 °C. The furnace heating mode was linear (dT/dt = 10), and the reference substance was calcined Al_2O_3. The sample weighed strictly 500 mg, with the sensitivity of the balance being 200 mg on the scale of measurement. The analysis survey was carried out within the following limits of the measuring systems of the device: DTA = 250 μV, DTG = 500 μV, TG = 500 μV, and T = 500 μV. The results of the analysis were compared with the data from the thermal curves of minerals and rocks given in the atlases and were compared with the descriptions of the thermal behavior of samples set out in other reference sources and accumulated in the databank of the laboratory that conducted these studies.

X-ray phase determinations were carried out on an automated diffractometer DRON-3 with $Cu_{K\alpha}$ radiation and a β-filter (DRON-3, manufacturer: "Burevestnik", Saint-Peterburg, Russia). X-ray phase measurements were performed on a semiquantitative basis using powder sample diffraction patterns via the method of equal weights and artificial mixtures. Quantitative ratios of crystalline phases were determined. Diffraction patterns were interpreted using data from the ICDD file, the powder diffraction database PDF2 (Powder Diffraction File), and diffraction patterns of pure minerals. The content was calculated for the main phases. Possible impurities, the identification of which cannot be unambiguous due to a small number of contents and the presence of only 1–2 diffraction reflexes or poor crystallization, are indicated in the tables.

Along with the traditional classical methods of studying the composition of rocks, in some samples, their elemental and oxide compositions were determined using the method of radiometric microanalysis. The selected samples from different areas were taken with a Superprobe 733 electronic probe microanalyzer from JEOL (Tokyo, Japan). Elemental analysis of samples and photography with various types of radiation were performed using an INCA 8N8RGY energy-dispersive spectrometer from Oxford Instruments (Abingdon, England), which was installed on the above microanalyzer (Superprobe 733) with an acceleration voltage of 25 kV and a probe current of 25 mA.

The involvement of these methods in the study of the material composition of rocks from the pre-salt deposits of the LTCGD made it possible to identify the presence of calcite, dolomite, magnesite, and other minerals that quantitatively vary depending on the depth of occurrence of the rocks.

3. Results

3.1. Results of Macro- and Microscopic Studies

The carbonate reservoirs of the deposit are mainly organogenic-lumpy limestones, which are partially or completely recrystallized and dolomitized; dolomites are substituted for organogenic limestones. In limestones, the shaped elements (lumps and clots) with a size of 0.1–1 mm and an oval, rounded, elongated, or irregular shape are composed of pelitomorphic calcite and are cemented with carbonate material for better crystallization (Figures 4–6).

Figure 5. Inclined joint of tectonic separation in Famennian–Carboniferous rocks (detail of Figure 4(3)) with subsequent development of vugginess and new formations of carbonate minerals.

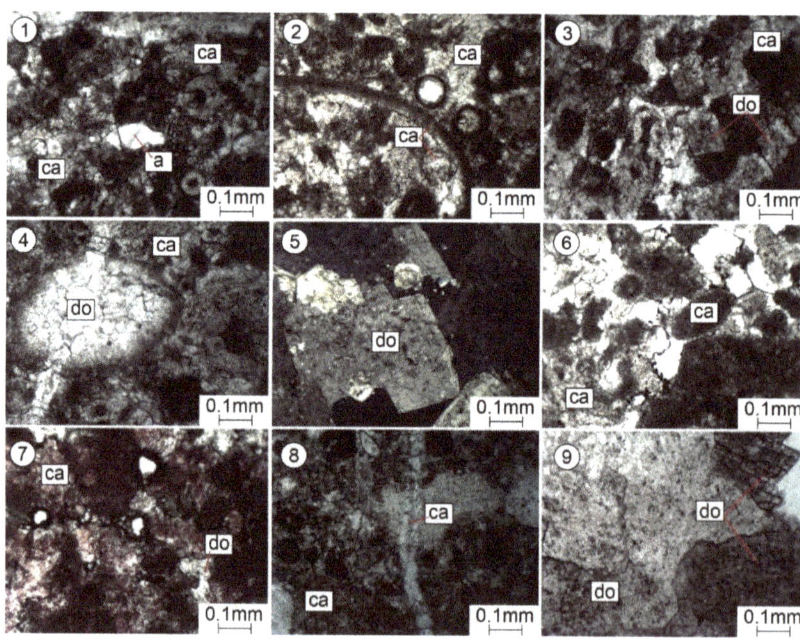

Figure 6. Pictures of thin sections. Epigenetic changes in Famennian–Tournaisian carbonate reservoirs. (**1–4**)—organogenic-lumpy limestones, partially recrystallized and dolomitized; depth, 4390 m; Nicole one. (**5**)—crystalline granular dolomite; depth, 4616–4621m; Nikoli +. (**6**)—organogenic-lumpy limestone; depth, 4380 m; Nicole one. (**7**)—organogenic-lumpy limestone (after staining with alizarin red); depth, 4380 m; Nicole one. (**8**)—organogenic-lumpy limestone, slightly recrystallized; depth, 4915 m; Nicole one. (**9**)—crystalline granular dolomite; depth, 4616–4621m; Nicole one.

Carbonate reservoirs are composed of calcite, dolomite, a small amount of barite, anhydrite, pyrite, and clay minerals, in addition to there being a finely dispersed impurity of organic matter, which is sometimes in the form of a film on the surface of carbonates or "effusions" in the core (Figure 4(6). Organogenic residues include ostracods, brachiopods, crinoids, and ammonites composed of fine-grained carbonates. The structure of rocks is organogenic-lumpy and crystal-grained, and the texture is weakly layered or erratic [4,5,12,13]. The rocks are porous, fractured, and cavernous (Figures 4 and 5). Lithogenetic and tectonic joints, as well as stylolites (according to morphology, serrated, and tubercular), are made of clay–organic substance. They are weakened zones along which rocks are split, forming an uneven surface (Figure 4(7,8)). Along the stylolites, there are new formations of large grains of carbonates (calcite, dolomite, and pyrite). These zones are favorable for the formation of open joints.

As a result of the microscopic studies, epigenetic processes of recrystallization, dolomitization, substitution, leaching, and secondary mineral formation were revealed. Sulphate mineralization was noted, which leads to the sealing of voids and thus reduces the porosity of carbonate reservoirs (Figure 6(1)).

In the studied rocks, there are voids formed as a result of the epigenetic leaching of calcite from formed formations (organogenic residues of Crinoidea, ostracods, etc.) (Figure 4(2,4)) of various sizes: from small pores to large caverns of an irregular shape (Figure 5). Pores and caverns are often encrusted with well-formed crystals of rhombohedral dolomite (Figure 6(3,5,9)).

Epigenetic dolomitization: dolomite crystals of a rhombohedral shape are found in the granular mass of calcite (Figure 6(3)) when it is relatively intense and multi-faced rhombohedrons of dolomite are in contact with either edges or vertices with faces (Figure 6(9)). Given such packaging, significant intergranular spaces comparable to grains remain free between them; as a result, the porosity of limestones increases significantly. The pores occupy about 10%, have an angular and irregular shape, are located between the large (up to 1 mm) rhombohedral grains of dolomite, and are mainly isolated (Figure 6(5)). Formations (organogenic fossils) are dolomitized, and some are crossed by a later vein, which indicates a later process of dolomitization (Figure 6(4)).

3.2. Results of Analytical Studies

When dynamically heating the 10-1 sample drawn from the depth of 4616–4621 m, the differential (DTA and DTG) curves showed pronounced endothermic manifestations caused by the reactions and the destruction of the thermally active components of the sample. The morphology of these curves, formed in the range of 495–850 °C, indicates the presence of dolomite, calcite, magnesite, and siderite in the sample, as shown in Figure 7 [14]. Peaks related to the dissociation of the compounds $MgCO_3$, $CaCO_3$, $CaMg(CO_3)_2$, and $FeCO_3$ were reconstructed from these curves based on the total thermal manifestations obtained during the firing process [15–18]. At the same time, the identified temperature intervals, in which the gradual removal of carbon dioxide (CO_2) into the atmosphere was carried out, were determined as weight losses corresponding to the values of 26.3%, 10.35%, 6.5%, and 2.6%. Taking into account these emissions and stoichiometric formulas of these carbonates, the percentage of their content in the sample was established, where the dolomite was 55.1%, calcite was 23.5%, magnesite was 12.5%, and siderite was 6.8% (Table 1).

The X-ray phase analysis of the sample resulted in a curve with a series of diffraction peaks, of which only one (D—2.8842) achieved an intensity acceptable for the identification of the phases present (Figure 7, Table 2). This peak indicates the presence of dolomite in the sample. The remaining diffraction reflexes left very poorly developed peaks on the curve, which may be due to the low content of substances in the sample that caused these dispersed spikes, or due to the presence of poorly crystallized impurity minerals in the rock.

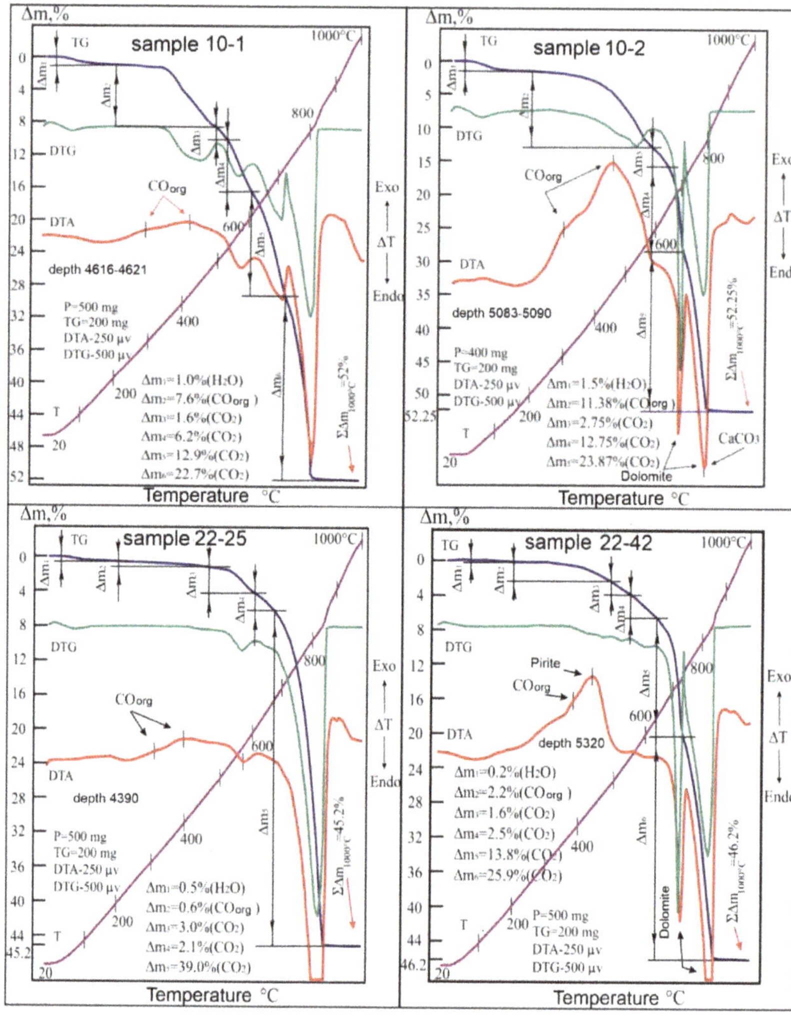

Figure 7. Derivatograms of samples: 10-1, 10-2, 22-25 and 22-42.

The results of the semiquantitative X-ray phase analysis of crystalline phases are shown in Figure 8, and the data of the interplanar distances and the phase composition of the studied sample are summarized in Table 2.

In contrast to the results of the semiquantitative radiometric measurements (Table 2), the thermal analysis found not only dolomite in the sample, but also calcite, magnesite, and siderite in noticeable amounts (Table 1). Thus, the absence of pronounced diffractive reflexes of calcite, magnesite, and siderite on the radiograph (Figure 8) does not mean their complete absence in the sample. These carbonates were not detected through X-ray structural analysis due to the poor perfection of the crystalline structures of their lattices. These minerals, even with significant defects in their structures, practically retain their inherent thermochemical properties and are easily diagnosed by DTA methods. According to the complex (XRD and DTA) measurements, the sample contains well-crystallized dolomite, as well as calcium, magnesium, and iron carbonates and crystalline lattices, which were underdeveloped.

Table 1. Summary table of mineral and material composition of the ChD rocks based on the results of thermal analysis.

Sample	Mineral Composition of Samples, wt %							Compounds Removed from Rocks and Minerals, wt %			LI, 1000 °C
	Depth, m	Dolomite	Calcite	Magnesite	Siderite	Miscellaneous Minerals	Thermally Inert Minerals (TIM)	CO_{org}	H_2O	CO_2	
10-1	4616–4621	55.1	23.5	12.5	6.8	–	2.1 (CaO, MgO)	7.6	1.0	43.4	52.0
10-2	5083–5090	56.0	28.4	2.5	–	Pyrite 1.5	0.5 (CaO, MgO)	11.3	1.5	39.58	52.2
22-25	4390	8.8	83.9	5.7	–	-	0.5 (CaO, MgO)	0.6	0.5	44.1	45.2
22-42	5320	59.5	29.3	4.8	–	Pyrite ~2	2.0 (CaO, MgO)	2.2	0.2	43.8	46.2

Note: LI—loss on ignition.

Table 2. Interplanar distances and sample phase composition of 10-1.

d, Å	I %	Mineral	d, Å	I %	Mineral
4.03153	0.5		2.19195	2.5	dolomite
3.69761	0.6		2.06467	0.4	
3.19480	2.4		2.01492	1.2	
2.97414	0.8	β quartz	1.84720	0.4	dolomite
2.88420	100.0		1.80371	3.1	dolomite
2.66722	2.4	dolomite	1.78564	2.3	
2.53819	1.0		1.78015	1.3	

The 10-2 sample showed a slightly different roasting behavior, which was removed from a deeper depth (5083–5090 m) than the 10-1 sample. However, the thermal degradation of its components is mostly similar to that of the latter, which indicates the similar identity of their material composition. Under the conditions of dynamic heating, the test sample showed on its curves the same bends that were traced on the lines of the above-considered derivatogram, according to Figure 7. Here, the leading role in the formation of analog information about the processes occurring in the system is played by the temperature and chronological parameters that cause the destruction of carbonate formations, and the quality of molecular bonds of H_2O, CO_2, and CO_{org} in the structures of minerals play a part as well. In this system, only one of these compounds, H_2O, when released into the atmosphere, does not carry any significant information about the state of the tested system due to its low content in the sample (1.5%), as there is an absence of links of this water to any crystalline inclusion of the sample.

Reactions traced at higher temperature intervals (645–825 °C) appeared on all derivatogram curves. According to the location of the peaks on the DTA and DTG curves in the test temperature range, and taking into account their amplitudes, which are responsible for the degree of intensity of thermal reactions, it was established that each sample decomposition process belongs to a particular structure.

Thus, according to the mineral accessories of the components of the complex, the differentiation of thermal manifestations was carried out. In the above temperature range, dolomite and calcite are usually decomposed. In dolomite ($CaMg(CO_3)_2$), the destruction process takes place in two stages. First, there is a dissociation of the magnesium component, in which 50% of the existing carbon dioxide is removed from the system; then, there is a decomposition of the calcium component, accompanied by the loss of the remaining amount of CO_2 (also 50%). Therefore, according to the thermogravimetric data of the sample (Table 3), the weight loss ($\Delta m_3 + \Delta m_5$) caused by the destruction of the specified carbonate in the range of 585–645 and 645–825 °C was 26.9%, and, taking into account its stoichiometric formula, the amount of dolomite in the sample composition corresponds to 56.0% (Table 1).

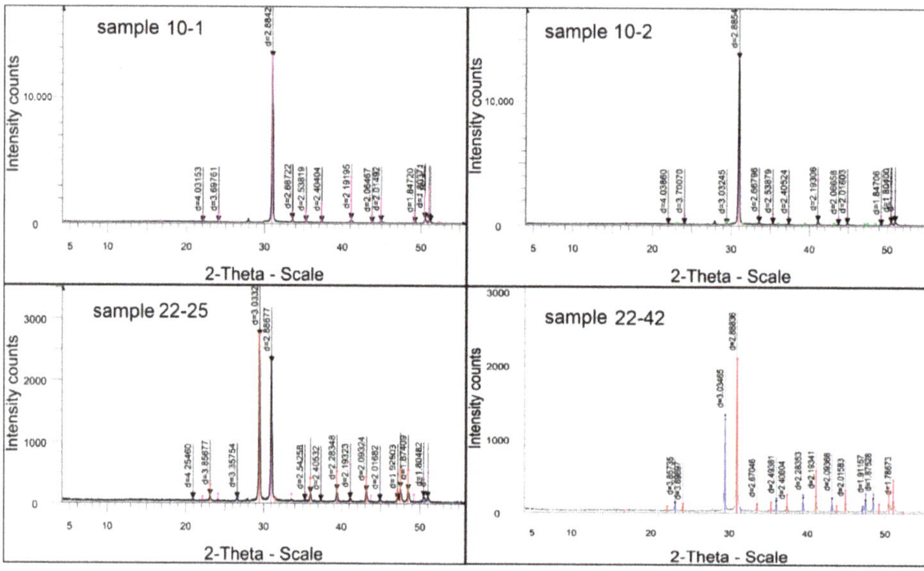

Figure 8. Sample diffraction pattern results of semiquantitative X-ray phase analysis of crystalline phases: 10-1 sample: dolomite $CaMg(CO_3)_2$—99.0%, X-ray amorphous impurities—1%. 10-2 sample: dolomite $CaMg(CO_3)_2$—99.2%; calcite $Ca(CO_3)$—0.8%. 22-25 sample: $Ca(CO_3)$ calcite—53.9%; dolomite $CaMg(CO_3)_2$—44.9%; SiO_2 quartz—1.2%. 22-42 sample: dolomite $CaMg(CO_3)_2$—62.3%; calcite $Ca(CO_3)$—37.7%.

Table 3. Thermogravimetric readings of sample 10-2 in the range of 20–1000 °C.

Weight Loss Sequence	Amount of Weight Loss, %	Volatile Components of the Heated Sample	Decomposition Stage Temperature Range, °C
Δm_1	1.5	H_2O	20–200
Δm_2	11.375	$10(CO_2)_{org} + 1.375(CO_2)$	200–585
Δm_3	2.75	CO_2	585–645
Δm_4	12.75	CO_2	645–705
Δm_5	23.875	CO_2	705–825
$\sum\Delta m1000$ °C	52.25	H_2O, CO_{org}, CO_2	20–1000

The presence of calcite in the sample composition was traced by the endothermic reaction, the temperature limits of which coincided with the stage of decomposition of the calcium component of dolomite. The amount of released carbon dioxide during $CaCO_3$ dissociation was determined based on the difference between the weight loss recorded in the range of 705–825 °C and the weight loss of calcite in the same temperature interval [16]. After corrective procedures for ordering the values of the thermogravimetric parameters (Figure 6), the weight loss of calcite was indicated to be 12.5% (CO_2). Taking into account this value and according to the stoichiometric formula of CaCO3, the calcite content in the sample composition was 28.4% (Table 1).

The main feature explaining the genetic affiliation of the revealed calcite, dolomite, and magnesite is the thermal behavior of the sample containing these minerals. The investigated aggregate under dynamic heating, among other thermochemical properties, indicates the following features of its thermal behavior:

(1) The studied carbonate complex within the temperature range of 600–820 °C decomposes in two stages, leaving on the DTA curve two endothermic peaks of different areas at 700 °C and 800 °C, caused by the effects of destruction first of the magnesium part of dolomite ($MgCO_3$), then calcium ($CaCO_3$) (Figure 7, samples 10-2 and 22-42)

(2) The temperature–chronological parameters of the second stage of dolomite destruction completely coincide with the parameters of decomposition of calcite present in the rock. In this regard, the endothermic effect of the second stage of decomposition of the indicated carbonate complex is caused by the total emissions from the CO_2 system as a result of the decomposition of the calcium component ($CaCO_3$) of dolomite and the simultaneous decomposition of calcite, the mineral component of the studied rock. At the same time, on the DTA curve, the area of the second peak (S_2) exceeds the area of the first (S_1) by more than 1.7 times. The proportionality coefficient 1.7 is borrowed from the ratio of areas ($S_2 = 1.7\ S_1$) of endothermic peaks formed on the DTA curve as a result of the decomposition of pure pure dolomite [17]. Similar kinetics of thermal decomposition of dolomite–calcite formation are characteristic of the carbonate complex, in which the structure of calcite was formed mainly due to the decomposition products of dolomite, i.e., due to the replacement of the magnesium cation in this mineral by calcium.

(3) Thermal parameters of decomposition of carbonates indicated that magnesite and calcite in these associations are secondary minerals in relation to dolomite. The secondary nature of magnesite (samples 10-1 and 22-25) is confirmed by the close location of the temperatures of the endothermic reactions of the destruction of magnesite (mineral) and the magnesium component of dolomite, and the secondary nature of the origin of calcite (samples 10-2 and 22-42) is explained by the coincidence of temperature intervals and time of decomposition of calcite (mineral) and calcium component of dolomite.

(4) Since dolomite in the first stage of decomposition loses 50% of carbon dioxide (Δm_1), and during the passage of the second stage of decomposition, the remaining 50% of carbon dioxide (Δm_2) is removed, then the violation of this balance is in favor of an increase in the release of Δm_2 (CO_2) in the final second part of dissociation due to the presence of calcite in the system. The weight loss dm caused by the degradation of this carbonate is included in the second step of the TG curve, which corresponds to the second step of CO_2 release from the complex under consideration. Using this parameter, it is easy to determine the content of secondary calcite in the composition of the studied rock.

It should be noted that in such mineral associations, where dolomite is primary in relation to calcite, organic compounds (CO_{org}) are often included, the presence of which in the samples is easily established by the exothermic effect manifested in the range of 250–550 °C (Figure 7, samples 10-2 and 22-42). Based on our results, these carbonate associations are characterized by high micro-porosity and fracturing and can be used as criteria in the search for hydrocarbon deposits.

The thermal dissociation of magnesite takes place together with the combustion of the organic matter (OM) present in the sample; i.e., it takes place within the temperature interval that causes the weight to decrease, Δm_2. The proportion of CO_2 emitted by magnesite was 1.375%, which corresponds to the 2.6% presence of this carbonate in the sample. The remainder of the weight loss (10%) was caused by the release of carbon monoxide from the combustion system of the organic compound (Table 1).

Radiometric data of this sample showed that the rock under study mainly consists of dolomite (99.2%) and only 0.8% calcite (Figure 6). Such a discrepancy between the sample composition and the thermal analysis data, as is the case for the 10-1 sample, is due to the presence of poorly developed calcite in the sample. Recording such poorly crystallized calcite in the sample by using the XRD method is difficult (Table 4).

Table 4. Interplanar distances and phase composition of sample 10-2. Note: the test substance in the sample is poorly crystallized.

d, Å	I %	Mineral	d, Å	I %	Mineral
4.03860	0.5		2.40524	0.8	
3.70070	0.7		2.19306	2.5	dolomite
3.19505	2.4	β quartz	2.06658	0.5	
3.03245	0.4	calcite	2.01603	1.1	
2.97326	0.9	dolomite	1.84706	0.4	dolomite
2.88540	100.0		1.80400	1.8	dolomite
2.66796	1.8		1.78664	2.4	

All the given diffraction peaks of the 10-2 sample (Figure 8) belong to the phases listed in Table 4 only. In the X-ray phase analysis of the test sample, characteristic diffraction reflexes are given, allowing for the identification of the phases present.

To control the mineral and material composition of this rock that was identified via the methods of XRD, DTA, and TGA, data on the elemental composition of samples obtained from an electron-probe microanalyzer were used (Tables 5–7, Figure 9).

Table 5. Results of the elemental composition of sample 10-2 obtained from a probe microanalyzer.

Processing Parameters: All Items Analyzed (Normalized)							
All results in weight %							
Spectrum	O	Mg	Si	S	Ca	Fe	Total
Spectrum 1	56.91	14.89	0.22	0.47	27.31	0.20	100.00
Spectrum 2	57.86	14.64	0.24	0.50	26.51	0.25	100.00
Spectrum 3	56.51	14.62	0.34	0.41	27.88	0.24	100.00
Secondary	57.09	14.72	0.27	0.46	27.23	0.23	100.00

Table 6. The 10-2 sample processing parameters: carbon by difference.

All Results in Weight %								
Spectrum	O	Mg	Si	S	Ca	Fe	C	Total
Spectrum 1	44.53	9.20	0.13	0.30	18.19	0.13	27.52	100.00
Spectrum 2	45.79	9.17	0.15	0.32	17.84	0.17	26.56	100.00
Spectrum 3	42.08	8.30	0.19	0.24	17.29	0.15	31.75	100.00
Secondary	44.13	8.89	0.15	0.29	17.77	0.15	28.61	100.00

Table 7. Results of the oxide composition of the 10-2 sample obtained from a probe microanalyzer.

Processing Parameters: Oxygen by Stoichiometry (Normalized)						
All Results in Compounds %						
Spectrum	MgO	SiO2	SO3	CaO	FeO	Total
Spectrum 1	36.32	0.78	1.91	60.58	0.41	100.00
Spectrum 2	36.39	0.88	2.07	60.14	0.53	100.00
Spectrum 3	35.43	1.18	1.65	61.24	0.50	100.00
Secondary	36.05	0.94	1.88	60.65	0.48	100.00

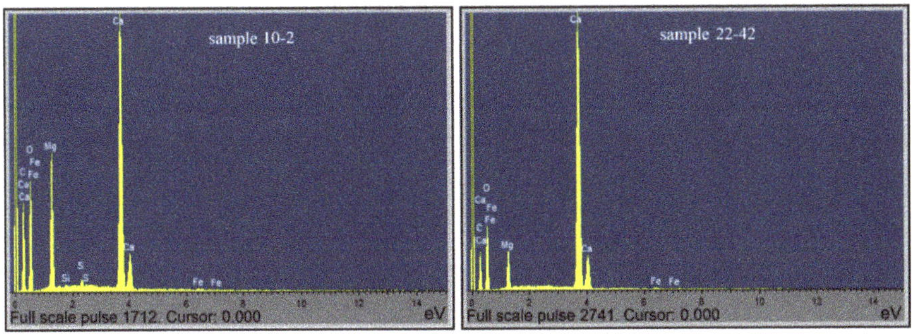

Figure 9. Elemental composition spectrum of samples: 10-2 and 22-42.

The given data of the electron-probe microanalysis on the elemental, oxide, and carboxylic content of the 10-2 sample rock correspond to the composition of minerals established by X-ray diffractometry and the thermal analysis complex (Figure 9). Within the limits of the capabilities of the used scientific equipment, the results obtained are quite acceptable for identifying the degree of perfection of the crystalline structures of carbonate minerals and can be used to determine the mechanism of accumulation and migration of hydrocarbon complexes in the sedimentary strata of the studied massif.

Limestone taken at the depth of 4390.9 m is an aggregate of calcite with impurities of magnesium carbonates. This is evidenced by the results of the thermal analysis of the powder sample from the studied core. Dynamic heating of this sample (22-25) led to the formation of many manifestations on the DTA, DTG, and TG curves due to the thermal destruction of calcite and, to a lesser extent, dolomite and magnesite (Figure 7). Along with these reactions, the thermal curves revealed weak emissions from the water system and organic carbon monoxide, which reduced the sample weight by 0.5 and 0.6%, respectively. Despite the similarity of the sample's composition with that of the 10-2 sample, the thermal behavior of the test sample was still somewhat different from the processes of decomposition of the compared core. This difference can be seen in the absence of an endothermic peak on the DTA curve of the studied system, which is responsible for the mode of destruction of the magnesium component ($MgCO_3$) of dolomite. Instead of said peak on the DTA and DTG curves, a slight deflection was recorded within 600–675 °C, indicating the development of a weak decomposition reaction of said magnesium component of dolomite in the sample. In this case, the thermogravimetric (TG) curve in the interval of the noted temperatures recorded a drop in the sample weight (Δm_4) by 2.1%. The same weight reduction of the sample's 2.1% dolomite was carried out in the range of 675–710 °C. Calcite decomposes simultaneously in the temperature range of 675–860 °C with dolomite (Figure 7). Using the stoichiometric formula of $CaMg(CO_3)_2$ and the amount of carbon dioxide emitted into the atmosphere (2 × 21.1%), it is not difficult to calculate the dolomite content in the system. The amount of it in the rock corresponded to 8.8%. Based on the results of the sample weight loss during the endothermic reactions in the regions of 460 °C and 830 °C, the presence of 5.7% magnesite and 83.9% calcite in the sample was calculated. The content of water and OM in the sample composition was determined based on the stages of weight loss in the temperature ranges of 20–200 °C and 200–460 °C, respectively (Figure 7, Table 1).

The high concentration of calcite (83.9%) in the sample composition made it possible to determine the kinetic constants of thermal dissociation of this carbonate. Using the technique from [18–20], the activation energy (E_A) of the emissions from the carbon dioxide system was calculated. In the process of $Ca(CO_3)$ destruction, the specified parameter was 250 kJ/mol, which corresponds to a high degree of perfection in the crystalline structure of this calcite.

The kinetic calculations also established that the modular value of EA of the studied carbonates is directly dependent on the value of free energy of their structures. Taking into account this regularity and inequality of $E_{A(calcite)} > E_{A(dolomite)}$, it follows that the direction of mineral transformation in calcite–dolomite systems largely depends on the energy levels of their crystalline structures, and then on the geochemical environment of the surrounding space. A sample analysis confirmed this inequality. In turn, the energy parameter of dolomite E_A is composed of the activation energy of the calcium component ($E_{A1}(CaCO_3)$) and its magnesium component ($E_{A1}(MgCO_3)$), i.e., the equality of $E_A(dolomite) = E_{A1}(CaCO_3) + E_{A1}(MgCO_3)$ is met.

It can be seen here that the energy level of the crystalline lattice of mono-mineral calcite is higher than the level at which the dolomite is located. In this regard, according to the Ostwald rule on the energy preference of the transition of one mineral system to another [19,20], the conversion of calcite to dolomite is more likely to occur than the conversion occurring in the opposite direction. This means that during the development of the new phase, the partial substitution of the Ca cation with Mg in the transformable calcite is less difficult than the substitution of magnesium dolomite with calcium. This conclusion is also true for the corresponding carbonates of the LTCGD. The process of converting calcite to dolomite in sediments, together with the introduction of clay minerals into carbonaceous complexes, can be attributed to one of the important stages of the formation of reservoirs and pore spaces (in carbonate strata) with high capacitive and filtration properties [21]. Further, it follows that the (mineral) destruction of $CaCO_3$ requires more energy consumption than the destruction of the dolomite structure. Such a conclusion, which is based on the direction of formation of mineral phases in the calcite of dolomite strata, can be used to identify the dependence of the productive properties of reservoirs not only on the composition of carbonate–clay complexes, but also on the degree of conversion of calcite to dolomite in these associations.

Table 8 lists the unit cell parameters: interplanar distances of the formation under study, d, Å, intensities of reflection, I%, and phase composition. In Figure 8, these parameters correspond to the positions of the peaks and the magnitude of their heights, indicating the composition and amount in the sample. This sample is characterized by the appearance of lines of quartz, β-quartz, dolomite, and calcite in the diffraction pattern.

Table 8. Interplanar distances and phase composition of the 22-25 sample.

d, Å	I %	Mineral	d, Å	I %	Mineral
4.25460	2.2	quartzite	2.40532	1.4	
3.85677	4.4		2.28348	7.7	
3.35754	2.6	β-quartz	2.19323	2.9	
3.19865	2.2	calcite	2.09324	7.1	
3.03324	100.0	dolomite	2.01682	1.9	dolomite
2.88677	84.1		1.92503	2.4	
2.84487	2.9		1.91098	9.4	
2.54258	1.8	dolomite	1.87409	7.9	
2.49272	5.9		1.80482	4.0	

It should be noted that the mineral composition in this rock, according to XRD data, somewhat differs from the material content presented by thermal analysis. Additionally, this is natural, since X-ray diffractometry gives the mineral content only from the crystalline part of the sample, whereas the thermogravimetric method used to determine the mineral composition is carried out relative to the entire mass of the substance under study, i.e., relative to the crystalline and amorphous parts of the sample. In addition, the presence of various salts and thermally inert disperse formations in the system under study can partially change the kinetics of the thermal decomposition of carbonates, which also leads to interference with the reliable determination of the quantitative content of mineral phases [18].

The 22-42 sample, which is from the lower stratigraphic level (5320 m), was heated via continuous temperature increase. This caused the 22-42 sample to leave a series of endo- and exothermic orientations on the thermal curves (Figure 7). The differential thermoanalytical (DTA) and differential thermogravimetric (DTG) curves in the range of 625–825 °C formed their trajectories in a W-shaped curve, which corresponds to the endothermic peaks of the thermal decomposition of dolomite. The depths of these peaks during $CaMg(CO_3)_2$ dissociation are usually the same. However, in the case under consideration, the value of the second peak (at 830 °C) was slightly higher than the value of the first one, which is due to the presence of calcite in the composition of the dolomite formation, the endothermic peak of which is also located at 830 °C. According to the thermogravimetric (TG) curve, it was found that when the dolomite dissociates into the atmosphere, a series of CO_2 molecules rush in (in two stages) with a mass of 28.4%, corresponding to the weight loss of the sample in the temperature interval under consideration. In turn, with the decomposition of calcite, the emission of carbon dioxide leads to a decrease in the weight of the sample by another 12.9%. Using these values to calculate the mineral content of the rock with TG determinations and stoichiometric formulas, the presence of dolomite (59.5%) and calcite (29.3%) was found in the sample.

According to the thermal effects recorded on the DTA curve in the range of 20–220 °C, the sample was found to be dehydrated with a weight loss of 0.2%, and for that in the range of 220–640 °C, the reaction of burning organic matter (CO_{org}) with a decrease in the sample weight of 2.2% was established. On the same curve, a pronounced exothermic peak with a vertex at 440 °C was noted. The effect was caused by the oxidation of pyrite along with the formation of hematite (α-Fe_2O_3) [17,18]. The amount of sulfide (FeS_2), counted in the peak area, corresponds to 2%. Finally, the presence of magnesite in the sample was detected from the weak endothermic dip of the DTA curve in the range of 550–640 °C. Within these temperatures, the TG line recorded a sample weight loss of 2.5%, which corresponds to a magnesite content of 4.8%. The parametric data of these reactions and information on the mineral and material composition of the rock are given in Table 1.

The results of the X-ray structural analysis are presented in Table 9, and the diffractometric reflexes of the test sample are reflected in the diffractogram in Figure 6. All diffraction peaks listed here belong to the above phases only. Characteristic diffraction reflexes are also noted, allowing for the identification of the phases present.

Table 9. Interplanar distances and phase composition of sample 22-42.

d, Å	I %	Mineral	d, Å	I %	Mineral
3.85735	5.4		2.19341	9.7	
3.69897	2.2		2.09368	8.6	calcite
3.36081	1.8	calcite dolomite	2.06347	1.8	
3.19855	3.0		2.01583	4.8	
3.03465	62.7		1.92592	3.5	
2.88836	100	calcite	1.91157	7.4	
2.67046	43		1.87528	8.5	
2.54064	2.9		1.84784	1.6	
2.49381	6.7		1.80471	7.1	
2.40604	3.0		1.78673	6.1.	dolomite

The control results of the elemental composition of the 22-42 sample, obtained by using the electron-probe microanalyzer, also revealed the correspondence of the mineral composition of the sample obtained via the X-ray and DTA methods (Tables 10–12, Figure 8).

Table 10. Electron-probe microanalysis results, Sample 22-42.

Processing Parameters: All Items Analyzed (Normalized)					
All results in weight %					
Spectrum	O	Mg	Ca	Fe	Total
Spectrum 1	57.58	6.43	35.52	0.47	100.00
Spectrum 2	56.98	5.28	37.49	0.25	100.00
Spectrum 3	57.49	6.71	35.45	0.35	100.00
Secondary	57.35	6.14	36.15	0.36	100.00

Table 11. Results of the oxide composition of sample 22-42 made on a probe microanalyzer.

Processing Parameters: Oxygen by Stoichiometry (Normalized)				
All Results in Compounds %				
Spectrum	MgO	CaO	FeO	Total
Spectrum 1	16.96	82.02	1.02	100.00
Spectrum 2	13.83	85.62	0.55	100.00
Spectrum 3	17.61	81.62	0.77	100.00
Secondary	16.13	83.09	0.78	100.00

Table 12. Processing parameters: carbon by difference.

All Results in Compounds %						
Spectrum	O	Mg	Ca	Fe	C	Total
Spectrum 1	47.61	4.59	27.30	0.36	20.14	100.00
Spectrum 2	45.46	3.57	27.75	0.19	23.02	100.00
Spectrum 3	48.17	4.90	27.72	0.28	18.94	100.00
Secondary	47.08	4.35	27.59	0.27	20.70	100.00

It should be noted that out of all the diversity within the material composition of the rocks, which is represented by carbonates, clays, siliceous formations, and other impurities, OM was also found in the studied wells. In limestones, where the amount of calcite exceeded 96%, the content of the organic compound varied within 0.3%–0.7%. In other rocks, where dolomite, magnesite, and siderite were present along with calcite, the concentration of OM reached 13 percent or more. The increased content of organic matter in these rocks is caused by the presence of a wider range of carbonate–clay minerals and silicon oxides necessary for the formation of reservoirs and pore spaces in the strata.

4. Discussion

The giant Karachaganak oil and gas condensate field located in the northern border zone of the Caspian Basin lies within the Lower Permian–Carboniferous–Upper Devonian carbonate bioherm platform and is oriented in the latitudinal direction with a length of 30 km; the width in the meridional direction varies along the strike from 15 to 5 km and a height of 1.7 km. The reservoir is a massive gas condensate. The productive horizons are composed of biohermic and biomorphic–detritic limestones and dolomites, which have transient differences. Reservoirs of pore and pore-cavern types are identified here, whereas reservoirs of the fracture type are insufficiently studied [1,2,4–8]. The Kashagan carbonate platform, located in the southern part of the Caspian Basin, was formed between the Visean and Bashkirian stages. The porosity in the interior of the platform is heterogeneous due to the superimposition of dissolution, grouting, and compaction processes. Complex diagenetic evolution led to the formation of mature matrix porosity within the platform and heterogeneous porosity due to the development of large, cavernous pores and joints [12].

An analysis of the literary sources on some oil- and gas-bearing areas within the ancient cratons of the world [22–28] makes it possible to carry out a comparative description of the areas of the LTCGD within the northern marginal zone of the Caspian Syneclise of the East European Craton.

In the Tarim and Sichuan Basins of China, similar diagenetic transformations and karst processes are present in the carbonate rocks, and the presence of fracturing has also been established [23–27]. The results of these works are argued and substantiated through modern methods of petrophysical studies. The development and evolution of porosity in dolomite reservoirs primarily reflect the complex effect of the dissolution and precipitation of minerals in the process of dolomitization [22,25].

In the northeastern part of the African Craton, the area within the Gulf of Suez and the area of the Nile Delta of Egypt, the effective porosity, matrix gas permeability, and reservoir quality parameters were measured in various oil and gas fields [26].

In the Gulf of Suez, at the Oktyabrskoye field (African Craton), a carbonate reservoir has been established [27]. On the basis of a facies analysis, the conditions of sedimentation and petrophysical characteristics of the Turonian Wata Formation were established.

In the South American Craton, the São Francisco do Sul area of Brazil, diagenetic processes control a zone of high permeability in carbonate rocks [28].

In all the above cases, a different instrumental–analytical research base was used.

The capabilities of the proposed integrated method for macro- and microscopic analysis of thin sections and core, X-ray diffraction analysis, DTA, TGA, and microprobe were not used. This technique has shown its effectiveness in studying the digenetic transformations of the reservoir carbonate reservoirs of the Famennian–Tournaisian reservoirs of the Northern marginal zone of the Caspian Syneclise.

The LTCGD are located above the protrusion of the Proterozoic basement [5,8]. Two genetic types of reservoirs were identified: (1) porous and pore-cavernous types and (2) joints (shear zones). It has been established that organogenic limestones are characterized by processes of recrystallization, dolomitization, substitution, leaching, and secondary mineral formation. The photos of the thin sections show various types of organogenic–lumpy limestones that are recrystallized and dolomitized to varying degrees (Figure 6). Often, as a result of post-sedimentation transformations, the rock completely loses its organogenic structure and acquires a crystalline structure, becoming relict–organogenic. In such cases, the primary structure is determined by relics and contours of organic fossils or by the relative arrangement of crystals of different sizes and orientations. Various stages of recrystallization can be traced in the thin sections of organogenic limestones, the photos of which are shown in Figure 6, where the cement and skeleton of the remains of organisms underwent recrystallization but retained their shape. Along with further recrystallization, the relics of the primary organogenic–lumpy structure are preserved in the form of a characteristic distribution of crystals of various sizes.

The staging of post-sedimentation processes is shown in Figure 6, where it is expressed via the following sequence: first, the incomplete recrystallization of the primary calcite mass, then uneven dolomitization, then leaching of the unchanged residual calcite and filling in the intercrystalline pores with anhydrite, and somewhat later, selective leaching of the rock with the formation of voids of several centimeters. The intensive manifestation of the dissolution and leaching processes led to the formation of large pore–cavernous cavities that are several centimeters in size.

Epigenetic mineral, texture-structural transformations of carbonate rocks, expressed through the recrystallization of carbonates and the emergence of a secondary joint–pore space in rocks, can be traced in Famennian–Carboniferous carbonates. The main rock-forming minerals are calcite, dolomite, and magnesite. Minor minerals are represented by siderite, rhodochrosite, manganocalcite, clay formations (kaolinite, hydromica, and mixed-laminated minerals), iron sulfides, magnesium, silicon, and calcium oxides.

5. Conclusions

The final formation of the reservoir properties of carbonate rocks occurred as a result of the development of tectonic fracturing and the diagenetic processes of leaching and secondary mineral formation. According to the data of the microscopic investigation, fracturing, stylolitization, and leaching increase the permeability and porosity of carbonate rocks. Epigenetic recrystallization, dolomitization, and sulfatization have a significant impact, worsening the reservoir properties of rocks in this case. On the contrary, tectonic fracturing contributes to an increase in pore space.

Based on the results of X-ray diffractometry (X-ray), thermal analysis (DTA and TGA), probe microanalysis, and complex microscopic study of Famennian–Tournaisian carbonate rocks of the LTCGD, their mineral and material composition was determined. The rocks were composed mainly of forms of calcite with crystalline lattices at different degrees of perfection, dolomites with moderately ordered structures, and magnesites with underdeveloped bonds of carbon dioxide in the intermolecular space.

Among these instrumental methods, a special place was given to thermal analysis. It was used to study the kinetics of thermal destruction of the $Ca(CO_3)$ and $CaMg(CO_3)_2$ structures, on the basis of which the scheme for the formation of dolomite due to the decomposition products of calcite and vice versa was established. It was found that most of the samples from dolomite–calcite associations contain calcite, the destruction of which requires the expenditure of higher activation energies (E_A) than is seen in the structures of dolomites. Thus, according to the Ostwald rule, the energy level of crystal lattices of similar calcites is higher than that of dolomite structures. In this regard, calcites in these associations are easily transformed into dolomites by partial replacement of the calcium cation with magnesium. Vice versa, if the energy level of the calcite crystal lattice is equal to or lower than the level of dolomite, then the transformation goes in the opposite direction: from the dolomite phase to the formation of the calcium carbonate structure. The interpretation of these data showed that in the studied carbonate sequences, the transformation of calcite into dolomite is more preferable than the reverse transformation.

As a result of comparing the activation energy of the ($E_{A1}(CaCO_3)$) decomposition of the calcite part of dolomite with the activation energy of $E_{A2}(CaCO_3)$ as it related to the dissociation of calcite proper, it was found that the value of $E_{A2}(CaCO_3)$ exceeds the values of $E_{A1}(CaCO_3)$. The calculations also showed that the activation energy of calcium decomposition of the containing component ($E_{A1}(CaCO_3)$) of dolomite was higher than the activation energy of the magnesium component of dolomite.

The interpretation of these data showed that in the studied carbonate strata, the transformation of calcite into dolomite is preferable to the reverse transformation. The results of X-ray, DTA, and TGA measurements and data from the probe microanalysis made it possible to identify thermally inert formations in the composition of carbonates as oxides of calcium, magnesium, silicon, iron, and other compounds accumulated during the conversion of carbonates.

On the basis of the conducted studies, an assessment of epigenetic changes in carbonate rocks was made. A comprehensive method for studying the properties of carbonate reservoirs has been proposed and justified, which opens up the possibility of using the applied methods for subsequent studies.

When studying hypergeneous transformations of nickel-bearing weathering crusts of the Kepirsay ultramafic massif, we applied a research technique (X-ray diffractometry for the solid phase of rocks, DTA, TGA), which showed its effectiveness [20]. When studying the Famennian–Tournaisian reservoirs of the northern marginal zone of the Caspian Syneclise in combination with macro/microscopy and a probe microanalyzer, the material composition and nature of epigenetic changes were determined, and data on the void space of reservoir carbonate reservoirs were obtained.

Our applied methodology is the scientific novelty of this work. We substantiated a rational, relatively inexpensive express method via the testing of samples from carbonate reservoirs.

Author Contributions: Conceptualization, V.K., A.C., Z.T., I.S. and Y.D.; methodology, V.K.; investigation, V.K., I.S. and Y.D.; resources, V.K., I.S. and Y.D.; writing—original draft preparation, V.K., I.S. and Y.D.; writing—review and editing, V.K., A.C., Z.T., I.S. and Y.D.; visualization, A.C., Z.T., I.S. and Y.D.; supervision, V.K.; project administration, V.K.; funding acquisition, V.K., A.C., Z.T., I.S. and Y.D. All authors have read and agreed to the published version of the manuscript.

Funding: This research was funded by the Ministry of Education and Science of the Republic of Kazakhstan (project IRN, grant number AP09260097, contract number no. 177/36-21-23, dated 15 April 2021) and actively supported by the leadership of the Kazakh–British Technical University (Almaty, Kazakhstan). The authors of this article are very grateful for the support provided.

Data Availability Statement: The data presented in this study are available on request from the corresponding author.

Acknowledgments: This research was supported by the leadership of the Kazakh–British Technical University (Almaty, Kazakhstan). The authors thank the reviewers for valuable comments and are grateful to the editor for careful editing.

Conflicts of Interest: The authors declare no conflict of interest.

References

1. Akchulakov, U.; Zholtayev, G.; Iskaziyev, K.O.; Kovrizhnyh, P.N.; Kuandykov, B.M.; Ogay, Y.K. *Atlas of Oil and Gas Bearing and Prospective Sedimentary Basins of the Republic of Kazakhstan*; Kazakshan Institute of Oil and Gas: Almaty, Kazakhstan, 2015; p. 97. (In Russian)
2. Vocalevskiy, E.S.; Bulekbayev, Z.E.; Iskuzhiyev, B.A.; Kamalov, S.M.; Korstyshevsky, M.N.; Kuandykov, B.M.; Kuantayev, N.E.; Marchenko, O.N.; Shudayev, K.S.; Matloshinsky, N.G.; et al. *Oil and Gas Field. Kazakhstan. Reference book*, 3rd ed.; Institute of Geological Sciences Named after K.I. Satpayev: Almaty, Kazakhstan, 2016; p. 409. (In Russian)
3. Bosikov, I.I.; Klyuev, R.V.; Tekiev, M.V. Analysis of spatial distribution of chemical elements of apatite-stafelite ore. *Geol. Geoph. Russ. South* **2021**, *11*, 137–157. (In Russian) [CrossRef]
4. Bagrintseva, K.; Dmitrievsky, A.; Bochko, R. *Atlas of Carbonate Reservoirs of Oil and Gas Fields of East European and Siberian Platforms*; Moscow: Moscow, Russia, 2003; p. 264. Available online: https://www.geokniga.org/books/14853 (accessed on 2 October 2022). (In Russian)
5. Fortunatova, N.K.; Kartseva, O.A.; Baranova, A.V.; Agafonova, G.V.; Offman, I.P. *Atlas of Structural Components of Carbonate Rocks*; VNIGNI: Moscow, Russia, 2005; p. 440. (In Russian)
6. Fortunatova, N.K.; Varlamov, A.I.; Kanev, A.S.; Poroskun, V.I.; Baranova, A.V.; Bushuyeva, M.A. Structure and assessment of the oil potential of carbonaceous carbonate-siliceous Domanik deposits in the Volga–Ural oil and gas province. *Rus. Geol. Geoph.* **2021**, *62*, 929–946. [CrossRef]
7. Daukeyev, S.Z.; Abdullin, A.A.; Bespayev, K.A.; Vocalevskiy, E.S. *Forecast Map of Oil and Gas Potential of Kazakhstan. M 1:2500000*; Ministry of Energy and Mineral Resources of the Republic of Kazakhstan: Almaty, Kazakhstan, 2000. (In Russian)
8. Daukeyev, S.Z.; Vocalevskiy, E.S.; Shlygin, D.A.; Piliphosov, V.M.; Paragulgov, K.K.; Kolomiyic, V.P.; Komarova, V.P. *Deep Structure and Mineral Resources of Kazakhstan (Part 2—Eastern Kazakhstan), Oil and Gas*; National Academy of the Sciences of the Reublic of Kazakhstan: Almaty, Kazakhstan, 2002; p. 248. Available online: https://www.geokniga.org/books/11437 (accessed on 1 November 2022).
9. Iskaziyev, K.O. *Strategy for the Development of Oil and Gas Resources in the Subsalt Deposits of the North of the Caspian Syneclise*; Abstract of the dissertation for a Doctor of Geological and Mineral Sciences; Gubkin Russian State University of Oil and Gas (National Research University): Moscow, Russia, 2021.
10. Zholtayev, G.Z.; Nikitina, O.I.; Zhaymina, V.Y.; Seitmuratova, Y.Y.; Pirogov, T.Y.; Ivanova, N.I.; Fazylov, Y.M.; Musina, Y.S.; Nigmatova, S.A.; Baishashov, B.U. *Modernization of the Phanerozoic Stratigraphic Schemes of Kazakhstan Based on the International Chronostratigraphic Scale—2016–2021*; Institute of Geological Sciences Named after K.I. Satpayev: Almaty, Kazakshan, 2021; p. 139.
11. Mitchell, J. A note on a method of staining to distinguish between calcite and dolomite. In *Colonial Geology and Mineral Resources*; H.M. Stationery Office: London, UK, 1956; p. 182.
12. Ronchi, P.; Ortenzi, A.; Borromeo, O.; Claps, M.; Zempolich, W.G. Depositional setting and diagenetic processes and their impact on the reservoir quality in the late Visean–Bashkirian Kashagan carbonate platform (Pre-Caspian Basin, Kazakhstan). *AAPG Bull.* **2010**, *94*, 1313–1348. [CrossRef]

13. Demeyeva, M.S. Features of formation of carbonate reservoirs at the Chinarevskoye field. *Eng. Sol. Oil Gas. Ind. Kaz.* **2020**, *6*, 130–135.
14. Ivanova, V.P.; Kasatov, B.K.; Krasavina, T.N.; Rozinova, Y.L. *Thermal Analysis of Minerals and Rocks*; Nedra: Leningrad, Russia, 1974; p. 398.
15. Feklichev, V.G. *Diagnostic Constants of Minerals*; Nedra: Moscow, Russia, 1989; p. 479.
16. Tsvetkov, A.I.; Vallashikhina, E.P.; Piloyan, G.O. *Differential Thermal Analysis of Carbonate Minerals*; Nauka: Moscow, Russia, 1964; p. 167.
17. Bondarenko, I.I.; Samatov, I.B. New in the study of siderites. *Izv. Akad. Nauk. KazSSR, Geol. Ser.* **1982**, *1*, 59–62.
18. Samatov, I.B.; Urumbaev, B.U. Thermochemical features of calcite-dolomite formations (on the example of Central Kazakhstan). *Geol. Kaz. Ser. Geol.* **1997**, *2*, 49–56.
19. Coats, A.P.; Redfern, J.P. Kinetic parameters of thermogravime data. *Nature* **1964**, *201*, 68. [CrossRef]
20. Korobkin, V.; Samatov, I.; Chaklikov, A.; Tulemissova, Z. Peculiarities of dynamics of hypergenic mineral transformation of nickel weathering crusts of ultramafic rocks of the Kempirsay group of deposits in Western Kazakhstan. *Minerals* **2022**, *12*, 650. [CrossRef]
21. Putnis, A.; McConnell, J.D.C. *Principles of Mineral Behaviour*; Elsevier: Amsterdam, The Netherlands, 1980; Available online: https://books.google.kz/books/about/Principles_of_Mineral_Behaviour.html?id=v9CzAAAAIAAJ&redir_esc=y (accessed on 4 November 2022).
22. Jiu, B.; Huang, W.; Mu, N.; Hao, R. Petrology, mineralogy and geochemistry of Ordovician rocks in the southwest of Tarim Basin, implications for genetic mechanism and evolution model of the hydrothermal reformed-paleokarst carbonate reservoir. *Mar. Pet. Geol.* **2022**, *140*, 105687. [CrossRef]
23. Chen, L.; Zhang, H.; Cai, Z.; Hao, F.; Xue, Y.; Zhao, W. Petrographic, mineralogical and geochemical constraints on the fluid origin and multistage karstification of the Middle-Lower Ordovician carbonate reservoir, NW Tarim Basin, China. *J. Pet. Sci. Eng.* **2022**, *208*, 109561. [CrossRef]
24. Shi, J.; Zhao, X.; Pan, R.; Zeng, L.; Luo, W. Natural fractures in the deep Sinian carbonates of the central Sichuan Basin, China: Implications for reservoir quality. *J. Pet. Sci. Eng.* **2022**, *216*, 110829. [CrossRef]
25. Yang, L.; Yu, L.; Chen, D.; Liu, K.; Yang, P.; Li, X. Effects of dolomitization on porosity during various sedimentation-diagenesis processes in carbonate reservoirs. *Minerals* **2020**, *10*, 574. [CrossRef]
26. Abuamarah, B.A.; Nabawy, B.S. A proposed classification for the reservoir quality assessment of hydrocarbon-bearing sandstone and carbonate reservoirs: A correlative study based on different assessment petrophysical procedures. *J. Nat. Gas Sci. Eng.* **2021**, *88*, 103807. [CrossRef]
27. Kassem, A.A.; Osman, O.A.; Nabawy, B.S.; Baghdady, A.R.; Shehata, A.A. Microfacies analysis and reservoir discrimination of channelized carbonate platform systems: An example from the Turonian Wata Formation, Gulf of Suez, Egypt. *J. Pet. Sci. Eng.* **2022**, *212*, 110272. [CrossRef]
28. La Bruna, V.; Bezerra, F.H.R.; Souza, V.H.P.; Maia, R.P.; Auler, A.S.; Araujo, R.E.B.; Cazarin, C.L.; Rodrigues, M.A.F.; Vieira, L.C.; Sousa, M.O.L. High-permeability zones in folded and faulted silicified carbonate rocks—Implications for karstified carbonate reservoirs. *Mar. Pet. Geol.* **2021**, *128*, 105046. [CrossRef]

minerals MDPI

Article

Early Diagenesis in the Lacustrine Ostracods from the Songliao Basin 91.35 Million Years Ago and Its Geological Implications

Zhenwu Liu [1,2], Yuke Liu [2], Xuejia Du [3], Dan Lyu [1], Huaichun Wu [1,*] and Huajian Wang [2,*]

1 School of Ocean Sciences, China University of Geosciences (Beijing), Beijing 100083, China
2 Research Institute of Petroleum Exploration and Development, Beijing 100083, China
3 Department of Petroleum Engineering, University of Houston, Houston, TX 77023, USA
* Correspondence: wanghuajian@petrochina.com.cn (H.W.); whcgeo@cugb.edu.cn (H.W.); Tel.: +86-010-8359-5538 (Huajian Wang)

Abstract: Diagenesis is a double-edged sword of geochemical recordings. It makes us always doubt about the representativeness of many geochemical indicators, especially the isotope and mineral related. It also provides a window to explore the biogeochemical processes at the water–rock interface, which are related to the interactions between the hydrosphere, biosphere, and lithosphere. In this study, we identified microbial early diagenesis in lacustrine ostracods from the Songliao Basin 91.35 million years ago by using in situ mineralogical and carbon isotope analytical methods. Our results suggest multiple biological early-diagenesis processes and the formation of a ferric and methane transition zone (FMTZ) in the sulfate-poor pore water, which are conducive to the formation of dolomite and ankerite. These secondary carbonate minerals related to dissimilatory iron reduction and methanogenesis have heavier carbon isotopic compositions than the calcified ostracod shell in the water column and might bring interferences to the geochemical parameters of ostracods.

Keywords: ankerite; carbon isotope; dissimilatory iron reduction; dissolved inorganic carbon; early diagenesis; ferric and methane transition zone; methanogenesis; ostracod

Citation: Liu, Z.; Liu, Y.; Du, X.; Lyu, D.; Wu, H.; Wang, H. Early Diagenesis in the Lacustrine Ostracods from the Songliao Basin 91.35 Million Years Ago and Its Geological Implications. *Minerals* **2023**, *13*, 5. https://doi.org/10.3390/min13010005

Academic Editors: Hamzeh Mehrabi and Vahid Tavakoli

Received: 13 November 2022
Revised: 14 December 2022
Accepted: 15 December 2022
Published: 20 December 2022

1. Introduction

Reconstruction of the water conditions of paleo-oceans and -lakes is of great importance in understanding the living environment of ancient life. Due to the ubiquitous diagenesis, many important geochemical parameters obtained from whole rock analysis, such as inorganic carbon and oxygen isotopes of carbonate ($\delta^{13}C_{carb}$ and $\delta^{18}O_{carb}$), strontium isotope ($^{87}Sr/^{86}Sr$), anomalies of rare earth elements (REE), etc., need to be carefully used [1–3]. The emergence and evolution of shelled organisms since the terminal of the Neoproterozoic Era [4], provide an effective archive for the primary geochemical information of the paleo-waters in their skeletonized shells. Geochemical compositions of some calcified zooplankton shells, such as foraminifera and ostracoda, are widely used in the reconstruction of paleo-climatic, -oceanographic and -ecological changes [5,6].

Carbon isotope of calcified zooplankton shells ($\delta^{13}C_{carb}$) is one of the most representative proxies to reveal the environments of paleo- oceans and -lakes, as well as the atmospheric carbon dioxide (CO_2) concentration [7–9]. Long-time scale and high-precision $\delta^{13}C_{carb}$ data play an important role in the stratigraphic correlation, paleoenvironment reconstruction, and deep-time carbon cycle modeling. For example, due to the $\delta^{13}C_{carb}$ of benthic foraminiferal and the micro-plankton criteria, the Abiod Formation of north-central Tunisia is restricted to the Late Campanian age, and the Campanian/Maastrichtian boundary is placed within the lowermost part of the El Haria Formation [10]. A high-resolution, benthic foraminiferal $\delta^{13}C_{carb}$ curve from the northwest Pacific ODP Site 1209 spanning 44 to 56 million years ago (Ma) with five thousand years (kyr) resolution, presents several excursions corresponding in timing and magnitude to hyperthermal layers previously described elsewhere [11].

Ostracods are one class of the oldest known microfauna, containing both marine and non-marine forms [12]. The first fossil representatives are from marine sediments 485 Ma [13] and have molecular clock evidence 600 Ma or earlier [14]. The earliest freshwater ostracoda species came from coal-forming swamps, ponds, and streams of the Pennsylvanian age (318~299 Ma) [15]. Preservation of ostracod fossil is attributed to the calcified shell present in most species, which is a valuable source of calcite. Carbon isotope analyses of ostracod shells have concentrated on lacustrine environments, from which other carbonate microfossils are largely absent [5]. The specimens' $\delta^{13}C_{carb}$ values are mainly controlled by the calcification completion of shell, the equilibrium with dissolved inorganic carbon (DIC) in the water column and pore water, as well as the seasonal and vertical variations of the carbon isotopic compositions of DIC ($\delta^{13}C_{DIC}$) [16]. In most cases, the impact of the first two factors is assumed to be negligible or in equilibrium. Then, the variation of ostracods' $\delta^{13}C_{carb}$ values were used to represent the climate and biological activities forced changes of $\delta^{13}C_{DIC}$. For example, the ostracods' $\delta^{13}C_{cal}$ values of the Late Cretaceous sediments from the Songke1s and Songke1n cores in the Songliao Basin, were considered to reflect changes in both global climate and regional basin evolution [17].

However, during the burial and early diagenetic stage, a series of heterotrophic microbial processes occurs to degrade organic matters and generate secondary carbonate minerals, such as dolomite, calcite, ankerite, and siderite [18–21]. As the bicarbonate (HCO_3^-) forming these secondary minerals is mainly from pore water rather than water column [19,22], the minerals are encased in ostracod shells and may be difficult to effectively remove, thereby causing geochemical contamination. Therefore, it is necessary to identify and evaluate the early diagenesis in the ostracods.

In this study, in situ mineralogical and carbon isotopic analyses of lacustrine ostracods from the Songliao Basin were performed by using quantitative evaluation of minerals by scanning electron microscopy (QemScan) and laser ablation isotope ratio mass spectrometry (LA-IRMS), respectively. Abundant ankerite, chlorite, and scarce pyrite in the studied ostracods indicated the occurrence of early diagenesis and the formation of a ferric-methane transition zone which generated ^{13}C-enriched ankerite. Our results support more strict treatment and in situ high-precision analysis to improve the reliability of ostracod-based geochemical data.

2. Sample Preparation and Analytical Methods

2.1. Sample Preparation

The ostracod samples are from Songke1s core with a depth of 1734 m in the Songliao Basin, northeast of China (Figure 1a). Three high-precision CA-ID-TIMS U–Pb zircon ages of 91.886 $\pm$ 0.11 Ma, 90.974 $\pm$ 0.12 Ma and 90.536 $\pm$ 0.12 Ma from the bentonites at 1780 m, 1705 m and 1673 m of the Songke1s core [23] and the critical astrochronological time scale [24,25] together constrain the samples' age to be 91.35 Ma (Figure 1b). The Songliao Basin is a large lacustrine basin filled with the Mesozoic and Cenozoic sediments [26]. Two sets of organic-rich black shales, Qingshankou and Nenjiang formations, mostly developed in the deep-water phases and became the main source rocks of the Songliao Basin [27]. The salinity of paleo-Songliao lake is interpreted as predominantly freshwater to oligohaline with multiple and small-scale seawater incursion events [28]. Although fourteen non-marine ostracod assemblage zones have been identified from the Member 1 of the Qingshankou Formation (K_2qn^1) to the Member 2 of the Nenjiang Formation (K_2n^2), only one assemblage zone coincides in the K_2qn^1 and is dominated by *Triangulicypris torsuosus* Netchaeva with a smooth shell and *Triangulicypris torsuosus* Netchaeva. *nota* Ten with tubercles [29]. These ostracods are rare in the black shales but are enriched in several individual layers. The studied samples are handpicked from the middle of K_2qn^1, with a lithological characteristic of centimeter-thick ostracod layers interbedded with micrometer-thick organic-rich laminae (Figure 1c).

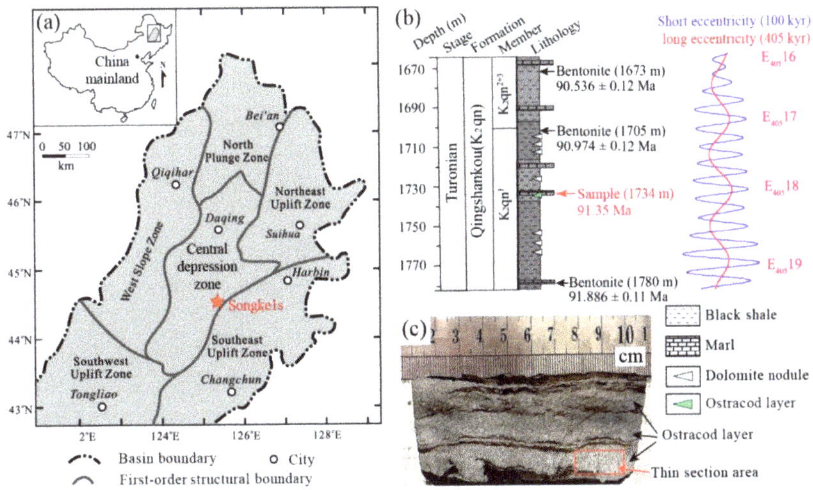

Figure 1. (**a**) Location of the Songke1s core in the Songliao Basin, northeast of China. (**b**) Stratigraphy, geochronology, and astrochronology of the Songke1s core in the 1782–1667 m range. The sample age (91.35 Ma) in this study was calculated from the high-precision U-Pb zircon ages [23] and the critical astronomical time scale tune by short eccentricity [24,25]. (**c**) Hand specimen characteristic of the studied ostracod layer. Thin sections were prepared from the area without organic-rich lamina.

Several standard thin sections with a 1.5 cm × 3 cm area were prepared. The thickness of lamina used for optical observation and LA-IRMS analysis was 30 μm and 200 μm, respectively, as the latter one met the requirement to provide sufficient laser ablated samples from micro-areas (at least 150-μm-thickness). Carbon-coated fragments with new fractures were prepared for scanning electron microscope (SEM) observation. Carbon-coated fragments with argon ion polished surface were prepared for QemScan analyses. These sections and fragments are from the same depth range to ensure the data comparability between different analyses (Figure 1c).

2.2. Optical Observation and Mineralogical Analyses

Optical observation of ostracods on sections was investigated by using optical microscope (Olympus 4500P, Olympus Company, Tokyo, Japan) under transmission light at the key laboratory of petroleum geochemistry of China National Petroleum Corporation (CNPC, Beijing, China), Research Institute of Petroleum Exploration and Development (RIPED). The carbon-coated fragments were examined by using a Apreo SEM (FEI Company, Hillsboro, OR, USA) equipped with an energy dispersive X-ray spectroscopy (EDS) for the semi quantitative element determination. An integrated high-speed detector (Bruker Company, Billerica, MA, USA) was used for backscattered electron (BSE) image and secondary electron image (SEI). The beam accelerating voltage was 15 kV, and the emission current was 71 mA.

Elemental imaging and mineral identification were performed at the key laboratory of hydrocarbon reservoir of CNPC, RIPED, by using a QemScan 650F (FEI Company, Hillsboro, OR, USA) with a pixel size of 1 μm. This instrument emits an X-ray spectrum and provides information on the content of elements at each measured point. The elemental contents were obtained by combining the BSE, image gray, and X-ray intensity, and then were converted into mineral phases. Individual mineral was identified by referring to a comprehensive mineral database incorporated into the QemScan software iExplorer. Several important elements, including aluminum (Al), calcium (Ca), iron (Fe), magnesium (Mg), silicon (Si), and sulfur (S) were selected for in situ imaging.

2.3. Laser Ablation Isotope Ratio Mass Spectrometry Analysis

In situ carbon isotope analysis was carried out in the key laboratory of carbonate reservoir of CNPC, RIPED. This experiment was performed on a coupled system of a Nd-1064 Nd:YAG laser ablation (Norla Institute of Technical Physics, Chengdu, Sichuan, China), a self-assembled elemental analyzer, and an isotope ratio mass spectrometer (Thermo, DELTA V Advantage, Waltham, MA, USA) (Figure 2). This system contains a movable sample stage that can be monitored by a computer-controlled video microscope (20 to 80×) and an internal light-emitting diode (LED) illumination. Due to the small amount of solid gas produced by laser ablation, the sample amount generated by a single ablation cannot generate sufficient CO_2 for mass spectrometry analysis. Therefore, we added a cold trap between LA and IRMS to capture and enrich the gas generated by multiple ablations (Figure 2). All laser carbonate systems to isotope operate on the principle of thermal decrepitation during laser heating by the following reaction (Equation (1)) [30].

$$M^{2+}CO_3 \rightarrow M^{2+}O + CO_2 \quad (1)$$

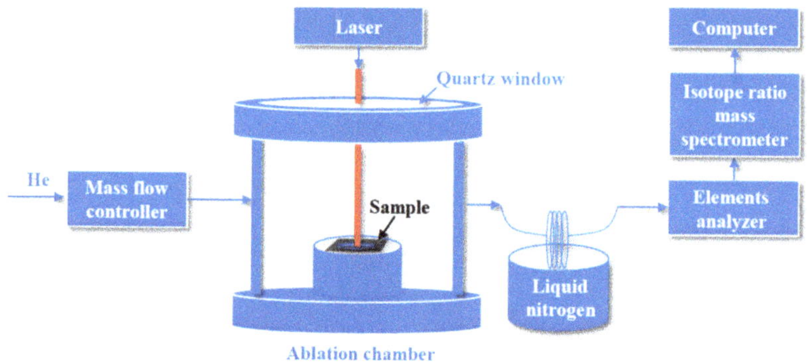

Figure 2. Diagram of the laser ablation isotope ratio mass spectrometry analysis.

After laser ablation, the generated CO_2 was frozen on a 0.5 mm i.d. fused silica capillary immersed in a liquid nitrogen bath. The trapping began a few seconds before ablation and continued for a total of 60 s. In most cases, the trapping continued for at least 30 s after sample ablation. Following the trapping step, we switched the carrier gas flow through the trap, reducing the high flow rate (10 mL/min) to a low flow rate (1 mL/min). Once the carrier gas flow rate was low enough, the capillary from the liquid nitrogen bath was removed. The CO_2 rapidly melted at room temperature and was transferred by the carrier gas to pass through a column in the element analyzer, which was filled with magnesium perchlorate ($Mg(ClO_4)_2$) to remove the organic matter and water and then passed through a quartz capillary to separate other possible debris.

Following the element analyzer, the CO_2 pulse passed into the IRMS instrument via an open split. The duration for each analysis, from ablation to isotope analysis, was about 8 min. High purity helium (with a volume fraction over 99.999%) was used as carrier gas to reduce potential CO_2 contamination from the atmosphere and improve transmission efficiency. For blank runs following the same method without any laser ablation, the CO_2 peaks were too small to be noticed for sample analysis. The national standard material GBW04405 with a uniform $\delta^{13}C$ value of $0.57 \pm 0.03‰$ was selected as the standard sample to optimize the operating parameters. With the selected and optimized operating conditions (Table 1), we can obtain stable $\delta^{13}C$ values of the standard samples with an uncertainty lower than $\pm 0.2‰$ (1σ) over 10 times tests. All $\delta^{13}C$ values were reported relative to the Vienna Pee Dee Belemnite (VPDB), with standard deviation (SD) of each sample being lower than $\pm 0.2‰$ based on three times analyses.

Table 1. Measurement parameters of LA-IRMS.

Laser Ablation System (LA)		Isotope Ratio Mass Spectrometry (IRMS)	
Wavelength	1064 nm	High voltage	3.0 KV
Pulse frequency	10 Hz	Box emission	0.8 mA
Pulse energy	66 μJ	Trap emission	0.7 mA
Output of energy	50%	Carrier gas (He)	1–10 mL/min
Energy density	7 J/cm^2	Data acquisition	Thermo Isodat 3.0
Spot size	50 μm	Temperature	26 °C
Scanning mode	Point	Humidity	52% RH
Pause time	1 s	Elemental isotope	^{13}C

3. Results

3.1. Petrographic and Mineralogical Features

The ostracod layer is mainly composed of ostracod shells, detrital quartz, and clay minerals. Most ostracods are well preserved in oval features with 0.7–1.2 mm for the long axis and 0.3–0.8 mm for the short axis (Figure 3). Their body sizes are similar to the previously reported *Triangulicypris torsuosus* Netchaeva and *Triangulicypris torsuosus* Netchaeva. *nota* Ten occurred in the K_2qn^1 [29]. As shown in Figure 3, most ostracods are filled with carbonate minerals and/or clay minerals, accounting for about half-and-half in observed area.

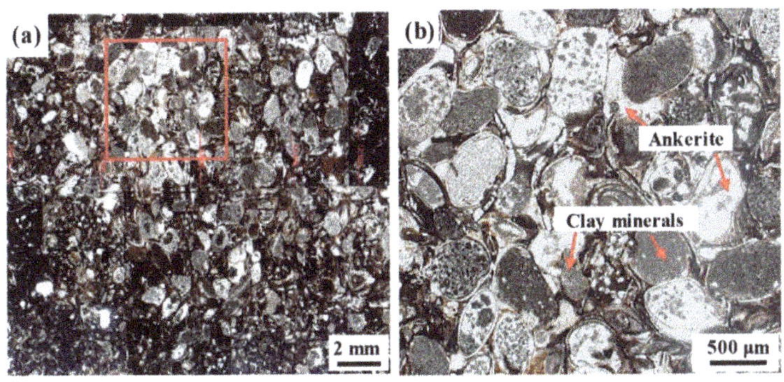

Figure 3. Petrographic features of the studied ostracod layer under transmission light. (**b**) is the enlargement of marked square area in (**a**). The white and gray fillings in the ostracods under transmission light are ankerite and clay minerals, respectively.

The well-preserved ostracods cluster together more clearly under SEM (Figure 4a). Both *Triangulicypris torsuosus* Netchaeva and *Triangulicypris torsuosus* Netchaeva. *nota* Ten are identified and presented in Figure 4b,c. The EDS data showed that the ostracod shell is composed of highly purified calcite ($CaCO_3$) with a sum atomic percentage of calcium (Ca), carbon (C), and oxygen (O) over 97% (Figure 4c, Table 2). The randomly oriented clay mineral with high atomic percentages of Fe (7.9%), Al (6.1%), Si (6.9%), and Mg (3.5%) (Figure 4d, Table 2) may be chlorite [$Y_3[Z_4O_{10}](OH)_2 \cdot Y_3(OH)_6$, Y = Fe, Al, Mg; Z = Si, Al]. The C-axis of this mineral can take any rotation angle on surface (Figure 4d). The rhombohedron of euhedral-subhedral shape carbonate mineral with a diameter of several to tens of micrometers and a higher atomic percentage of Fe (6.7%) than Mg (5.5%) (Figure 4e,f, Table 2) may be ankerite [$Ca(Mg_x, Fe_y)(CO_3)_2$, $x < y$]. Ankerite with a similar occurrence also can be found in the ostracod filled with chlorite (Figure 4d), further indicating the extensive distribution of ankerite.

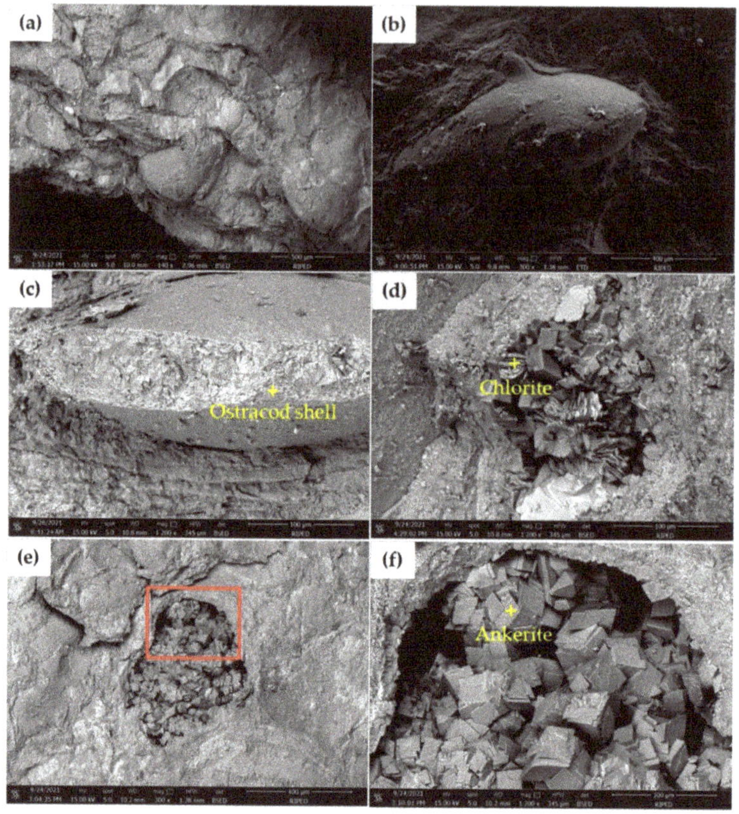

Figure 4. Scanning electron microscope photographs of the studied ostracods. (**a**,**b**) are appearances of well-preserved ostracod shells. (**b**) is *Triangulicypris torsuosus* Netchaeva. *nota* Ten with a tubercle. (**c**) is *Triangulicypris torsuosus* Netchaeva with a smooth shell. (**d**) is a broken ostracod filled with chlorite and ankerite. (**e**) is a broken ostracod filled with ankerite. (**f**) is the enlargement of marked area in (**e**). Energy dispersive X-ray spectroscopy (EDS) analysis was performed on the points with yellow cross forks in panels (**c**,**d**,**f**). The EDS data are shown in Table 2.

Table 2. Elemental contents of different points on sample based on SEM-EDS method.

Points	Units	Al	C	Ca	Fe	Mg	O	S	Si
Calcified ostracod shell (Figure 4c)	Un-normalized count (wt%)	0.5	8.0	24	0.8	0.4	26	0	0.7
	Normalized count (wt%)	0.9	13.2	39	1.3	0.7	43	0	1.2
	Atomic count (at%)	0.7	22.4	20	0.5	0.6	55	0	0.9
Chlorite (Figure 4d)	Un-normalized count (wt%)	7.3	5.2	6.0	17	3.3	37	0.1	8.2
	Normalized count (wt%)	8.7	6.2	7.1	20	4.0	44	0.1	9.7
	Atomic count (at%)	6.1	11.1	3.8	7.9	3.5	59	0.1	6.9
Ankerite (Figure 4f)	Un-normalized count (wt%)	0.4	5.1	24	11	3.8	23	0	0.3
	Normalized count (wt%)	0.6	7.6	35	16	5.7	35	0.1	0.5
	Atomic count (at%)	0.5	14.9	21	6.7	5.5	51	0	0.4

Elemental and mineralogical distributions of two types of ostracods filled with chlorite and ankerite were further investigated by using QemScan. Both of them show concentric mineralogical distribution inside (Figures 5 and 6). The co-enrichment of Fe, Ca, and Mg represents ankerite. The co-enrichment of Al, Si, Mg, and Fe is consistent with chlorite. Pyrite (FeS_2) is characterized by the co-enrichment of Fe and S. Quartz (SiO_2) is characterized by extreme enrichment of Si. The QemScan data give more refined mineral distributions and are consistent with the optical observation (Figures 3 and 4). For the ostracod filled with ankerite, the chlorite inside is mainly adjacent to the shell (Figure 5). Conversely, when the ostracod is filled with chlorite, ankerite is the dominating secondary mineral and distributes along the shell edge (Figure 6). A small amount of siderite was also identified in this type of ostracod (Figure 6). Small amounts quartz with poor roundness can be identified on the calcified shells (Figures 5 and 6). It should be noted that pyrite is rare in both two types of ostracods and is mainly associated with ankerite (Figures 5 and 6).

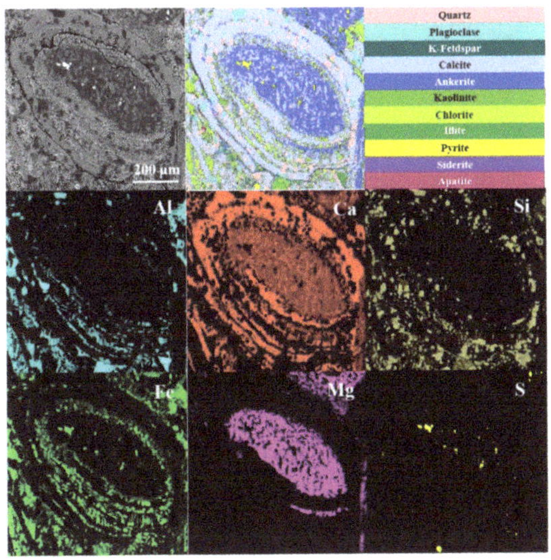

Figure 5. Scanning electron microscope photo and QemScan images of an ostracod filled with ankerite. Color brightness in each map represents relative abundance of its corresponding element. Brighter color indicates higher element content.

Figure 6. Scanning electron microscope photo and QemScan images of an ostracod filled with chlorite and ankerite. Color brightness in each map represents relative abundance of its corresponding element. Brighter color indicates higher element content.

3.2. Carbon Isotopic Compisitions

In situ $\delta^{13}C$ analyses of these two types of ostracods were performed on calcified shells and filling ankerites. The $\delta^{13}C$ values of calcified shells are from 0.8‰ to 1.1‰, with a mean value of 0.9‰ (Figure 7). For the ostracods filled with ankerite, the $\delta^{13}C$ values of ankerites are from 4.0‰ to 4.5‰, with a mean value of 4.2‰ (Figure 7a,b). However, low abundance of ankerite and interference from chlorite make it difficult to obtain authentic $\delta^{13}C$ values from a single ablation point. Thus, we used the liquid nitrogen freezing method to collect samples from three or four ablation points to obtain a mixed $\delta^{13}C$ value. The $\delta^{13}C$ values of ankerite in this type ostracods are from 1.9‰ to 3.0‰, with a mean value of 2.3‰ (Figure 7c,d).

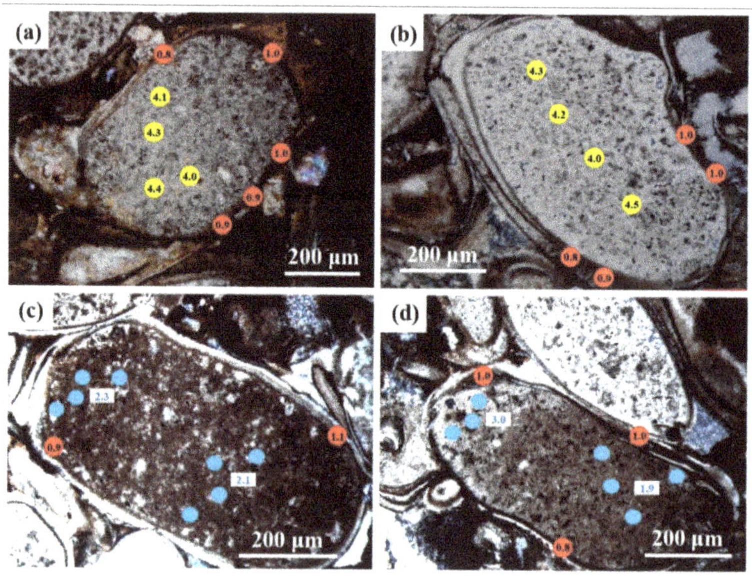

Figure 7. In situ $\delta^{13}C$ values of calcified shells and filling ankerites determined by using LA-IRMS. (**a**,**b**) are the ostracods filled with ankerite. (**c**,**d**) are the ostracods filled with chlorite. The red, yellow, and blue dots are ablation points in the experiment. Data in red and yellow dots are $\delta^{13}C$ values of calcified shells and ankerites, respectively. Data beside the blue dots are mixed $\delta^{13}C$ values of ankerite from neighbor ablation points.

4. Discussions

4.1. Microbial Early Diagenesis in the Ostracods

Three types of depositional pattern of ostracod-bearing beds have been recognized from the Songliao Basin, including mixed siliciclastic–ostracod deposits in the delta front, sheeted ostracod deposits, and dotted ostracod deposits in shallow or semi-deep lacustrine settings [31]. Here, our concerned sheeted ostracod layers with interbedded black organic-rich shales were deposited in a semi-deep lacustrine setting and were from the cementation and diagenesis of ostracods and detritus. A variety of authigenic minerals (e.g., ankerite, chlorite, pyrite, siderite) were identified in the ostracods, demonstrating complex early diagenesis during the burial of ostracods. Formation of these secondary minerals needs a relatively closed environment, which can be easily satisfied in pore water during burial time. The most important is the supply of required anions and cations, including the ferrous ion (Fe^{2+}) and hydrogen sulfide (H_2S) that are related with the water redox condition [32] and ammonia (NH_4^+) and HCO_3^- that are related with the water alkalinity [33]. The two most abundant authigenic minerals in our studied ostracods are ankerite and chlorite, both of which contain high proportion of Fe^{2+}. Previous studies showed that the comparatively large (>8 μm) and wide size distribution (8–30 μm) of the framboid diameter between 1725 and 1750 m of the Songke1s core, indicating a weakly oxic to anoxic depositional water environment [34]. Enrichments of the redox-sensitive elements supported the anoxic and non-euxinic deep water with low sulfate concentration [35]. Although Fe^{2+} is soluble in the anoxic water, it is difficult to form ankerite directly for the large enthalpies of hydration of Mg^{2+} and Ca^{2+} [36]. Transfer of Fe from water to sediment and pore water mainly depends on biotic and/or abiotic oxidation to $Fe(OH)_3$ or iron-manganese oxides, then reduced to Fe^{2+} through dissimilatory iron reduction (DIR) accompanied with the oxidation of organic matters (Equation (2), Figure 8a) [37]. This reaction will generate HCO_3^- (Equation (2), Figure 8b), which can elevate the alkalinity of pore water and is more conducive to the formation of siderite. However, the dominating carbonate mineral in our studied ostracods is ankerite rather than siderite, indicating non-dominant Fe^{2+} concentration and breakthrough of the kinetic and thermodynamic barriers of Mg^{2+} and Ca^{2+}. Experiments and geological evidence have proved that microbial mediation can realize the formation of dolomite in a low-temperature environment [38–40]. Thus, the ankerite might be from the isomorphic replacement of Mg^{2+} by Fe^{2+} in pore water [41].

$$4Fe(OH)_3 + CH_2O + 7H^+ \rightarrow Fe^{2+} + HCO_3^- + 10H_2O \quad (2)$$

The association of pyrite (even in low abundance) and ankerite proves the occurrence of bacterial sulfate reduction (BSR) and generation of H_2S (Equation (3)), which is prone to combine with Fe^{2+} to produce iron monosulfide phases (amorphous FeS) [42]. The latter will be further oxidized to pyrite (FeS_2) by elemental sulfur or other oxidants [42]. Low abundance of pyrite indicates limited sulfate supply and a narrow sulfate reduction window without free H_2S (Figure 8a) and is consistent with the previous study [35]. Bicarbonate generated through DIR and BSR processes has light $\delta^{13}C$ values (Figure 8c) for its organic origin (Equations (2) and (3)). This will produce ^{13}C-depleted carbonate minerals, e.g., the siderite with $\delta^{13}C$ values ranged from −27‰ to −8‰ from the Mesoproterozoic Xiamaling Formation [22,43] and the dolomite with $\delta^{13}C$ values lower than −10‰ from the Ediacaran Shuram Formation [1].

$$SO_4^{2-} + 2CH_2O \rightarrow H_2S + 2HCO_3^- \quad (3)$$

However, the $\delta^{13}C$ values of our studied secondary ankerites are from 1.9‰ to 4.5‰, heavier than those of the calcified shells (from 0.8‰ to 1.1‰, mean value 0.9‰). Furthermore, carbon in the ostracod shells is from the bicarbonate captured in the surface water, which usually has a heavier $\delta^{13}C$ value than that of the bottom and pore water. This can be proved by the difference of over −2‰ between the carbonate rocks deposited in deep-water and shallow facies in the same basin [44]. Variation of the $\delta^{13}C$ values of ostracods (from −6.4‰ to 4.4‰) from K_2qn^1 to the Mingshui Formation in the Songliao Basin were interpreted as the joint influence of both local basin evolution and global carbon cycle [17]. This process is mainly realized through the production and degradation of organic matter. Therefore, these seemingly contradictory measured data indicate that organic matter should have experienced further diagenetic degradation to generate ^{13}C-enriched HCO_3^-.

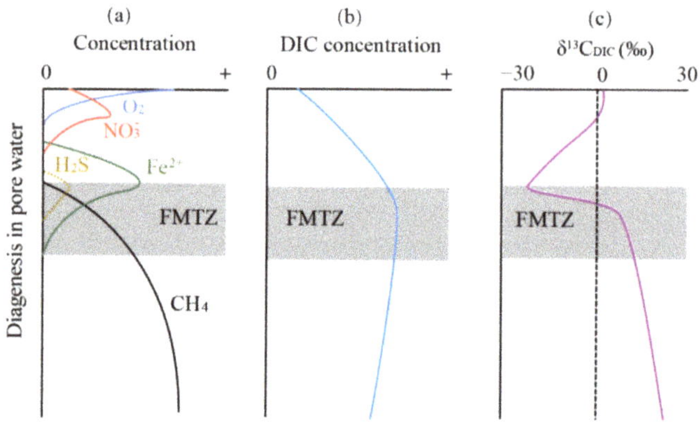

Figure 8. Schematic model of the early-diagenesis (**a**) and ferric-methane transition zone (FMTZ) controlled concentrations of dissolved inorganic carbon ((**b**), DIC) and carbon isotopic composition of DIC ((**c**), $\delta^{13}C_{DIC}$) in the pore water. Oxygen (O_2) and nitrate (NO_3^-) curves in (**a**) represent the aerobic respiration and nitrate reduction before dissimilatory iron reduction [45]. The dotted line of hydrogen sulfide (H_2S) represents that free H_2S might not exist in a sulfate-poor environment but be quickly combined with Fe^{2+} to form sulfide. The FMTZ occurs due to a lack of sulfate in pore water which impairs the sulfate reduction reaction and thus does not inhibit methanogenesis. The ^{13}C-enriched CO_2 generated through methanogenesis significantly increases the $\delta^{13}C_{DIC}$ value in the pore water (**c**) and makes an alkaline environment conducive to dolomite/ankerite precipitation and then decreases the DIC's concentration in the pore water (**b**).

Methanogenesis is the most possible process when explaining this finding (Figure 8), due to the synchronously produced extreme ^{13}C-depleted methane (CH_4, lower than −60‰) [46]. Furthermore, the mole ratio of CH_4: CO_2 is 4:1 when dissimilated for every mole C assimilated into biomass (Equation (4)) [46]. According to the isotope mass balance, if we assume the carbon isotopic value of organic matter ($\delta^{13}C_{org}$) to be −30‰ [35] and the mole ratio of methanogenic genetic HCO_3^- to pore water HCO_3^- (with an assumed $\delta^{13}C$ value of −8‰) to be higher than 1:5, it is entirely possible to generate HCO_3^- with a $\delta^{13}C$ value heavier than 8.3‰. This will cause a significant turning point of carbon isotopic composition of DIC, with a $\delta^{13}C_{DIC}$ value of pore water heavier than that of the bottom water (Figure 8c). Huge generation of HCO_3^- is also conducive to the formation of dolomite, which may result in the DIC concentration decrease after reaching supersaturation (Figure 8b).

$$5CH_2O + 6H_2 \rightarrow 4CH_4 + CO_2 + 3H_2O \quad (4)$$

In the Junggar Basin, the maximum $\delta^{13}C$ value of methanogenic genetic dolomite in the Permian Lucaogou Formation can reach 20‰ [18]. In the Ordos Basin, the carbonate concretion related with methanogenesis arising from bacterial activities in the fermentation zone have $\delta^{13}C$ values ranged from 10.0‰ to 14.2‰ [47]. Furthermore, the release of CH_4 may also promote the prosperity of methanotroph with biomarkers of ^{13}C-depleted 3β-methylhopanes [48]. Extended 3β-methylhopanes up to C_{45} identified from the sediments of K_2qn^1 in the Songliao Basin support robust methanogenesis during the early diagenetic stage [49]. Highly depleted $\delta^{13}C_{org}$ values (minimum −32.4‰) and increased C/N values (maximum 25.6) of K_2qn^1 were interpreted as influences from methanogenesis and methanotroph [35]. The $\delta^{13}C$ values of ankerite associated with high content chlorite are from 1.9‰ to 3.0‰, slightly lower than the more aggregated ankerite but still heavier than those of the ostracod shells (Figure 7). Differences of the $\delta^{13}C$ values of ankerites from these two types of ostracods might be from the degree of shell closure and the proportion of clays entering into. The precipitation of chlorite prior to ankerite was another important factor, resulting in slightly weaker methanogenesis, and Fe^{2+} generated through DIR were bound within chlorite.

Therefore, the identified early diagenesis of two types of ostracods recorded in our study involved DIR, BSR, and methanogenesis through the production of Fe^{2+}, H_2S, CH_4 and HCO_3^- (Figure 8). Although chlorite, pyrite, and ankerite in these ostracods were formed in the ferric reduction zone, sulfate reduction zone, and methanogenic zone, respectively, low abundance of sulfate makes the other two redox processes more important. High amounts of ankerite with slightly heavier $\delta^{13}C$ values and low contents of pyrite also suggest a ferric-methane transition zone (FMTZ) without a sulfate reduction zone between them (Figure 8). This transition zone might be of great importance in the sulfate-poor and ferruginous water dominated Precambrian ocean [50] and might induce methane release to maintain long-term warm-house [51]. This also can be used to explain the massive ^{13}C-enriched carbonate rock deposits in Precambrian without large-scale black shale deposits [52,53]. However, it should be noted that the ankerite found in ostracods might be different from the dolomite nodules and layers found in the K_2qn^1, which were formed in a dynamically varied diagenetic water redox conditions [21]. The dolomite nodule was suggested to have a primary form as dolomite and a gradual transformation into ankerite as well as manganese-ankerite [21]. Rare earth elements and Sr isotope values also provide further evidence for the formation of dolomite nodules and layers resulting from the seawater intrusion events of the Songliao Basin [53,54].

4.2. Implications for the Geochemical Analyses of Ostracod

In situ $\delta^{13}C$ analysis revealed obvious differences between calcified shells and filled early diagenetic ankerites of ostracods (Figure 7), indicating certain interferences from secondary minerals on the primary DIC signals recorded in the calcified shells. In fact, before conducting the geochemical analysis with ostracod fossils, fine treatment is necessary, such as the visual examination for possible secondary carbonate minerals and the following brushing [17] and roasting for a certain time at high temperature in vacuum to remove organic matter [55]. However, some uncertainties still exist from extraneous aspects, such as the selection of cleaning methods, specimen ontogenetic stage, and shell preservation. Based on cleaning tests, clay minerals adhering to the shells have been considered as major contaminants in specimens [56]. Although the diagenetic dolomite and ankerite wrapped in ostracod are difficult to find and remove, they should be considered another noteworthy contaminant. Thus, we suggest to use strict treatment and high-resolution analysis to improve the reliability of ostracod-based geochemical data. Furthermore, detailed works are necessary to obtain a high-resolution $\delta^{13}C_{DIC}$ curve of the paleo-Songliao lake if we want to better understand the carbon cycle model in the Songliao Basin between oceanic anoxic event 2 (OAE2) and OAE3 [35].

5. Conclusions

Diagenesis keeps us in doubt about the representativeness of many geochemical indicators, especially the isotope and mineral related, forcing us to choose the carrier that formed in the water column and remained stable during diagenesis. Ostracoda are a good geochemical carrier of paleo-water and paleo-climate information and are widely used in the paleo-environment reconstruction. We identified DIR, BSR, and methanogenesis in the ostracods during their burial in the Songliao Basin 91.35 Ma. For the scarcity of sulfate in the pore water, a ferric-methane transition zone is suggested to control the formation of ^{13}C-enriched authigenic ankerite, which may be of great importance in the sulfate-poor but ferruginous water dominated Precambrian ocean. We also suggested strict treatment and high-resolution analysis to improve the reliability of ostracod-based geochemical data.

Author Contributions: Conceptualization, Z.L. and H.W. (Huajian Wang); methodology, Z.L., Y.L. and H.W. (Huajian Wang); investigation, Z.L., Y.L., D.L. and H.W. (Huajian Wang); resources, H.W. (Huajian Wang) and H.W. (Huaichun Wu); data curation, Z.L., Y.L. and X.D.; writing—original draft preparation, Z.L. and Y.L.; writing—review and editing, Z.L., Y.L., X.D. and H.W. (Huajian Wang);

supervision, H.W. (Huaichun Wu) and H.W. (Huajian Wang); project administration, H.W. (Huaichun Wu) and H.W. (Huajian Wang); funding acquisition, H.W. (Huajian Wang) and H.W. (Huaichun Wu). All authors have read and agreed to the published version of the manuscript.

Funding: This research was funded by the National Key R&D Program of China (grant no. 2019YFC0 605403), the National Natural Science Foundation of China (grant no. 41872125, 42002158, 41790451) and the scientific and technological projects of RIPED (grant no. 2021ycq01, yjkt2019-3).

Acknowledgments: We thank Wang Yongsheng for his help on the LA-IRMS analysis.

Conflicts of Interest: The authors declare no conflict of interest.

References

1. Derry, L.A. A burial diagenesis origin for the Ediacaran Shuram–Wonoka carbon isotope anomaly. *Earth Planet. Sci. Lett.* **2010**, *294*, 152–162. [CrossRef]
2. Shields, G.; Stille, P. Diagenetic constraints on the use of cerium anomalies as palaeoseawater redox proxies: An isotopic and REE study of Cambrian phosphorites. *Chem. Geol.* **2001**, *175*, 29–48. [CrossRef]
3. Phan, T.T.; Hakala, J.A.; Lopano, C.L.; Sharma, S. Rare earth elements and radiogenic strontium isotopes in carbonate minerals reveal diagenetic influence in shales and limestones in the Appalachian Basin. *Chem. Geol.* **2019**, *509*, 194–212. [CrossRef]
4. Maloof, A.C.; Rose, C.V.; Beach, R.; Samuels, B.M.; Calmet, C.C.; Erwin, D.H.; Poirier, G.R.; Yao, N.; Simons, F.J. Possible animal-body fossils in pre-Marinoan limestones from South Australia. *Nat. Geosci.* **2010**, *3*, 653–659. [CrossRef]
5. Holmes, J.A.; Chivas, A.R. Ostracod shell chemistry—Overview. In *The Ostracoda: Applications in Quaternary Research*; American Geophysical Union: Washington, DC, USA, 2002; pp. 185–204.
6. Katz, M.E.; Cramer, B.S.; Franzese, A.; Hönisch, B.; Miller, K.G.; Rosenthal, Y.; Wright, J.D. Traditional and emerging geochemical proxies in foraminifera. *J. Foramin. Res.* **2010**, *40*, 165–192. [CrossRef]
7. Doney, S.C.; Fabry, V.J.; Feely, R.A.; Kleypas, J.A. Ocean acidification: The other CO_2 problem. *Annu. Rev. Mar. Sci.* **2009**, *1*, 169–192. [CrossRef]
8. Spero, H.J.; Lerche, I.; Williams, D.F. Opening the carbon isotope "vital effect" black box, 2, Quantitative model for interpreting foraminiferal carbon isotope data. *Paleoceanography* **1991**, *6*, 639–655. [CrossRef]
9. Spero, H.; Williams, D. Opening the carbon isotope "vital effect" black box 1. Seasonal temperatures in the euphotic zone. *Paleoceanography* **1989**, *4*, 593–601. [CrossRef]
10. Farouk, S.; Faris, M.; Bazeen, Y.S.; Elamri, Z.; Ahmad, F. Upper Campanian-lower Maastrichtian integrated carbon isotope stratigraphy and calcareous microplankton biostratigraphy of North-central Tunisia. *Mar. Micropaleontol.* **2021**, *166*, 102003. [CrossRef]
11. Westerhold, T.; Röhl, U.; Donner, B.; Zachos, J.C. Global extent of early Eocene hyperthermal events: A new Pacific benthic foraminiferal isotope record from Shatsky Rise (ODP Site 1209). *Paleoceanogr. Paleoclimatol.* **2018**, *33*, 626–642. [CrossRef]
12. Rodriguez-Lazaro, J.; Ruiz-Muñoz, F. A general introduction to ostracods: Morphology, distribution, fossil record and applications. In *Developments in Quaternary Sciences*; Elsevier: Amsterdam, The Netherlands, 2012; pp. 1–14.
13. Salas, M.J.; Vannier, J.; Williams, M. Early Ordovician ostracods from Argentina: Their bearing on the origin of binodicope and palaeocope clades. *J. Paleontol.* **2007**, *81*, 1384–1395. [CrossRef]
14. Regier, J.C.; Shultz, J.W.; Kambic, R.E. Pancrustacean phylogeny: Hexapods are terrestrial crustaceans and maxillopods are not monophyletic. *Proc. R. Soc. B* **2005**, *272*, 395–401. [CrossRef]
15. Benson, R.H. Ecology of ostracode assemblages. In *Treatise on Invertebrate Paleontology: Part Q. Arthropoda 3-Crustacea/Ostracoda*; Geological Society of American: McLean County, IL, USA, 1961; pp. 56–63.
16. Decrouy, L.; Vennemann, T.W.; Ariztegui, D. Controls on ostracod valve geochemistry: Part 2. Carbon and oxygen isotope compositions. *Geochim. Cosmochim. Acta* **2011**, *75*, 7380–7399. [CrossRef]
17. Chamberlain, P.; Wan, X.; Graham, S.A.; Carroll, A.R.; Doebbert, A.C.; Sageman, B.B.; Blisniuk, P.; Corson, K.; Zhou, W.; Wang, C. Stable isotopic evidence for climate and basin evolution of the Late Cretaceous Songliao basin, China. *Palaeogeogr. Palaeocl. Palaeocl.* **2013**, *385*, 106–124. [CrossRef]
18. Sun, F.; Hu, W.; Wang, X.; Cao, J.; Fu, B.; Wu, H.; Yang, S. Methanogen microfossils and methanogenesis in Permian lake deposits. *Geology* **2021**, *49*, 13–18. [CrossRef]
19. Wang, H.; Ye, Y.; Deng, Y.; Liu, Y.; Lyu, Y.; Zhang, F.; Wang, X.; Zhang, S. Multi-element imaging of a 1.4 Ga authigenic siderite crystal. *Minerals* **2021**, *11*, 1395. [CrossRef]
20. Khan, D.; Qiu, L.; Liang, C.; Martizzi, P.; Mirza, K.; Liu, J. Tracing forming mechanism of the sparry calcite growth in the lacustrine shale of east China: A glimpse into the role of organic matter in calcite transformation. *Geol. J.* **2022**, *57*, 1820–1836. [CrossRef]
21. Liu, Y.; He, W.; Zhang, J.; Liu, Z.; Chen, F.; Wang, H.; Ye, Y.; Lyu, Y.; Gao, Z.; Yu, Z.; et al. Multi-element Imaging Reveals the Diagenetic Features and Varied Water Redox Conditions of a Lacustrine Dolomite Nodule. *Geofluids* **2022**, *2022*, 9019061.
22. Wang, H.J.; Ye, Y.T.; Deng, Y.; Wang, X.M.; Hammarlund, E.U.; Fan, H.F.; Canfield, D.E.; Zhang, S.C. Isotope evidence for the coupled iron and carbon cycles 1.4 billion years ago. *Geochem. Perspect. Lett.* **2022**, *21*, 1–6. [CrossRef]

23. Wang, T.T.; Ramezani, J.; Wang, C.S.; Wu, H.C.; He, H.Y.; Bowring, S.A. High-precision U–Pb geochronologic constraints on the Late Cretaceous terrestrial cyclostratigraphy and geomagnetic polarity from the Songliao Basin, Northeast China. *Earth Planet. Sci. Lett.* **2016**, *446*, 37–44. [CrossRef]
24. Wu, H.; Zhang, S.; Jiang, G.; Hinnov, L.; Yang, T.; Li, H.; Wan, X.; Wang, C. Astrochronology of the Early Turonian–Early Campanian terrestrial succession in the Songliao Basin, northeastern China and its implication for long-period behavior of the Solar System. *Palaeogeogr. Palaeocl. Palaeocl.* **2013**, *385*, 55–70. [CrossRef]
25. Wu, H.; Hinnov, L.A.; Zhang, S.; Jiang, G.; Yang, T.; Li, H.; Xi, D.; Ma, X.; Wang, C. Continental geological evidence for Solar System chaotic behavior in the Late Cretaceous. *GSA Bull.* 2022, *in press*. [CrossRef]
26. Feng, Z.Q.; Jia, C.Z.; Xie, X.N.; Zhang, S.; Feng, Z.H.; Cross, T.A. Tectonostratigraphic units and stratigraphic sequences of the nonmarine Songliao basin, northeast China. *Basin Res.* **2010**, *22*, 79–95.
27. Feng, Z.; Fang, W.; Li, Z.; Wang, X.; Huo, Q.; Huang, C.; Zhang, J.; Zeng, H. Depositional environment of terrestrial petroleum source rocks and geochemical indicators in the Songliao Basin. *Sci. China Earth Sci.* **2011**, *54*, 1304–1317. [CrossRef]
28. Wang, C.; Scott, R.W.; Wan, X.; Graham, S.A.; Huang, Y.; Wang, P.; Wu, H.; Dean, W.E.; Zhang, L. Late Cretaceous climate changes recorded in Eastern Asian lacustrine deposits and North American Epieric sea strata. *Earth Sci. Rev.* **2013**, *126*, 275–299. [CrossRef]
29. Wan, X.; Zhao, J.; Scott, R.W.; Wang, P.; Feng, Z.; Huang, Q.; Xi, D. Late Cretaceous stratigraphy, Songliao Basin, NE China: SK1 cores. *Palaeogeogr. Palaeocl. Palaeocl.* **2013**, *385*, 31–43. [CrossRef]
30. Sharp, Z.; Cerling, T. A laser GC-IRMS technique for in situ stable isotope analyses of carbonates and phosphates. *Geochim. Cosmochim. Acta* **1996**, *60*, 2909–2916. [CrossRef]
31. Pan, S.; Liang, S.; Ma, L.; Liu, C.; Liu, H. Genesis, distribution, and hydrocarbon implications of ostracod-bearing beds in the Songliao Basin (NE China). *Aust. J. Earth Sci.* **2017**, *64*, 793–806. [CrossRef]
32. Canfield, D.E. Reactive iron in marine sediments. *Geochim. Cosmochim. Acta* **1989**, *53*, 619–632. [CrossRef]
33. Slaughter, M.; Hill, R.J. The influence of organic matter in organogenic dolomitization. *J. Sediment. Res.* **1991**, *61*, 296–303. [CrossRef]
34. Wang, P.; Huang, Y.; Wang, C.; Feng, Z.; Huang, Q. Pyrite morphology in the first member of the Late Cretaceous Qingshankou formation, Songliao Basin, northeast China. *Palaeogeogr. Palaeocl. Palaeocl.* **2013**, *385*, 125–136. [CrossRef]
35. Jones, M.M.; Ibarra, D.E.; Gao, Y.; Sageman, B.B.; Selby, D.; Chamberlain, C.P.; Graham, S.A. Evaluating Late Cretaceous OAEs and the influence of marine incursions on organic carbon burial in an expansive East Asian paleo-lake. *Earth Planet. Sci. Lett.* **2018**, *484*, 41–52. [CrossRef]
36. Hardie, L.A. Dolomitization; a critical view of some current views. *J. Sediment. Res.* **1987**, *57*, 166–183. [CrossRef]
37. Bekker, A.; Slack, J.F.; Planavsky, N.; Krapež, B.; Hofmann, A.; Konhauser, K.O.; Rouxel, O.J. Iron formation: The sedimentary product of a complex interplay among mantle, tectonic, oceanic, and biospheric processes. *Econ. Geol.* **2010**, *105*, 467–508. [CrossRef]
38. Sánchez-Román, M.; Vasconcelos, C.; Schmid, T.; Dittrich, M.; McKenzie, J.A.; Zenobi, R.; Rivadeneyra, M.A. Aerobic microbial dolomite at the nanometer scale: Implications for the geologic record. *Geology* **2008**, *36*, 879–882. [CrossRef]
39. Vasconcelos, C.; McKenzie, J.A.; Bernasconi, S.; Grujic, D.; Tiens, A.J. Microbial mediation as a possible mechanism for natural dolomite formation at low temperatures. *Nature* **1995**, *377*, 220–222. [CrossRef]
40. Gregg, J.M.; Bish, D.L.; Kaczmarek, S.E.; Machel, H.G. Mineralogy, nucleation and growth of dolomite in the laboratory and sedimentary environment: A review. *Sedimentology* **2015**, *62*, 1749–1769. [CrossRef]
41. Matsumoto, R.; Iijima, A. Origin and diagenetic evolution of Ca–Mg–Fe carbonates in some coalfields of Japan. *Sedimentology* **1981**, *28*, 239–259. [CrossRef]
42. Canfield, D.E.; Raiswell, R. Pyrite formation and fossil preservation. In *Taphonomy: Releasing the Data Locked in the Fossil Record*; Plenum Press: New York, NY, USA, 1991; pp. 337–387.
43. Canfield, D.E.; Zhang, S.C.; Wang, H.J.; Wang, X.M.; Zhao, W.Z.; Su, J.; Bjerrum, C.; Haxen, E.; Hammarlund, E. A Mesoproterozoic Iron Formation. *Proc. Nat. Acad. Sci. USA* **2018**, *115*, E3895–E3904. [CrossRef]
44. Jiang, G.; Shi, X.; Zhang, S.; Wang, Y.; Xiao, S. Stratigraphy and paleogeography of the Ediacaran Doushantuo Formation (ca. 635–551Ma) in South China. *Gondwana Res.* **2011**, *19*, 831–849. [CrossRef]
45. Canfield, D.E.; Thamdrup, B. Towards a consistent classification scheme for geochemical environments, or, why we wish the term 'suboxic' would go away. *Geobiology* **2009**, *7*, 385–392. [CrossRef] [PubMed]
46. Summons, R.E.; Franzmann, P.D.; Nichols, P.D. Carbon isotopic fractionation associated with methylotrophic methanogenesis. *Org. Geochem.* **1998**, *28*, 465–475. [CrossRef]
47. Zhu, R.; Cui, J.; Luo, Z.; Li, S.; Mao, Z.; Xi, K.; Su, L. Isotopic geochemical characteristics of two types of carbonate concretions of Chang 7 member in the middle-upper Triassic Yanchang Formation, Ordos Basin, Central China. *Mar. Petrol. Geol.* **2020**, *116*, 104312. [CrossRef]
48. Summons, R.E.; Jahnke, L.L.; Roksandic, Z. Carbon isotopic fractionation in lipids from methanotrophic bacteria: Relevance for interpretation of the geochemical record of biomarkers. *Geochim. Cosmochim. Acta* **1994**, *58*, 2853–2863. [CrossRef] [PubMed]
49. Zhu, C.; Cui, X.; He, Y.; Kong, L.; Sun, Y. Extended 3β-methylhopanes up to C_{45} in source rocks from the Upper Cretaceous Qingshankou Formation, Songliao Basin, northeast China. *Org. Geochem.* **2020**, *142*, 103998. [CrossRef]
50. Fakhraee, M.; Hancisse, O.; Canfield, D.E.; Crowe, S.A.; Katsev, S. Proterozoic seawater sulfate scarcity and the evolution of ocean–atmosphere chemistry. *Nat. Geosci.* **2019**, *12*, 375–381. [CrossRef]

51. Pavlov, A.A.; Hurtgen, M.T.; Kasting, J.F.; Arthur, M.A. Methane-rich Proterozoic atmosphere? *Geology* **2003**, *31*, 87–90. [CrossRef]
52. Cui, H.; Warren, L.V.; Uhlein, G.J.; Okubo, J.; Liu, X.-M.; Plummer, R.E.; Baele, J.-M.; Goderis, S.; Claeys, P.; Li, F. Global or regional? Constraining the origins of the middle Bambuí carbon cycle anomaly in Brazil. *Precambrian Res.* **2020**, *348*, 105861. [CrossRef]
53. Cadeau, P.; Jézéquel, D.; Leboulanger, C.; Fouilland, É.; Floc'h, L.; Chaduteau, C.; Milesi, V.; Guélard, J.; Sarazin, G.; Katz, A. Carbon isotope evidence for large methane emissions to the Proterozoic atmosphere. *Sci. Rep.* **2020**, *10*, 1–13. [CrossRef]
54. Liu, Y.; Wang, H.; Zhang, J.; Liu, Z.; Chen, F.; Wang, X.; Zhang, S.; Liu, H. Rare earth elemental and Sr isotopic evidence for seawater intrusion event of the Songliao Basin 91 million years ago. *Petrol. Sci.* 2022, *in press*. [CrossRef]
55. Schwalb, A.; Lister, G.S.; Kelts, K. Ostracode carbonate δ18O-and δ13C-signature of hydrological and climatic changes affecting Lake Neuchâtel, Switzerland, since the latest Pleistocene. *J. Paleolimnol.* **1994**, *11*, 3–17. [CrossRef]
56. Rodríguez, M.; De Baere, B.; François, R.; Hong, Y.; Yasuhara, M.; Not, C. An evaluation of cleaning methods, preservation and specimen stages on trace elements in modern shallow marine ostracod shells of Sinocytheridea impressa and their implications as proxies. *Chem. Geol.* **2021**, *579*, 120316. [CrossRef]

MDPI
St. Alban-Anlage 66
4052 Basel
Switzerland
www.mdpi.com

Minerals Editorial Office
E-mail: minerals@mdpi.com
www.mdpi.com/journal/minerals

www.ingramcontent.com/pod-product-compliance
Lightning Source LLC
Chambersburg PA
CBHW040840250926
51850CB00007BA/1364
* 9 7 8 3 7 2 5 8 0 6 8 8 1 *